**An Artificial
History of
Natural
Intelligence**

An Artificial History of Natural Intelligence

THINKING WITH MACHINES FROM DESCARTES TO THE DIGITAL AGE

David W. Bates

The University of Chicago Press
Chicago and London

The University of Chicago Press, Chicago 60637
The University of Chicago Press, Ltd., London
© 2024 by The University of Chicago
All rights reserved. No part of this book may be used or reproduced in any manner whatsoever without written permission, except in the case of brief quotations in critical articles and reviews. For more information, contact the University of Chicago Press, 1427 E. 60th St., Chicago, IL 60637.
Published 2024
Printed in the United States of America

33 32 31 30 29 28 27 26 25 24   1 2 3 4 5

ISBN-13: 978-0-226-83210-4 (cloth)
ISBN-13: 978-0-226-83211-1 (e-book)
DOI: https://doi.org/10.7208/chicago/9780226832111.001.0001

Library of Congress Cataloging-in-Publication Data

Names: Bates, David William, author.
Title: An artificial history of natural intelligence : thinking with machines from Descartes to the digital age / David W. Bates.
Description: Chicago : The University of Chicago Press, 2024. | Includes bibliographical references and index.
Identifiers: LCCN 2023033201 | ISBN 9780226832104 (cloth) | ISBN 9780226832111 (e-book)
Subjects: LCSH: Thought and thinking. | Artificial intelligence. | Intellect.
Classification: LCC BF441.B335 2024 | DDC 153.4/2—dc23/eng/20230802
LC record available at https://lccn.loc.gov/2023033201

♾ This paper meets the requirements of ANSI/NISO Z39.48-1992 (Permanence of Paper).

Dédié à l'esprit de Bernard Stiegler

Contents

FRAME | 1

1 Autonomy and Automaticity: On the Contemporary Question of Intelligence | 2

Part One: The Automatic Life of Reason in Early Modern Thought | 15

2 Integration and Interruption: The Cartesian Thinking Machine | 16
3 Spiritual Automata: From Hobbes to Spinoza | 33
4 Spiritual Automata Revisited: Leibniz and Automatic Harmony | 49
5 Hume's Enlightened Nervous System | 65

THRESHOLD: KANT'S CRITIQUE OF AUTOMATIC REASON | 79

6 The Machinery of Cognition in the First Critique | 80
7 The Pathology of Spontaneity: The *Critique of Judgment* and Beyond | 95

Part Two: Embodied Logics of the Industrial Age | 107

8 Babbage, Lovelace, and the Unexpected | 108
9 Psychophysics: On the Physio-Technology of Automatic Reason | 116
10 Singularities of the Thermodynamic Mind | 125
11 The Dynamic Brain | 131
12 Prehistoric Humans and the Technical Evolution of Reason | 141
13 Creative Life and the Emergence of Technical Intelligence | 152

PROPHECY: THE FUTURE OF EXTENDED MINDS | 161

14 Technology Is Not the Liberation of the Human but Its Transformation... | 162

Part Three: Crises of Order: Thinking Biology and Technology between the Wars | 163

15 Techniques of Insight | 164
16 Brains in Crisis, Psychic Emergencies | 181
17 Bio-Technicity in Von Uexküll | 188
18 Lotka on the Evolution of Technical Humanity | 194
19 Thinking Machines | 201
20 A Typology of Machines | 205
21 Philosophical Anthropology: The Human as Technical Exteriorization | 209

HINGE: PROSTHETICS OF THOUGHT | 227

22 Wittgenstein on the Immateriality of Thinking | 228

Part Four: Thinking Outside the Body | 233

23 Cybernetic Machines and Organisms | 234
24 Automatic Plasticity and Pathological Machines | 245
25 Turing and the Spirit of Error | 260
26 Epistemologies of the Exosomatic | 289
27 Leroi-Gourhan on the Technical Origin of the Exteriorized Mind | 311

THE BEGINNING OF AN END | 327

28 Technogenesis in the Networked Age | 328
29 Failures of Anticipation: The Future of Intelligence in the Era of Machine Learning | 345

Acknowledgments | 353
Notes | 355
Index | 391

We form our division of natural history upon the three-fold state and condition of nature; which is, 1) either free, proceeding in her ordinary course, without molestation; or 2) obstructed by some stubborn and less common matters, and thence put out of her course, as in the production of monsters; or 3) bound and wrought upon by human means, for the production of things artificial.

Let all natural history, therefore, be divided into the history of generations, præter-generations, and arts; the first to consider nature at liberty; the second, nature in her errors; and the third, nature in constraint.

Francis Bacon, *The Advancement of Learning* (1605)

Not much can be achieved by the naked hand or by the unaided intellect. Tasks are carried through by tools and helps, and the intellect needs them as much as the hand does. And just as the hand's tools either give motion or guide it, so—in a comparable way—the mind's tools either point the intellect in the direction it should go or offer warnings.

Francis Bacon, *The New Organon* (1620)

Frame

1
Autonomy and Automaticity

On the Contemporary Question of Intelligence

The historical evolution and development of artificial intelligence (AI) has long been tied to the consolidation of cognitive science and the neurosciences. There has been, from the start of the digital age, a complex and mutually constitutive mirroring of the *brain*, the *mind*, and the *computer*.[1] If to be a thinking person in the contemporary moment is to be a "brain,"[2] it is also true that the brain is, in the dominant paradigms of current neuroscientific practice, essentially a computer, a processor of *information*. Or, just as easily, the computer itself can become a brain, as the development of neuromorphic chip designs and the emergence of "cognitive computing" aligns with the deep learning era of AI, where neural networks interpret and predict the world on the basis of vast quantities of data.[3] These disciplines and technologies align as well with a dominant strand of evolutionary theory that explains the emergence of human intelligence as the production of various neural functions and apparatuses.[4]

In a way, this is a strange moment, when two powerful philosophies of the human coexist despite their radical divergence. For in the world of social science theory, science and technology studies, and the critical humanities, the dominant framework of analysis has emphasized the *historicity* and cultural plurality of the "human," and has, over the past few decades, moved more and more to a consensus that humans are just one part of distributed, historically structured networks and systems that subject individuals to various forms of control and development. We are, that is, functions in systems (political, economic, social, moral, environmental, etc.) that seem so familiar and almost natural but can be relentlessly critiqued and historicized.

On the other hand, we have a conceptual and disciplinary line that has increasingly understood human beings as essentially driven by unconscious and automatic neural processes that can be modeled in terms of information processing of various kinds, and the brain is the most complex network mediating these various processes. The result, to borrow the title from a cognitive science paper, is a new condition, namely, "the unbearable automaticity of being."[5] For the cognitive scientist, the human will is demonstrably an illusion, appearing milliseconds after the brain has *already* decided in controlled experimental conditions.[6] Consciousness, while still a philosophical problem, is understood as just another evolutionary function, linked now to attention mechanisms that can prompt responses from the unconscious space of operations. Whether we are thinking "fast" or "slow," to use Daniel Kahneman's terms, the *system* of human cognition as a whole is encompassed by the brain as the automatic—and autonomous—technology of thinking.[7] What else could thought be in the contemporary scientific moment? As one psychologist observed a while ago, "Any scientific theory of the mind has to treat it as an automaton."[8] If the mind "works" at all, it has to work on known principles, which means, essentially, the principles of a materially embodied process of neural processing. Steven Pinker, whom humanists love to hate (often for good reason), has put it bluntly: "Beliefs are a kind of information, thinking a kind of computation, and emotions, motives, and desires are a kind of feedback mechanism."[9] However crude the formulation, the overarching principle at work here is important. Cognitive science and neuroscience, along with myriad AI and robotic models related to these disciplines, cannot introduce what might be called a *spiritual* or transcendental element into their conceptualizations. Even consciousness, however troubling it may be, can be effectively displaced, marked as something that will eventually be understood as a result of physiological organization but that in the meantime can be studied like any other aspect of the mind. As the philosopher Andy Clark claims, a key contemporary philosophical issue is *automaticity*: "The zombie challenge is based on an amazing wealth of findings in recent cognitive science that demonstrate the surprising ways in which our everyday behavior is controlled by automatic processes that unfold in the complete absence of consciousness."[10]

Much as we may not want to admit it, Yuval Harari, of *Sapiens* fame, is probably right about the current moment, in at least one crucial way. As he says, "we" (cognitive scientists, that is) have now "hacked" humans, have found out why they behave the way they do, and have replicated (and in the process vastly *improved*) these cognitive behaviors in various artificial technologies.

> In the last few decades research in areas such as neuroscience and behavioural economics allowed scientists to hack humans, and in particular to gain a much better understanding of how humans make decisions. It turned out that our choices of everything from food to mates result not from some mysterious free will, but rather from billions of neurons calculating probabilities within a split second. Vaunted "human intuition" is in reality "pattern recognition."[11]

While we (rightly) rail against the substitution of human decision making, in judicial, financial, or other contexts, by algorithms, according to the new sciences of decision,[12] there is nothing more going on in the human brain, and to be fair, it isn't like humans were not exemplifying bias before the age of AI. As we know, the development of algorithmic sentencing, for example, was motivated by the desire to avoid the subjectivity and variability of human judgments.

In any case, we have to recognize that Harari is channeling the mainstream of science and technology on the question of the human: since we are, so to speak, "no more than biochemical algorithms, there is no reason why computers cannot decipher these algorithms—and do so far better than any *Homo sapiens*."[13] Hence the appearance of recent books with such horrifying titles as *Algorithms to Live By: The Computer Science of Human Decisions*, which helpfully introduces readers to concepts from computing that can improve their day-to-day lives,[14] and *Noise: A Flaw in Human Judgment*, which advises us humans to imitate the process of clear, algorithmic objectivity.[15] But my main point is that Harari reveals the contemporary crisis very clearly: it is a crisis of *decision*. "Computer algorithms," unlike human neural ones, "have not been shaped by natural selection, and they have neither emotions nor gut instincts. Hence in moments of crisis they could follow ethical guidelines much better than humans—provided we find a way to code ethics in precise numbers and statistics."[16] Computers will make better and more consistent decisions because they are *not* decisions in crisis but applications of the rule to the situation, objectively considered.

The backlash against this vision of AI, however well intentioned, has often been driven by just the kind of platitudes about the "human" that humanists and social science scholars have been dismantling for decades (if not centuries). New centers for moral or ethical or human-compatible AI and robotics assert a natural "human" meaning or capacity that the technology must serve—usually couched in the new language of "inclusion," "equity," and "fairness," as if those concepts have not emerged in historically specific ways or have not been contested in deadly conflicts (in civil wars, for example). As the home page for Stanford's Center for Human-Centered Computing proclaims, "Artificial Intelligence has the potential to help us realize our shared dream of a better future for all of humanity."[17] As we might respond: so did communism and Western neoliberal democratic capitalism.

But what do the critical scholars have to offer? At the moment, it seems that there is a loose collaboration that is hardly viable for the long term. One can critique technical systems and their political and ideological currents pretty effectively, and in the past years much brilliant work on media and technology has defamiliarized the image of "tech" and its easy "solutionism" with research on the labor, material infrastructures, environmental effects, and political undercurrents of our digital age.

And yet: What can we say in any *substantial* or positive sense about what can oppose the "new human" of our automatic age? What will ground a new organization or animate a new *decision* on the future? "Inclusion," for example, is not a political term—or maybe more accurately, it is *only* a political, that is, polemical, term. The challenge, obviously, is that the consensus among critical thinkers of the academy is that there *is* no "one true" human, or one way of organizing a society, a polity, or a global configuration. However, lurking in much contemporary critique is a kind of latent trust in an "automatic" harmony that will emerge once critique has ended—a version of Saint-Just's legitimation of terror in the French Revolution.

We are facing then a crisis of decision that must paradoxically be *decided*, but the ground of decision has been dismantled; every decision is just an expression of the system that produces it, whether that is a brain system, a computer network, or a Foucauldian disciplinary matrix. Is it even possible to imagine an actor-network system "deciding" anything? When we have undercut the privilege of the human, where is the point of beginning for a new command of technology, one that isn't just a vacuous affirmation of "multiplicity" or diversity against the Singularity? Or a defense of human "values" against technical determination?

I want to suggest that the current crisis demands a rethinking of the human in this context, the evolution of two philosophies that seek to dissolve the priority of decision itself. This cannot be a regressive move, to recuperate human freedom or institutions that cultivate that freedom. We must, I think, pay attention to the singular nature of *automaticity* as it now appears in the present era, across the two philosophies. The goal of this project has been to rethink automaticity, to recuperate what we can call *autonomy* from within the historical and philosophical and scientific establishment of the automatic age.[18] What I offer here is not a history of automaticity, or a history of AI, or a history of anything. There is no history of AI, although there are many histories that could be constructed to explain or track the current configuration of technologies that come under that umbrella. But this is also not a "history of the present," or a genealogy, that tries to defamiliarize the present moment to produce a critical examination of its supposed necessity through an analysis of its contingent development. There has been much good work in this area, but at the same time, the conceptual or methodological principle is hardly surprising. We (critical humanists) always know *in advance* that the historical unraveling will reveal, say, the importance of money, or political and institutional support, or exclusions in establishing what is always contingent.

Critique and Crisis in the Automatic Age

The historian Reinhart Koselleck published his postwar classic, *Critique and Crisis: Enlightenment and the Pathogenesis of Modern Society*, in 1959, as the Cold War emerged as a new epoch in world history.[19] Koselleck tied this new era to the foundation of a new technical apparatus. "History has overflowed the banks of tradition and inundated all boundaries," he claimed. "The technology of communications on the infinite surface of the globe has made all powers omnipresent, subjecting all to each and each to all."[20] With the appearance in our own day of the digital revolution, which is only accelerating in its expanse and reach into the operations and lives of the globe, this statement seems all too relevant—although we are not so sure, as one might have been in 1959, what these omnipresent "powers" really are today, although we do know that technologies from automated drones to algorithmic sentencing are surely vehicles of political and social control. Koselleck's goal in *Critique and Crisis* was to examine how the world had come to the point where two political blocs not just opposed one another, but actively *excluded* the legitimacy of the other, preparing the way for a potentially annihilistic war. The roots of the crisis were to be found in the kind of historical concepts inherited from the Enlightenment: "They are the philosophies of history that correspond to the crisis and in whose name we seek to anticipate the decision, to influence it, to steer it, or, catastrophically, to prevent it."[21]

As Koselleck would explain, here and then in more detail in his conceptual history of the term, "crisis" was already a question of *decision*. The crisis was the moment when history could not predict the outcome, and there was no automatic resolution. Crisis meant decision, for to recognize a crisis is to know that a decision must be made.[22] This is why preventing the decision in a critical time may well be catastrophic. Other forces will define the future or maybe even destroy it, at least for human beings. Koselleck's *Critique and Crisis* was indebted to the work of the German legal scholar Carl Schmitt, whose theorization of sovereignty as decision in a time of crisis was a warning to liberal democracies in the Weimar era, as well as a kind of perverse "reason of state" principle for the Third Reich, during which Schmitt continued to work and publish. After the war, however, Koselleck's invocation of the Schmittian decision in the context of new global technical infrastructures animating military machines of apocalyptic scale was hardly unproblematic.

Only a few years earlier, Schmitt himself had reflected on the new era, supplementing his 1950 book, *Nomos of the Earth*, with an analysis of technology. In a set of dialogues on power and the state, published in 1954, Schmitt gave an astonishing rebuttal, in a way, to the idea that the crisis of the moment (the threat of *unlimited* warfare) demanded a decision. The decision, he proposed, had been taken over, assimilated to the technical system that now uses the human being as instrument rather than the other way around. As Schmitt wrote, "The human arm that holds the atom bomb, the human brain that innervates the muscles of the human arm is, in the decisive moment less an appendage of the individual isolated human than a prosthesis, a part of the technical and social apparatus that produces the atom bomb and deploys it."[23] As Schmitt argued in this period, the question was no longer the decision on the crisis but a *crisis of decision*. The new challenge of technology was its unprecedented independence: "The one who manages to restrain the unencumbered technology, to bind it and to lead it into a concrete order has given more of an answer than the one who, by means of modern technology, seeks to land on the moon or on Mars."[24]

Today, technology, in particular, digital technology and the power of artificial intelligence, is raising the same kind of questions. No longer is it assumed (if it ever really was) that the process of technology would be "beneficial" to humanity, and the backlash has begun, with figures as prominent as Elon Musk, Bill Gates, and Stephen Hawking warning the world of the impending dangers of AI, for example. But how to meet the challenge? Musk, fulfilling Schmitt's prophecy, has suggested that starting afresh on Mars might be a good idea. The libertarian Peter Thiel for some time championed the idea of independent city-states flourishing offshore in a techno-anarchic paradise. (Thiel, by the way, actually knew Schmitt's work, through his interest in Leo Strauss, and famously invested in Facebook because he recognized at work in social media anthropological principles explicated by his teacher René Girard at Stanford.)[25] But what would it mean to take back technology, to make the *decision* for humanity and against the acceleration of automation and automatic governmentality, to borrow Michel Foucault's term?[26]

What I want to do here is prepare the way for facing this crisis of decision. The aim is not to resurrect old concepts of liberty to counter the scientific understanding of *automaticity*. Rather, we must confront the intimate (and tangled) historical and philosophical connection between autonomy and automaticity from within the very heart of a tradition that is understood to be the very source of the pathology that is *instrumental reason*. This is not, therefore, a history of a technology and research program ("artificial intelligence"), nor is it an intellectual history of the concept of mind and body in Western thought, though it intersects with these themes. In an important sense, this is not a history at all. What I am tracing is an *entanglement* in modern thought, one that begins in a specific historical moment (the emergence of a certain scientific worldview and method in the seventeenth century) and the opening up of the possibility of a total mechanistic understanding of nature, including organic nature and our own living bodies. At the same time, this modern scientific perspective allowed for, perhaps even *required*, the persistence of a divine order, and this proved to be the space for thinking anew what we can call the exception that is the human—part of nature, yet forever outside of the natural. At the heart of this entanglement was the machinic body and the nervous system as control mechanism, for no longer was it enough to connect physiology with sensory experience. Cognition itself would be reorganized around the living brain, and here the exception of the human could be attacked or at least *normalized* in terms of scientific methods of explication. This much we know from the history of psychology, a discipline that emerged in a new form in this period.[27] The other thread in this entanglement is *technology*. As we know from much work in the history of science on mechanistic models and metaphors in this period of the "clockwork universe," there was an unstable interplay between artificial machines and the order and organization of nature, of the cosmos itself. More recent work on the history of automata reveals the new importance of artificial robotic beings for thinking the body and for setting the stage for a total replication of human action—including cognition and rationality itself. Again, the seventeenth-century concept of the human is never stabilized due to the intricate entangling of ideas about the brain and nervous system, the theory and practice of human technologies, and the philosophical reflections on the capacities of the human mind. Descartes is our beginning point, since his work so clearly elucidates the topology of this modern entanglement, the shifting boundaries that link artifice, nature, automaticity, and human *autonomy*.

What follows is not a history per se but instead an attempt to track the multiple, evolving lines of thought that begin in this early modern moment, lines of thought that move from body to mind to nature to technology, thereby weaving new entanglements as certain ideas and concepts solidify and come to the fore. I also try to show how new ideas and experiences (e.g, industrialization, evolutionary theory) reconfigured the ways in which the automaticity of the body could be linked with technical systems, while at the same time the mind, as the inventive power *of technology*, could still create the space for autonomy and the possibility of an exception from nature itself.

To be clear, the trajectories I am tracing here are resolutely Euro-American and indeed, with rare exceptions, the domain of white male minds. Normally, critics of the contemporary computational, algorithmic regimes of surveillance and asymmetric legal justice can trace their origins to a central line of thought deep in the Western tradition, one that centers "reason" in the sovereign subject and aligns that reason with an essential technical organization of the world. If this tradition begins with figures such as Descartes, there is no doubt that the emergence of computational and cybernetic forms of rationality in the twentieth century accelerated this historical movement. My goal here is to delve into what Achille Mbembe, echoing Nick Land, calls the "Dark Enlightenment" of contemporary computational regimes,[28] to rediscover within this new *heart of darkness* that is Western rationality lines of thought that conceptualized a different form of reason, and different forms of epistemology, not through a rejection of technology but rather with an intense reflection on the essential technical dimension of human thought itself. While the figures participating in these lines of thought are no doubt among the most privileged in intellectual history, my argument is that this tradition harbors resources for an internal critique of what has been spawned by "modernity"—in all its worst guises.

We are witnessing today a moment when, to quote Mbembe, there is "the very distinct possibility that human beings will be transformed into animate things made up of coded digital data." And in an extraordinary comment, Mbembe warns of a loss of self-determination on a massive scale, going so far as to write, "This new fungibility, this solubility, institutionalized as a new norm of existence and expanded to the entire planet, is what I call the *Becoming Black of the world*."[29] For Mbembe, reason is no longer the (unequally shared) faculty that defined the human as such: "The computational reproduction of reason has made it such that reason is no longer, or is a bit more than, just the domain of human species. We now share it with various other agents. Reality itself is increasingly construed via statistics, metadata, modelling, mathematics."[30] My artificial history of reason is an attempt to recuperate models of human thought that preserve an exceptional space for human autonomy *despite* the very real infiltration of technical supplements into our own nervous systems. Cybernetics was not the end of thinking (pace Heidegger) but in fact the continuation of a complex history that fueled both the automatization of the social and political world and new concepts of human autonomy appropriate to this new condition. This history is of course one that marginalized certain groups and individuals and produced the kind of technologized world implicated in all the worst excesses of Euro-American hegemony. But as I try to demonstrate, there is at work here conceptions of the human that rely on notions such as plasticity, error, interruption, and so on, concepts that have been effective in the many critiques of Western thought, especially in the domain of media and technology.

The first part of the book tracks the ways in which some of the major thinkers of "mind" in the seventeenth and eighteenth centuries met with the challenge raised first by Descartes, namely, how the intellect relates to a complex organismic body armed with sophisticated sensory organs and an integrating brain and nervous system. With Hobbes and Spinoza, we will see how reason and cognition was, in part, the result of an artificial regimen of training, linked to a nervous body that was capable of formation and re-formation. For Spinoza, this insight offered a path to thinking anew the ways in which human thought were connected to materiality, at the site of the body but also at the very site of God. A crucial element here will be the figure of artifice. I then take up Leibniz, whose infamous doctrine of pre-established harmony will be reframed as *automatic harmony*. Again, I tease out different lines of thought to see how Leibniz deploys order and organization across different fields—the body, the mind, and the natural cosmos writ large—with an eye to how the creative capacity of the mind can help us understand how the human can maintain its exceptional character.

This sets up the analysis of Hume and Kant, in which that status is relentlessly dismantled in favor of an analysis of the human mind that emphasizes internal processes and laws of regulation. If Hume set out to dismantle early modern pretensions through a refiguring of the "animal spirits" and the emergence of reason in the midst of passionate activity, Kant thoroughly systematized the plurality of cognitive operations while speculating on the peculiarity of organismic causality. My goal in these chapters is to isolate the challenge of autonomy and creative intelligence as it appears in accounts of the mind that rigorously articulate the automaticity of cognition itself. Kant is a threshold to the modern neuroscientific worldview, one that resolutely embodies cognition and perception in the structures of automatic neural machinery.

The second part of the book ranges more widely, through new psychologies, thermodynamics, evolutionary thinking, and so on, to see how in the period of industrialization in Europe thinkers and scientists reframed the body as both within and outside the artificial regulations and organizations produced and demanded by new circuits of manufacturing and economic ordering. The lines of thought move through the new brain sciences to emerging experimental psychology and refigured philosophy, where we can see technology itself emerge as the marker of the human exception—in evolutionary time but also in terms of individual human development. The goal of this part of the book is to show that in the midst of automatic machinery, intellectual figures from a wide variety of domains understood the mind to be a space of possibility for *interrupting* automaticity, using newly available concepts and language that were to be found in disciplines such as thermodynamics, evolutionary theory, or neuroscience.

The third part explores, in what can only be a preliminary way, the rich territory of interwar thought, to see how radically new concepts of the integrative nervous system, alongside new philosophical approaches to mind and body spurred by the physiology and psychology of both humans and animals, drew on—while simultaneously influencing—a new and intense interest in the rise of automatic technologies, technologies controlled not by human operators but by complex new informational systems. This sets up part 4, where the very idea of "artificial intelligence" emerges with the development of the digital computer during World War II. Having followed the often-errant paths of several lines of thought, we will see in the early disciplines of cybernetics and AI a continuity with earlier concepts and problems, now filtered through the most radically automatic technology ever invented, namely, the digital computer. In this moment when brain, mind, and computer were first becoming fused in certain disciplinary frameworks, a host of other possibilities were in play, and the argument here emphasizes how the radical automaticity of the computer did not inevitably lead to the kind of reductive cognitivism dominant in contemporary sciences of the mind and body but in fact provoked significant and sophisticated rethinking of the nature of technology itself, and its relationship to the human mind. The final part of the book tracks this last line of thought—a series of concepts and frameworks that cross disciplines but are linked by the key philosophical issue of what I call *technogenesis*, to use the term employed by Bernard Stiegler. An analysis of an example of contemporary neurocognitive science, the theory of predictive processing, I argue, offers a critique from within cognitive science itself, as contemporary researchers and theorists struggle with the entanglement of ideas that must be understood across this longer historical time axis.

This project of providing an "artificial history" of human intelligence is of course a massive one. I offer here only a failed version of this more grandiose vision. The lines of thought I trace and the concepts I sift out are fragmentary, selective, and very limited, hampered by the constraint of time, the contingencies of research, my specific abilities and languages, my own idiosyncratic interests and psychic challenges. There are, of course, many other lines of thought that point in a dizzying plurality of different directions, and while I do think it is fair to say that the core concepts and issues inherent in my "artificial history" do emerge historically from within a particular (and let it be said, *peculiar*) European constellation, the network of intersection, opposition, and juxtaposition would only get richer as the threads from different contexts and zones of thinking are confronted in an extended time and space.

Still, with respect to our current crisis of decision, I hope at least to make the case that there is a critique of the automatic era that is possible from within the domain of technology and from within the domain of *the automatic* in particular. Autonomy can be rethought as the *foundation* of one's own norms: the artificial history of intelligence reveals a kaleidoscopic variety of examples of how the decision on norms is always dependent on a certain openness to automaticity, to the automatic regulation of human life on many different planes. What grounds critique in this space is the special character of human beings—poised between the *automaticity* of the organic and physical world and the *automaticity* of its own technical being. The human is not outside natural or artificial life. But thinking, I argue—by tracing intersecting and divergent lines of thought in neurology, philosophy, biology, technology, and psychology, from Descartes to deep learning—is not possible except in that *gap* between the two. There is no natural intelligence. All intelligence is artificial. And so, we might say, there is no *artificial intelligence*, at least as we usually think of it, because machines are not living and are (unlike us) *only artificial*. Hence this is not a history. It is a conceptual trajectory that aims to release from contemporary ideas historical traces of the question of intelligence as it emerges as the very mark of the artificial.

PART ONE

The Automatic Life of Reason in Early Modern Thought

2
Integration and Interruption
The Cartesian Thinking Machine

> I thought too, how the same man, with the same mind, if brought up from infancy among the French or Germans, develops otherwise than he would if he had always lived among the Chinese or cannibals.
>
> Descartes, *Discours sur le méthode*[1]

It is no surprise that a prominent cognitive scientist like Antonio Damasio would locate René Descartes's fundamental "error" in the philosopher's insistence on the "abyssal separation between mind and body."[2] For the program of cognitive science is arguably the total reduction of the mind to its neurobiological foundation, and this foundation, as Bernard Stiegler for one has pointed out, is essentially machinic in origin, given the intertwined histories of computing technology and artificial intelligence research, which gave rise to cognitive science itself as a discipline.[3]

Of course, we could just as easily celebrate Descartes as the first cognitive scientist.[4] As most scholars now recognize, Descartes was intensely interested in the physiological foundations of cognition and emotion, elaborating a complex theory of the nervous system and brain[5] while developing a sophisticated medical philosophy.[6] Descartes was the first intellectual to explore systematically the ramifications of the new mechanical philosophy for thinking about embodied human experience. As he wrote in a letter of 1632, "I am now dissecting the heads of various animals, in order to explain what imagination, memory, etc. consist in."[7] And yet Descartes is still chastised by so many (in so many disciplines) for holding onto some immaterial, spiritual "substance" as the ground of the "Cartesian subject."

I would like to zero in on the intersection of these two domains—pure intellect and the body as responsive automaton—to ask the deeper question of how to think historically and conceptually about the more fundamental relationship linking humanity with its technology, which is what I will be tracking across early modern thought and beyond. The history of "artificial intelligence" cannot be the genealogy of a technology, since the first early modern concepts of machine cognition were inextricably entwined with concepts of intelligence that veered uneasily between the artificial and the natural.

Descartes, I argue, was interested in mapping systematically the complexities of somatic machinery, not so much to "reduce" aspects of thinking to the actions of that body, but instead to reveal the ways our minds were constantly being shaped and organized by these automatic material processes even as they *resisted* total determination—as the interventions of what he called "pure intellect" attest.

We must begin, then, with a Descartes seldom encountered in philosophy or critical theory, the proto-cybernetic theorist of automata. Descartes in fact recognized the crucial importance of a form of *information* within the physiological mechanism that operated as a competing logic within the organization of the body. The threshold notion of information is what will connect the rigorous materialism of Descartes with his equally persistent spiritualism—the body and mind, in his system, although these terms fail to do justice to the way Descartes understood cognition and its organismic function.[8]

If we look closely at what I call here *Cartesian robotics*, we can glimpse a novel concept of the human emerging in the seventeenth century. For Descartes, the human body was a robotic, even cybernetic information machine that steered itself, yet it was also one that was capable of *interrupting itself*. This will be the key contribution. Indeed, Descartes's depiction of an intellect capable of interfering with the sensory machinery of the body can only be understood if we realize just how intimately bound the soul was to the organs and structures of a *living* body. With this supplemental capacity, the complex technical and informatic machinery of the human body became radically open in a new way and thereby became capable of the most radical transformations and unprecedented reorganizations. The Cartesian robot was, in essence, a plastic being.

Living Machines

In adopting the mechanical philosophy as a foundational starting point of his scientific investigations, Descartes banished any notion resembling the Aristotelian idea of "soul" to explain natural phenomena, and that included living beings, the natural forms that most resisted mechanical explanations.[9] His most notorious claim was perhaps his denial of any soul in the animal. Descartes was committed to a physiological theory that depended on purely mechanical explanation; there was, in the end, no way that he could explain what he knew to be the free and open nature of the mind. This is usually read as the beginning point of Descartes's problematic "dualism." More important to note here is that the dualistic approach was predicated on a prior, revolutionary *redescription* of both animal and human bodies as mechanically organized entities, yes, but automatic mechanisms that were also *self-governing*. This project of Cartesian robotics reveals (in a negative fashion) the key role that the soul will play in his effort to understand the exceptional nature of human identity as something distinct from, while still embodied in, the explicitly technological understanding of animal-human automata.

We can begin with Descartes's infamous claim that the animal was simply a machine—no experience, no feeling, no emotion. Descartes, like his early modern contemporaries, were very familiar with automata, and indeed, robotic machines had been a part of academic and even religious culture for some time.[10] In his *Discours sur la méthode* of 1637, Descartes imagined that if someone built a robotic monkey we would not be able to recognize a real creature when confronted with this mechanical version at the same time. And this was for a simple reason: the real creature was itself a robot according to Descartes, an "automaton," or self-moving machine. Defending this conjecture in a letter the following year, he presented a more elaborate take on this robotic imitation game.

> Suppose that a man had been brought up all his life in some place where he had never seen any animals except men; and suppose that he was very devoted to the study of mechanics, and had made, or helped to make, various automatons shaped like a man, a horse, a dog, a bird, and so on, which walked and ate, and breathed, and so far as possible imitated all the other actions of the animals they resembled, including the signs we use to express our passions, like crying when struck and running away when subjected to a loud noise.[11]

Descartes claims that if this mechanical genius was transported to our own world, he would instantly recognize our animals for what they really are: intricate automata that were just incomparably more accomplished than any of those he had previously made himself. He would be struck, that is, by the structural resemblance between the real dogs and horses and his own mechanical constructions. As Descartes explained in his physiological works, as well as numerous letters in the 1630s, since all animal behaviors could be perfectly explained in purely mechanical terms, there was absolutely no need to introduce the hypothesis of an animal soul: "Since art copies nature, and people can make certain automatons [*varia fabricare automata*] which move without thought, it seems reasonable that nature should even produce their own automatons, which are more splendid than artificial ones—namely all the animals." It was much more astonishing, Descartes claimed—and this is what we need to focus on—that the human body, which was in essence one of these "natural" works of art, turns out to have a soul.[12]

But what about these human automata? Would our imaginary roboticist be fooled into thinking our fellow citizens were merely machines when he arrived in our midst? "Suppose that sometimes he found it impossible to tell the difference between the real men and those which had only the shape of men." Perhaps initially fooled by his own walking, laughing, crying human robots he would have eventually

> learnt by experience that there are only two ways of telling them apart[,] ... first, that such automatons never answer in word or sign, except by chance, to questions put to them; and secondly, that though their movements are often more regular and certain than those of the wisest men, yet in many things which they would have to do to imitate us, they fail more disastrously than the greatest fools.[13]

In this critique of "expert systems" *avant la lettre*, Descartes implies that the automaton would inevitably confront a situation for which it was not programmed, so to speak, to handle. But, as he had already noted in the *Discours*, genuine humans arrange their words differently in response to inquiries, and crucially they can think their way out of challenging circumstances despite the lack of precedents. "It is unimaginable," he writes, "for a machine to have enough different organs to make it act in all the contingencies of life in the way in which our reason makes us act." Humans reveal themselves by their essential *flexibility*, their adaptability and their *creative* capacity: "Reason is a universal instrument which can be used in all kinds of situations."[14]

The Nervous System as Information Machine

... the substance of the brain being soft and pliant ...
Descartes, *Traité de l'homme*[15]

It is important to keep in mind that Descartes was never really interested in the traditional philosophical division between mind and body that we now associate with his name but rather a more ephemeral transition point between what might be called forms of "corporeal cognition" produced by the nervous system and brain of the body and the kind of *pure* intellection that could be performed only by the soul.[16]

To understand the importance of this liminal space, we must read Descartes's foray into conjectural human robotics, the *Traité de l'homme*, written around 1630 but never published in his lifetime. Descartes's conceit here is that he will, like his imaginary counterpart, construct—virtually—a human automaton, a machine made up only of physical matter (the conjectural method deployed for Descartes's theory of the formation of the universe). After building the automaton, he will then show that this robotic creature would be able to imitate its real human counterpart in almost every way, demonstrating that the bodies we possess must be essentially machines—albeit of divine origin. (Descartes's implicit argument will be that any action *not* explained by this virtual robotic simulation must be ascribed to the soul and not to our bodies.)

Descartes was not only dissecting animals regularly himself, but he was also clearly well versed in the medical and anatomical tradition.[17] He was of course not the first to offer a theorization of the nervous system (in fact, he borrows heavily here from Galen's standard, if by then outdated, work, not to mention the more recent anatomical investigations of Andreas Vesalius and especially Caspar Bauhin),[18] nor was he the first to speculate about how certain mental operations could be localized in specific parts of the brain.[19] However, Descartes took the terminology and concepts of earlier medical and psychological theories and reoccupied them, replacing their sometimes ephemeral notions of order and organization with precise, and purely mechanistic, explications.

One of the main aims of the *Traité* as an exercise in virtual robot construction was to discover the mechanisms of "self-movement" in the human body, the *control systems*, in other words, that make possible the continuing integration of the bodily organs and maintain the process of life. The key locus of explanation is the nervous system. (Figure 1.1.) Descartes will detail how "animal spirits," defined as the most rarified form of particulate

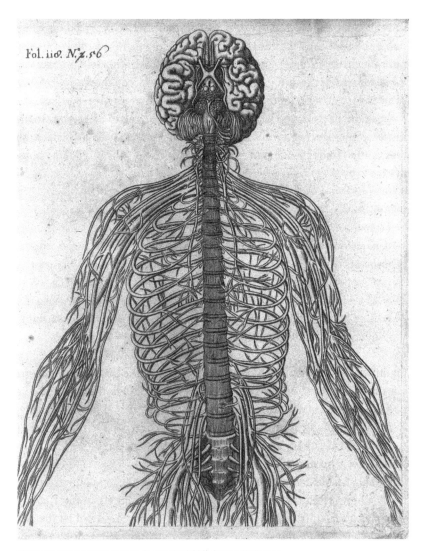

Figure 1.1. From René Descartes, *De Homine* (1662). Source: Wellcome Collection. CC BY 4.0 / https://creativecommons.org/licenses/by/4.0/legalcode.

matter, what Descartes calls a distillation, or "fine wind" (the term itself can be traced to Galen),[20] flowing through exceedingly small and narrow passages in the nervous system and brain, could explain a diversity of rather complex animal and human actions. In adopting the mechanistic stance here, Descartes does away completely with the Aristotelian concepts of the sensitive or vegetative soul—as that which gives form and unity or life itself to matter—thereby opening up both a new way to think about the organization of living bodies and, perhaps more importantly, paving the way for a radically new approach to the function of what used to be called the rational soul.[21] For Descartes, the rational intellect had to be linked to—but also radically distinct from—the wholly material organization and process taking place within the automaton.[22] The rational soul could not "direct" (like some sovereign figure) the activity of the automatic corporeal systems.

In a famous passage, Descartes likens the mechanism of the body to the intricate engineering animating the moving statues in the artificial grottoes at the famous royal gardens at Saint-Germain, which operated automatically by means of complicated waterworks.

> And truly one can well compare the nerves of the machine that I am describing to the tubes of the mechanisms of these fountains, its muscles and tendons to divers other engines and springs which serve to move these mechanisms, its animal spirits to the water which drives them, of which the heart is the source and brain's cavities the water main. Moreover, breathing and other such actions which are ordinary and natural to it, and which depend on the flow of the spirits, are like the movements of a clock or mill which the ordinary flow of water can render continuous.[23]

Significantly, these automata could even react to the presence of visitors via external sensory devices. For example, a visitor unwittingly steps on a particular special stone in order to better glimpse Diana at her bath, she retreats, and suddenly Neptune appears, wielding his trident.

For Descartes, these automata were essentially *cybernetic systems*, functioning not according to the rigid, serial logic of the clock but rather following from a flow of *information* within the system understood now as a totality. The automaticity of the reactive mechanism required an information processing system. That is, the "outside" world was converted by the system into an internal coding of sorts, which could then set in motion various kinds of bodily activity. The act of sensing can be best understood here as a *perturbation* of the system, a provocation that sparks a reorganization and then action in response to this flow of internal information.

Descartes clearly goes much further than his mechanical and hydraulic analogies would suggest. The body can, he imagined, perform a kind of thinking that greatly exceeded the relatively straightforward (if complex) mechanistic activity of these early modern waterworks. Because the sense organs are made of exceptionally pliant material, they are physically "imprinted" with the movements generated by the objects of external world—like wax imprinted with a seal. The animal spirits (unlike the flow of water in the earlier analogy, which was mostly a physical force) can in fact encode real information as they respond to, then transmit, the configurations or textures of the physical environment. While this is somewhat murky in Descartes's texts, external objects have what I would call a certain *structural topology* relating their parts, and different organs "sense" different organized topologies by virtue of the possibility that this network of *relationships* can be imprinted on the open, elastic surface of the sense organs.[24] The *pattern* of organization and not the physical collision per se is what is transmitted from the "screen" of the organ to the animal spirits flowing in the nerves that are attached to those organs.

These configurations—information patterns—therefore embody any number of "sensible" qualities: figure, position, size, distance, but also colors, odors, titillation, and other passions.[25] "There is a code of the senses, antecedent to that of the sensations of the soul united to the body," as Jean-Pierre Séris concisely puts it.[26] Somewhat cryptically, Descartes even depicted this coding of the nervous system in the *Traité* as a series of lines forming endlessly complex geometric figures. The important point is that the code does not need to "represent," if by that we mean *resemble*, the external object in order to transmit these qualities, which are after all dependent only on topological configurations of matter themselves.[27] Each sensory organ transmits (or better, transduces) a particular modality or topology, which is then integrated with other partial configurations in the common sense. Only later will these material codes be "experienced" by the soul as actual sensations. (And in animals, of course, there will never be such an "experience," only the machinic transduction into nerve patterns in the ongoing organization and reorganization of the brain.)

As Descartes describes, the coded information flowing from the various organs of sense is eventually inscribed on the "common sense," that venerable cognitive function that was now located with anatomical precision (and thereby newly *materialized*) in the infamous pineal gland, deep within the structure of the brain.[28] There, Descartes suggests, the information can be "read" (or better, "felt") by the intellectual soul as a single unified experience of the world. Because sensory information is transmitted as a wave through the animal spirits instantaneously through the nerves, Descartes shows how the state of the sensory organs would be immediately *doubled* within the pineal gland, which is the center of all the nerve channels leading into the brain.[29] (Figure 1.2.) Even at this first level of neural organization the information system has its own internal economy. The brain manipulates and *reorders* this information to effect certain physical activities. Reflex action—the body moving away from the fire, the hands positioned to protect against a sudden fall—is just the result of a movement of infor-

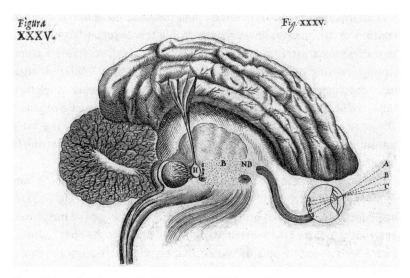

Figure 1.2. From René Descartes, *Traité de l'homme* (1664). Source: Wellcome Collection. CC BY 4.0 / https://creativecommons.org/licenses/by/4.0/legalcode.

mation through the nerves to the brain and the pineal gland and back down to the muscles, where the action finally occurs—"without any mental volition, just as it would be produced in a machine."[30]

The Memory System

Cognition can first be understood by Descartes as largely a physical manipulation of this virtual reality, organized and reorganized by the imagination—strictly redefined as a purely *corporeal* faculty. Memory introduces a new complexity. Indeed, memory truly distinguishes this information machine—a cybernetic body with a nervous system—from merely reactive automata such as the royal waterworks. With memory (first described by Descartes in the *Traité de l'homme* as actual patterns formed by physical "holes" in the brain, then in later works as structural "folds" in the brain)[31] the body becomes capable of ever more complex actions, because it is in effect responding not just to *present* stimuli but also to patterns of *past* experiences, at the very same time.[32] In an almost Pavlovian mode, Descartes (around the time he was writing the *Traité*) writes to Mersenne, "If you whipped a dog five or six times to the sound of a violin, it would begin to howl and run away as it heard that music again."[33]

This all leads to a rather startling admission by Descartes: if the automaton is outfitted with a memory system, he says in the *Traité de l'homme*, "without there being any soul in this machine, it can be naturally disposed to imitate all the movements that real men (or many other, similar machines) will make when the soul is present."[34] For Descartes the delayed effects of the memory structure, the *persistence* and circulation of information even after the original physical movements have dissipated, make possible a "natural" simulation of the interventional capacity of the soul.[35] Memory disrupts the flow of information in the system, allowing the robotic machine to act (at least in the view of an observer) in a seemingly flexible and adaptive manner.

What exactly is it, then, that distinguishes the work of the soul from the complex cognitive functioning of the temporalized nervous system? Or to put it another way, what possible advantage does the soul offer the body? One thing seems clear: the soul (from the robot's perspective, that is) must enable a *new* form of response to the environment, reactions that exceed the capacity of a corporeal cognitive system—what Descartes calls memory and imagination.

Nonautomatic Cognition

The rational soul must operate both inside and outside the automatic system: it will have "its principal seat in the brain, and reside there like the fountain-keeper who must be stationed at the tanks to which the fountains' pipes return if he wants to produce, or prevent, or change their movements in some way."[36] This raises the question of how that corporeal space of information can be radically disrupted (i.e., *interrupted*) from *within* its very own economy and logic.

This question haunts one of Descartes's earliest works on thinking, the unfinished *Rules for the Direction of the Intellect* (ca. 1628), where he rejects traditional rhetorical and logical methods of discovery, methods that were essentially discursive and often formally syllogistic, and moves toward a new cognitive model of understanding that privileged the immediacy of what he called, somewhat mysteriously, *intuition*.[37] For Descartes, intuition was self-grounded; that is, it produced knowledge that was not derived from any other source. "By 'intuition,'" he writes, "I do not mean the fluctuating testimony of the senses or the deceptive judgment of the imagination as it botches things together."[38] Intuition was that immediate grasp of connection between ideas. The common sense might synthesize or associate those separate experiences, but it will never remark the rational, indubitable *relationship* between these experiences.

The model of thinking in this early text locates intelligence in the grasping and reconfiguration of relations and proportions as they suddenly appear within the field of experience generated by the physiological systems of sensation, imagination, and memory. Descartes can say that the soul represents a spark of "the divine" within the corporeal system, because there is a structural analogy between divine creativity and the unprecedented intuition that disrupts the regular economy of the embodied information system. The divine soul marks an exception to that causality.

Intellectual Judgment and the Moment of Insight

From the perspective of Cartesian robotics, the *Meditations* is a demonstration of just how difficult it is to isolate pure thinking within a complex cognitive realm dominated by automatic and corporeal forms of mental organization and reorganization. "My habitual opinions keep coming back, and, despite my wishes, they capture my belief, which is as it were bound over to them as a result of long occupation and the law of custom."[39] We should note here that Descartes is referring to the power of the animal spirits. Philosophy is literally a counterbalancing force that keeps at bay the corporeal cognitive flow.

In famous passages from the Second Meditation, Descartes locates pure intellection in the act of judgment, and we now gain insight into the specific domain of *l'esprit*—the ability to see something that is not at all present in our senses. We might see, touch, even hear the innumerable changes in, for example, a piece of malleable wax; we can even imagine (in the corporeal sense of that word) changes that have not taken place yet—endless new shapes, for example. However, only the intellect can "see" wax itself. But here "wax" marks that *invisible form of identity* that persists throughout these perceptual changes.[40] The pliant wax—like the pliant brain or the mind itself—can take on many different forms, but only the intellect can perceive the underlying identity that is itself never accessible to sensory perception, for *this* form of organization (substantial unity in and of itself) can literally make *no impression* on the nervous system. The intellectual judgment is no doubt parasitic on the corporeal cognition generated by sensation and its processing in the common sense,[41] but it is not, Descartes demonstrates here, identical with it. The mind, like the wax, is a foundational unity that escapes direct perception but can be located between, so to speak, its variety of modes. "Therefore this insight [*comprehensio*] is not achieved by the faculty of the imagination." The perception of the identity is an "inspection of the mind alone [*solius mentis inspectio*]."[42]

But what does the intellect "see"? The judgment is more than an artificial synthesis of corporeal experiences. Descartes gives us another, brief (but revealing) example of this intellectual ability to judge beyond perception, one that nicely ties together the *Meditations* and the robotic writings. He writes:

> But then if I look out my window and see men crossing the square, as I just happen to have done, I normally say I see the men themselves, just as I say I see the wax. Yet do I see any more than hats and coats which could conceal automata? I *judge* that they are men. And so something which I thought I was seeing with my eyes is in fact grasped solely by the faculty of judgment which is in my mind.[43]

And yet, why should we judge them to be men? This judgment may well be in error (think *Blade Runner*), for in order to recognize (or better, *cognize*) a genuine human, we would need to perceive the signs of "pure intellect" itself, this strange capacity to see what is not really there.

The difference between the robot and the human is not attributable to any substantial content of the soul's being, something new that is "added" to the robotic organization. Rather, the soul *intervenes*—it is the cut into the system that opens up a new form of action. The intervention is an *interruption* in the radical sense of the word. The soul does not construct higher unity so much as it interferes with the automatic integrations of sensory information by making and remaking the order of relations that emerge in the common sense of the brain.

Unitary Systems

Descartes's last published work, the *Passions de l'âme* (1649), returned to the peculiar relationship between the active soul and the automatic robotic body. The key trope was *unity*. Descartes is concerned with both the self-enclosed automatic unity of the body and a much stranger unity that *twists* intellect and body together in the unstable figure of the human. The passions will be studied closely, since they constitute an essential connection between the sensing body and the purely intellectual functions of the soul. Descartes confronts a liminal space of connection in which some kind of unity brings together, while keeping absolutely separate, body and mind. (To condemn Cartesian "dualism," one must, of course, ignore this key question of the required ontological plane that can allow their unification.)

First, Descartes's theory of the organism constitutes an independent *existential* logic. The body's internal economy, its organization and motion, is not at all dependent on the soul for its unity: "Death never occurs through the absence of the soul, but only because one of the principal parts of the body decays."[44] *Life* and *death* mark the distinction between functioning and damaged physiological systems—the success and failure of *unity*.[45] After a brief overview of his theory of the gross anatomy and physiology of the body, Descartes turns to the nervous system, explaining it in some detail, repeating ideas developed earlier in the *Traité de l'homme* and other texts—but with a new inflection. He emphasizes here, first, the absolute integrity of the animal or human body mechanism. Every movement that is not willed "occurs in the same way as the movement of a watch is produced merely by the strength of its spring and the configuration of its wheels."[46] In other words, the soul is here absolutely excluded from the logic of the body's own operations. But the soul does inhabit this space in a special way: "The various perceptions or modes of knowledge present in us may be called its passions . . . for it is often not our soul which makes them such as they are, and the soul always receives them from the things that are represented by them."[47]

The most critical point, as we know, is that the soul can initiate "actions that terminate in our body"; that is, the soul *can will* particular actions. This special relationship is obviously important, but we cannot help but remark on its profound, almost perverse, strangeness. Elsewhere, Descartes claimed that if an "angelic intelligence" were to inhabit the body, to position itself, as it were, inside the pineal gland, "it would not sense the world as we do." Instead, the angel would simply observe "the motions which are caused by external objects" as they pass through the animal spirits; it would perceive, that is, only the raw "coding" of the information flowing through the nervous system. This angel spliced into a human automaton, therefore, "would differ from a real man."[48] Our human souls are so deeply intertwined with our bodies that we actually *feel* the sensory information as subjective experience. As he says repeatedly in the *Passions*, these relations (i.e., between soul and nervous system) are themselves "ordained by nature."

The *Passions* is from one angle a protracted attempt to understand the *function* of this intimately bound soul, to understand it, that is, from the perspective of the corporeal economy. The passions emerge here as a supplement to corporeal memory, which was already staged as a supplement to sensory information and therefore capable of disrupting the linear sequence of sensation, action and reaction. The passions, though, unlike memory, operate more as a warning system for the body.[49] Pain, joy, sadness, courage, and so on, all are ways the body presents challenges or opportunities to itself, based on its needs.

The soul is provoked by these signals, according to Descartes, in moments of crisis and can constitute a new space of *decision* for the living body. When passion "impels us to actions which require an immediate decision [*résolution*], the will must devote itself mainly to considering and following reasons which are opposed to those presented by the passions, even if they appear less strong." The soul can interrupt the automaticity of the fear response, as well as a strong desire for honor, both of which arise in one single situation. The soul can turn *against* these responses, intellectually, by *attending* to their opposites. Descartes's example is an "unexpected attack of the enemy" when there is "no time for deliberation."[50] However, there is still time for decision, or we can say, the judgment that is decision *takes no time*. The conceptual difficulty is locating the *plane of connection* between soul and brain here.

First, we can say the body is a "unity which is in a sense indivisible because of the arrangement of its organs, these being so related to one another that the removal of any one of them renders the whole body defective."[51] Second, the soul has its own form of indivisibility, its own foundational unity—the *cogito*.

Descartes suggests that the point of contact between soul and body is negotiated precisely at the intersection of these two unities. The soul is joined to the whole body, or, more precisely, to the body's being where it appears *as wholeness*. The concrete physiological space where that unity is best expressed is the brain's own center of organization, namely, the pineal gland. That gland, as we saw, was automatically inscribed by the informational systems of the body simultaneous to their occurrence. Not surprisingly, then, the soul "exercises its functions" in relation to the body "more particularly" in the pineal than anywhere else. I rely here on Nima Bassiri's innovative reading of this fundamental point. "The pineal gland," he claims, "is *organizationally* reduplicating the body's sensory affections."[52] The connection between mind and body as parallel unities is indeed less mysterious now than the connection between the body's organizational wholeness (whatever status that has metaphysically) and its various complex parts. In any case, we can now understand that for Descartes the soul is in a sense able to influence the pineal gland via its intimate connection to the body *as a whole*, a body that itself can influence specific parts, though the medium is still not explained here by Descartes.

As some scholars are now asserting, the *Passions* introduces a tentative but powerfully novel conception of the human as a "single system" consisting of physiological organization *and* the functions of intellectual intervention.⁵³ I would supplement these persuasive interpretations. Descartes is setting the stage for the idea that the soul (whatever its own aspirations) is the entity that takes *responsibility* for acting on behalf of the body's own integral unity in the absence of any such existential reaction in the robotic system. In moments of real crisis, the virtual integration of the body's diversity in the brain does not constitute the most effective weapon.

As Descartes wrote in the *Meditations*, the soul is not at all like the pilot of the ship, who controls the vessel as if it were an organ—that is, a mere external instrument of its own desire, an *organon*. As in cybernetics, the discipline that took its name from the Greek word for "steering," *kyber*, Descartes here insists on seeing the organism as a functional whole, an *integration* of intellectual and bodily systems, of informational and material operations. That is, the body, even a human body, is a complex unity that steers itself as a reactive and adaptive automaton.

In the *Passions*, Descartes redescribes the soul's function in relation to the existential, vital demands of the body. The passion of "anxiety," for example, is raised in the soul when something "very strange and terrifying" is perceived. While one can imagine an animal fleeing automatically from this danger, the point that Descartes emphasizes is that the soul can make a decision, one that is not at all predetermined by the nature of the bodily response. The soul is co-opted by the body to serve its own existential drive: "The principal effect of all the human passions is that they move and dispose the soul to want the things for which they prepare the body."⁵⁴ The passions are excited in us by objects in the world "because of the various ways in which they may harm or benefit us."⁵⁵ The soul is affected by the passions, in that it *now desires* what the body already "desires" and thus may persist in a course of action despite the changing physiological conditions of the body or the environment, or, perhaps, the presence of *conflicting* impulsions.

Wonder: Epistemology of the Automaton

The key moment in the text is when Descartes points out a certain special kind of passion that is not like any of the others—an exception to the very logic of desire and fear that shape the movements in the brain and produces the basic passions he has enumerated. The singular passion of "wonder" (*admiration*) is unusual precisely because it has no relation at all to the immediate well-being of the body. This passion affects only the *brain*. Yet a pliant, plastic Cartesian brain can hardly have its own object of desire, or fear. This is, I argue, why wonder is something quite unusual—the "sudden surprise of the soul which brings it to consider with attention the objects that seem to it unusual and extraordinary."[56] Wonder is really a radical *disruption* of the body's operations, not an expression of vitality. With wonder, the body is alerting the soul to the presence of something fundamentally *unknown*. Now, it is exactly *this* state of wonder (*admiration*) that is impossible in the robotic animal, for it can react to something only if it can be understood in relation to its immediate or automatically anticipated needs.[57] The shock of the new within an animal-machine cannot produce any *specific form of behavior*.

So how does the machinic body go beyond itself, its own automaticity, to "alert" the soul to something *unknown*? Wonder is, Descartes tells us, an "impression of the brain, which represents the object as something unusual and worthy of special consideration."[58] The challenge is to explain how a material system can express its own *absence of information*. As Descartes will suggest, the novelty of the impression is marked by the fact that certain parts of the pliant brain that have not normally been affected in the past have been suddenly forced into new configurations. At this moment, the animal spirits flow to this site of novelty *and* to the sense organs, so that they remain fixed on the new object; therefore the soul will be forced, in a sense, to acknowledge it.

Whatever the physiological explanation, the functional point is absolutely clear. Wonder is the way that a body interrupts itself in the face of a novel and unprecedented situation, in order to alert the soul to initiate a response that is no longer aiming to duplicate or amplify an anticipated machinic response but rather act in a way wholly foreign to the body—but still in the service of organismic life.

All other passions identify what is "good" or "evil," of benefit or not to the bodily machine. Yet wonder is a state of surprise, the purest form of interruption—good and evil are no longer in play. So wonder, unlike the other passions, is agnostic with respect to life because its object is defined as *unknown*, as absence of knowledge of fear or desire. Therefore wonder will always provoke a genuine decision while simultaneously preparing the ground for new knowledge. Or at least, if the soul rises to the challenge and is not absorbed completely by novelty, to the point of immobility, the pathological state Descartes calls "astonishment."[59]

To sum up: The soul intervenes on behalf of the life of the organism, by forcing the automaton to act against its own automaticity. However slight the impact of the soul might be in the face of the passions, this is the zone where the soul realigns the organization altogether, beyond the logic of its own automaticity. The human is, therefore, no simple addition to the living mechanized body, a reasoning being trapped in a material existence, an awkward marriage in both philosophical and practical terms. The human as a Cartesian robot is a hybrid, defined by its material organization, yet open to both cultural and technical formation—and often inspired to use that very capacity of openness to remake itself freely in moments of true decision. (This is why this hybrid entity must always be, for Descartes, an ethical being, but that is another question.)

A genuinely autonomous robotic system must therefore itself be somewhat "autonomous" of its own subsystems if it is going to display the flexibility and adaptability of the human being.[60] The human automaton is not just capable of reorganizing itself; it is also endowed with a capacity for radical *self-interruption*.

Cartesian robotics demands that we interrogate early modern thought and its theorization of cognition as an immersion in the complex intersection of rapidly evolving conceptualizations of minds, vital organic bodies, material technologies, and cosmic speculation. In the wake of Descartes, the cognitive capacities that resisted simplistic models of *l'homme-machine* will also spur parallel investigations of the body and technology. The crucial line was between *artificial* and *natural*, and this line was drawn in surprising and complex ways. The concept of automaticity was inseparable from ideas concerning openness, plasticity, creativity, and indetermination—whether in physical science, emerging forms of biological thought, theories of technology, or the philosophy of mind. And underlying these complex strands of thought was Descartes's open question, the ontological plane that would function as a *common space* for the appearance of both automatic machines and spiritual intellects.

3
Spiritual Automata

From Hobbes to Spinoza

Man alone is a concrete Spirit.
> Jean-Baptiste van Helmont

Organisms and Machines: Hobbesian Rationality

For those seeking early modern roots of artificial intelligence, Hobbes is often cited for both his mechanistic "reduction" of cognition to the physics and physiology of matter in motion and his infamous remark that "by reasoning, I understand computation [*per ratiocinationem autem intelligo computationem*]."[1] What usually goes unremarked is the original distinction of the natural and the artificial in Hobbes's conceptualization of the human. The natural world of living creatures implies the existence of forms of organization and systemic integration that work automatically to produce complex and varied action. What distinguishes the human is its capacity for artificial forms of behavior. Indeed, for Hobbes intelligence is not *naturally* computation; rather, intelligence is dependent on the acquisition and deployment of a new technology of organization that allows for rigorous processing of information. Hobbes looks to machines such as the abacus, or "paper machines" like accounting ledgers,[2] to emphasize that it is precisely the artificial system that allows thinking to do more than it ever could in its natural state.

Reason is "attained by industry." Philosophy is therefore the disciplining of the forms of natural thinking inherent in human beings; it is a form of what Hobbes will describe as *cognitive agriculture*. As Hobbes put it in *De Corpore*, "All men can reason to some degree, and concerning some things: but where there is need of a long series of Reasons, there most men wander out of the way, and fall into Error for want of Method, as it were for want of sowing and planting, that is, of improving their Reason."[3] This reason—with its artificially linear furrows and precise, regular distribution of "seeds"—is, we find in *Leviathan*, a method of manipulating names, especially universal names, the most precious seeds of knowledge. Remember that for Hobbes the name itself is an *artificial unity*: a multitude of experiences is made *one* through the prosthetic of a material sign.[4]

Unity is a critical issue for Hobbes and not just because of his political argument for singular sovereign power. Hobbes's model of the human is *organismic* and not so much mechanistic, as we are so often led to believe.[5] As Hobbes explained in his late text *Decameron Physiologicum* (1678), "It is very hard to believe, that to produce male and female, and all that belongs thereto, as also the several and curious organs of sense and memory, could be the work of anything that had not understanding"[6]—in other words, the organic body resists comprehension because its order and *reproduction of order* cannot be deduced from the mere existence of given matter put in motion. Bodies have an organismic order with *their own logic of operation*, and they are therefore the result of an invention, a "product" of intelligence. This is perhaps why Hobbes struggles to explain how our naturally organized bodies could ever extricate themselves from the automaticity of natural cognition. How can intelligence produce itself from nothing? What is the theory of generation?

Hobbes emphasized that a natural body ("animal") is a collection of mechanical parts—the heart a spring, the nerves so many strings, the joints so many wheels—these parts are *organic*, and what is at stake here is not the materiality of the body but instead, as Hobbes writes in the introduction to *Leviathan*, how one gives "motion to the whole body, such as was intended by the artificer."[7] To imitate nature with automata, he implies, some insight into how natural bodies are *animated* is needed—which is also another way of saying we need to understand how natural bodies produce and maintain their unity.

The figure who coined the term "mechanical philosophy" in fact agreed with Hobbes on this score, underlining the importance of the question of organismic order in this period. Robert Boyle would admit that it was impossible to imagine, at the origin of the universe, that the "Great Mass of matter" having been "barely put into motion, and then left to itself" could ever produce the order of nature, in particular organic nature—"such curious Fabricks as the Bodies of men and perfect animals, and such yet more admirably Contriv'd parcels of matter, as the seeds of living Creatures." Boyle thought it necessary to add another principle, "an Architectonick

Principle" that would through "skilfull guidance of the motions of the small parts of matter" transform the confused Chaos into an orderly world.[8]

Without exactly denying this implication, Hobbes leverages the very ambivalence of the distinction between the automatic machine and the natural body in order to highlight the artificial character of political sovereignty, the real subject of *Leviathan*. The commonwealth is described as an "artificial man," created to protect and defend individuals. But *unlike* the automatic machine, the automaton that is the commonwealth has to *imitate* the organism by having that "life and motion" animating a "whole body." The trick, which is at the center of Hobbes's argument but easy to forget, is that this particular automaton will have, as he puts it, an "artificial soul"—the sovereign figure who will produce and maintain the unity of the commonwealth.

The obvious but often ignored truth that lies behind this claim is that an organic body has such a soul naturally, though this soul—we can really call this organization—has nothing to do with the Aristotelian version, according to Hobbes. In any case, successful imitation of the function of a natural soul entails a clear understanding of the organic version that comes before it. To be clear, nowhere does Hobbes argue that animal (or human) bodies are mere machines—though to be sure he often seems to be saying exactly that when he wants to highlight the materiality of human experience and action; in fact, he recognized that the organization of bodies requires a new form of explanation.

This distinction between the natural body and the artificial machine is in fact essential to Hobbes's body-politic analogy, given the importance in his work of the sovereign figure as the *substitute* for the automatic ordering principles that inhere in organisms but are absent in machines. In chapter 42 of *Leviathan*, for example, Hobbes critiques the argument that in both the "natural body" and the automaton-commonwealth, individual members "depend one of another." For Hobbes, it is important to recognize that *any* coherence in the state is dependent on the sovereign, the political "soul," such that in the absence of sovereignty (in a civil war, say) the lack of "common dependence" means the dissolution of coherence—and Hobbes pointedly notes that this is just like when the parts of a dying natural body "dissolve into earth for want of a soul to hold them together."[9]

Human beings are not naturally in agreement at all, hence any agreement is "by covenant only, which is artificial," and it is precisely this artificial construction that cannot sustain itself, be "constant and lasting," without intervention and coercion directing the parts making it up (120). On the other hand, the unitary "system" of systems (155) that is the commonwealth can learn to *become* "organic" to the point where it is even capable, Hobbes says, of artificial procreation—that is, the formation of colonial "clones," so to speak, that, like a child who has matured and become independent, can be emancipated from its parent metropolis (175–76).

The analogy between living unitary organisms and artificial commonwealths depends on an understanding of how biological tissues and organs—the subsystems—are integrated into an overall system that provides the unity and direction necessary for maintaining *life*. In *De Corpore* Hobbes draws on the mechanistic physics of his time, stating that "all Resistance is Endeavour opposite to another Endeavour, that is to say, Reaction."[10] To understand an organism, Hobbes realized, means to understand how organized arrangements of matter could produce specific kinds of "reactions" to events in the world. The living body is not made of particles but instead organs, and it is the "whole organ" that reacts to motions outside of it. The reaction of the organ is therefore conditioned wholly by its organization, "by reason of its own internal natural motion," as Hobbes puts it.[11]

The sense organ, in particular, conducts its reaction via the nerves to the brain, where the endeavor of the organ appears as "Phantasme or Idea."[12] But given the "endeavor" (like Descartes's "coded figures") is now separated from the organ, the endeavor "appears" to the mind as something external—outside the body, that is. Hobbes, like Descartes, understands that the motions of objects hitting the organs of the body produce inward reactions that are organized and directed due to the internal configuration of the organs themselves.

Furthermore, as Hobbes points out, if sense is nothing more than reaction, then the body must have some way of retaining within itself that reaction as motion. Again, this will depend on the precise configuration of the organs involved. He can imagine inanimate reaction producing what he calls a "Phantasme," but it would cease to exist as soon as the object was removed. As he writes, in *Leviathan*, "For unless those bodies had organs, as living creatures have, fit for the retaining of such motion as is made in them, their sense would be such as that they should never remember the same."[13]

Hobbes therefore constructs his physiologically based psychology on "organics" and not mechanics. He lays out, in *Leviathan*, a distinction between what we would call the "autonomic" operations of the body's internal regulatory functions and the "voluntary" motions that direct movements in an environment. The vital motions begin "in generation" and continue throughout life—circulation of blood, breathing, nutrition, and so on. What exceeds the automaticity of the motion of organs and reaction is the *retain-*

ing of sense in the organs of cognition. The important question: How exactly does the *unnatural* organization and reorganization of thought—the very technologies of language and reason—appear within a natural body?

The "Umwelt" of the Bloodworm: Spinoza on Reason

Each *Umwelt* [subjective environment] forms a closed unit in itself, which is governed, in all its parts, by the meaning it has for the subject. According to its meaning for the animal, the stage on which it plays its life-roles [*Lebensbühne*] embraces a wider or narrower space. This space is built up by the animal's sense organs, upon whose powers of resolution will depend the size and number of its localities.
          Jakob von Uexküll, *Theory of Meaning* (1940)[14]

Despite radical philosophical differences, Descartes, Hobbes, and other early modern figures were, we can say, positioning the mind as something simultaneously both inside and outside of the organismic body—a claim that opened up corporeal psychologies but also the possibility of a connection between cognition and the intelligent order of nature itself. This would of course lead to the some of the most troubling philosophical puzzles of this period. Spinoza's thought, I propose, marks a new approach altogether. He will reject Hobbes's attempt to ground the rational mind in the sensing body while refusing to displace the mind into some higher plane of existence. In a radical turn, Spinoza will confront for the first time the philosophical question of how to understand how our corporeal experience of the world *and* the ordered world that we experience are both parts of a system that exceeds and incorporates, so to speak, these two different domains simultaneously.

However, if Nature ("God") is *one*, how is it possible to claim that the human mind—as a fragment, however privileged, of that unity—could have true knowledge of this universal unity, given that it comprehends at one and the same time our cognitive experiences and what is emphatically *not* experience, namely, the external "what" of extended existence that the mind perceives in fragments?

To answer this question, we will begin not with God but instead a parasite—a microscopic being, namely a bloodworm,[15] that Spinoza, a lens maker by trade, speculates about in a letter to Oldenburg from 1665. Imagine a little being that lives in our veins, Spinoza proposes, a worm that is capable of distinguishing by sight the different particles of blood (lymph, chyle, etc.) that surround it. This "little worm," Spinoza explains, could "observe by reason how each particle, when it encounters another, either bounces back, or communicates a part of its motion" and therefore could navigate its way through the bloodstream.[16] However, what escapes the worm's experience is the the systematic quality of the blood understood as a *unity* with a "universal nature." These individual particles "adapt themselves to one another, in relation to their size and shape, that they completely agree with one another and they all constitute one fluid together."[17] What Spinoza is saying it that the worm recognizes the particles as "wholes" that interact with other similar entities according to regular laws but does not see that the blood is itself a whole, actively governing the behavior of its constituent parts. The point of this exercise is to remind us that we humans are in exactly the same position as the bloodworm. All bodies in nature, Spinoza explains, are like these particles of blood, determined in a specific and certain manner by other entities, their motions governed by a higher system of unity that comprehends them, and these systems—all systems—are ultimately regulated by the total unity that is the entire system of the universe.

Spinoza is here seizing on a lacuna in early modern thought. If all bodies—especially organic ones—are subject to the formal order that structures them, where can we draw the boundaries between unified systems and those that are what Hobbes called "subordinate systems," or organs? When we (or the bloodworm) perceive some body as a "whole," we understand that all the parts "cohere," in that they adapt themselves to each other and interfere with each other as little as possible. The threshold between part and whole is marked by the idea in our mind of agreement and disagreement only—that is, we grasp a whole by distinguishing it from other ideas that do not agree at all with it. However, that particular whole may well adapt itself to another whole and be understood as a "part" of some larger system. Spinoza's conclusion is that the human mind, or the bloodworm's, could never grasp the coherence of nature as a whole, since we only ever have access to very particular perspectives and cannot track how every demarcated unitary order related to every other in the entire cosmic order. The bloodworm would have to leave its *Umwelt* in order to understand how the blood as a system regulates the parts. Similarly, the human mind would have to get outside of its own specific experience—and ultimately outside of nature itself—in order to grasp cosmic order. Given that the human body and the human mind are themselves undeniably *parts* of nature, that escape is absolutely impossible. The part can never transcend itself to grasp the totality of a system of which it is part.[18] At best, we can experience disruptions of our reasoning that are based on the properties that we have access to—an action whose cause is the larger system that encompasses the parts we perceive and reason through.

Still, we must understand that the mind and the body are not the same kind of thing. The body is a part of Nature, a part or expression of the material and physical attribute of the universe, which is to say, it is a specific formation in the total organization of all substance. At the same time, the mind is a part of the *intellectual* Nature that represents the "whole" of Nature as extended and formed substance. The individual mind represents in ideas its own particular and limited natural environment, the one that the finite individual body reveals to that mind through the senses. This is a deflationary account. Even as Spinoza admits, with Descartes and Cudworth (and Hobbes at the edges of his argument), that the mind is capable of grasping relations and formal order in sensible experience, he follows the consequences of that argument to its radical endpoint, which is that the finite mind will never truly understand Nature at all. What we take to be "natural laws" are only tentative insights into the local relations that we observe as regulating certain bodies—in other words, completely contingent. The limitation of this form of inquiry (which Spinoza calls "reasoning") is that any observation of regularity (i.e., perceived relations of cause and effect) relies on the preliminary identification of the body itself, and yet whether or not a body in fact "coheres" with other bodies will always escape our observation, or at least, we have to recognize that it is impossible to observe all the ways in which bodies react to one another across all of time and space. Reason in this context is, for Spinoza, therefore just a local logical prediction tool and not an easy path to knowledge—because knowledge, in this early modern context, must be founded on *certainty*, and reason can draw only on our limited experience.[19] In other words, no matter how adequate the universal notions we gain by reason, the tracking of causation is still fundamentally limited, despite our adequate knowledge of, say, how bodies act and react to each other.

Spinoza's Organismic Rationality

I do not presume that I have discovered the best Philosophy; but I know that I understand the true one. Moreover, if you ask how I know this, I will reply: in the same way you know that the three Angles of a Triangle are equal to two right angles. No one will deny that this is enough, not if his brain is healthy and he is not dreaming of unclean spirits, who inspire in us false ideas which are like the true. For the true is the indicator both of itself and of the false.
Spinoza, letter to Albert Burgh (1675 or 1676)[20]

The path to knowledge must therefore bypass, rather than *transcend* in some way, the contingency of experience that limits reason's grasp. As

Spinoza put it in a letter to Simon de Vries, experience can only reveal to us things that are *not* necessary, such as modes, which cannot therefore be derived from the essential definition of a thing. That definition of an essence cannot be grasped through experience since the existence of an essence is *exactly the same* as the essence. (The mode, in other words, is not what it "is" completely, given that its definition relies on its status as a *modification* of something more essential.) However, we clearly do understand what an essence is—therefore, for Spinoza, the mind must be able to grasp the "essence" of a thing in an intellectual way, going beyond the identification of disciplined universal concepts, the work of reason.[21]

To track Spinoza's argument we need to pay attention to both the way he frames the body, as an exemplar of this object of mental experience, and then the way he positions our purely intellectual capacity in that context. This is not a question of the "mind-body" relation, because the real end of philosophy for Spinoza is an understanding of exactly what must comprehend them both. The *relation* between ideas and materiality will reveal his ontology that positions them as *modifications of one essential reality*.

We can begin with Spinoza's postulates concerning the human body as spelled out in the *Ethics*. This body is made up of many individual parts, each "extremely complex," and all of them distinct—some are hard, some soft, and some fluid. This body is, of course, affected by external bodies, but it also "stands in need for its preservation" of these external bodies, as a source of replenishment of matter, because the human body is constantly regenerating itself from within—a process that we can call, loosely, metabolism.[22] Finally, Spinoza makes two points. First, the "fluid" parts of the body, when impinged by foreign bodies, can actually alter the surface of the "soft" parts that fluids flow through, so that these plastic spaces are left with an "impression" of the external body—a kind of inscription in a new medium. Second, the human body can intervene, to move external bodies and "arrange them in a variety of ways" (2, Postulates 1-6, p. 128).[23]

Now to unpack the implications of these postulates. One key argument is that any compound body that can lose particular bodies and replace them with those of the same nature must still preserve its "nature" as a *unitary body*: there is no change in its "actuality" (*forma*). The body is defined by its union and not the substantial materiality that is brought together in that union, since that substance can be continually changed. As Hans Jonas first noted in a 1965 essay, Spinoza's concept of the organism locates identity in the spatial and dynamic pattern of this composite body, such that the whole is compatible with a change in its parts, which is essential to the life of any being. For Jonas, there is no machinic analogue to this kind of organized body, which is continually constituting itself through regeneration of its parts.[24] The body is a local zone of *auto-regulation*.

But as Spinoza insists, the operations of the body can be understood in terms of the strict causal relations that govern all of its material activity. In the *Ethics*, Spinoza mocks those that celebrate any discovery of "harmony" in the natural world—whether religious enthusiasts or more sober philosophers carefully observing the heavenly bodies. His response to this reaction could have come from Hobbes: "Each one has judged things according to the disposition of his brain." And Spinoza proceeds to quote an aphorism: "There are as many differences of brains as of palates" (1, Appendix, p. 114). What will seem ordered and harmonious to one brain will seem disordered and unpleasant to another. More to the point, Spinoza notes how so many people are "struck by a foolish wonder" when seeing the intricate structure of the human body; ignorant of how its organs function and interrelate, they leap to the conclusion that some kind of "supernatural art" has arranged the parts so they act harmoniously, through extramechanical means (1, Appendix, p. 113). The body is a machine, an automaton, capable of highly complex behavior, as the example of the somnambulist reveals (3p2s). But it is also a body that is capable of maintaining itself through a transformation of *external* bodies to replace any loss or damage to the *internal* bodies making up the tissues and organs of the physiological system. However, the real question does not concern the metaphysical status of the perceived "order" of the system. For Spinoza, the challenge is to explain the existence of the mind. What purpose does it serve in the autopoietic (self-organizing) economy of living bodies?

Given the undeniable reality of mental experience, Spinoza argues that the basic function of thought is to give representation of the body in the sphere of ideas. Of course, if that representation was merely passive, it would have absolutely no function. Which is not to say that the mind can be assumed to have some special "power" over the body. Just the opposite. Spinoza is concerned to locate the function of mental representation precisely on the seemingly abyssal threshold connecting but also dividing the body and the mind. Or to put it another way, the question is not to ask how the mind and the body are "connected" but rather to probe how the human being—which is at one and the same time both a material body governed by causal laws *and* a series of mental representations—operates as a single system. The mental experience is the direct analogue of the physiological, in that both material and mental activity constitute the modes of one being.

With this in mind, we can see how Spinoza will explain the valuable role of a mind for this being—by focusing on the crucial role of what we might call passionate affect.

Affective Interference

The advantage of having mental representations of the body's own causally determined responses to its internal and external environments hinges on the *separation* of the representation, which allows for the emergence of a psychic realm where ideas interact independent of their origin. This is not unlike Hobbes's argument in *Leviathan*. However, unlike Hobbes, Spinoza does not keep separate the passions of the body and the logic of sensory information. Like Descartes in his *Passions of the Soul*, Spinoza will argue that the intersection of psychic and organismic realms will have a specific effect on the body's capacity for survival and success. Take Spinoza's definition of "joy" and "sadness," for example. Joy is somewhat cryptically defined as the *idea* of something that increases the body's power of acting or increases the mind's power of thinking; sadness refers to the exact opposite (3p11s). The mind can have an idea of an external body only indirectly, so to speak, that is, through the impression made on our body's affection by that foreign entity—which for Spinoza means through the mind's representation of the body's own causally determined material transformation.

The next part of the argument is admittedly not easy to grasp: Spinoza is claiming that when the mind imagines (i.e., causes the representation to appear again in the mind) that external body as "present," our physiological systems will again be affected in the same way, as if that body was actually present again (3p12). The mental experience is not merely a repetition of the original experience. The being that is subject to *both* material and mental "expression" (to use Deleuze's term) mediates the two domains.[25] Mental representations do not "act" on the body. However, the body reacts, as a material expression of the unitary being, to *its* (i.e., the system's) transformation expressed in the mental representation. The logic of interaction is precisely *not* one of causality: the causal logics are distinct within the separate domains of the material dimension and the intellectual sphere.

The implication Spinoza draws out is this: the mind will strive to imagine those things that increase the capacity of the body to act, since that condition will also allow the mind a greater power to think. And vice versa: the mind will strive to imagine situations that do not involve those bodies that decrease the body's ability to act, since the mind would lose its capacity at the same time. But what does this mean exactly? One way to approach Spinoza's theory here is to focus on the organismic unity of the body alongside the unity that connects the material with the psychic. Only the mind can learn from the body's history of affection and feed back that knowledge in order to alter the automatic trajectory of the body, and it does this not by intervening or suspending mechanical causality but instead by clarifying its ideational representation of the physical states in the body.

The mind learns through the use of the sensible imagination (where ideas can be organized and reorganized) and through memory, defined in the *Treatise on the Emendation of the Intellect* as the "sensation of impressions on the brain" along with a sense of "duration." Spinoza is arguing that the function of the mind is to track, through its own system of ideas, the causal order of the material environment,to penetrate, that is, the "laws" that govern the interaction of the body and foreign entities, as well as the laws governing interaction between external bodies.[26]

How does the mind accomplish this? For Spinoza, the method of *reason* is the tool that the mind uses to comprehend the cause-and-effect relations that hold sway in the material domain. Reason is not at all a mere function of the brain, some kind of cognitive machinery. The value of the mind is its liberation from the logic of the body's own sensibility, the trace of the logic of extension that haunts the world of ideas. The task of the understanding is first to articulate the distinction between the fictitious array of representations that emerge from the accidents of origin in the sensory and nervous systems. The understanding must identify what Spinoza calls "true ideas," those that clearly show "how and why something is or has been made; and that its subjective effects in the soul correspond to the actual reality of its object." With true ideas in view, the mind will naturally "perceive" (rationally, that is) the causal relations that determine the behavior of the external world and the ways in which the body will respond to certain kinds of interactions with the world.

But if the soul will naturally (by virtue of its nature) reason and comprehend, for it to be successful it needs to be furnished with the proper kinds of ideas—true ideas, not those generated by the material logic of the brain, appearing in the mind as "imagination." And here Spinoza admits that a certain kind of tool—what he calls "method"—is important for the proper functioning of the mind. Method is the construction of a standard, the "laying down certain rules and aids," so that the mind can be saved "useless mental exertion." Method is an artificial *training of the rational mind*, through the adoption of rules that will generate the occasions for proper reasoning and comprehension. Human beings are not rational; they be-

come rational.²⁷ For Spinoza, the analogy is literally the origin and development of *technology*—the intellect uses "intellectual tools" to bootstrap its progress in the search for knowledge.²⁸

Method will literally make the mind into a *machine*: once so trained, the soul acts "according to fixed laws, like a spiritual automaton [*automa spirituale*]."²⁹ The mind is not *naturally* an automaton ("like a . . ."); neither is the mind *analogous* to a material automaton, if that means an ordering of parts that interact according to rigorous *physical* laws of cause and effect. The spiritual automaton is one that is radically self-constructed, and it operates therefore according to an independent psychic logic, namely that of *reason*.³⁰

The struggle that results from the creation of the spiritual automaton hinges on the coexistence, now, of *two* logics within the realm of ideas: first, the corporeal and contingent experiences of the senses and the imagination that arrive together in a somewhat ad hoc manner and therefore do not clearly reveal necessary relations of things in reality; and alongside that, second, the *logic of reason*, the development of a system of understanding that discerns the necessary laws of nature from the limited and often confused conjunctions of ideas that represent our corporeal affections. This struggle is perpetual. The persistence of "inadequate ideas" challenges the spiritual automaton to locate and resolve any new incoherence within thought. The artificial life of reason—an orderly and internally consistent life based on adequate, that is, clear, ideas—is much like the life of the *material* automaton that is the organism; however, the spiritual automaton is never fully in sync with its counterpart precisely because there is not, and there *cannot* be, any causal relationship between them. Moreover, there is no one-to-one parallel of ideas to extensive body: for Spinoza, the mind is constituted by a limited, fragmented number of ideas of bodily modification. The value of reason for the body is a heightened awareness of the necessary order of nature and hence a capacity to act on the basis of prediction, or better, anticipation (which is, according to Spinoza, resolutely more than the mere deployment of memory; Hobbes also argued this in his discussion of foresight).

The engineering of the spiritual automaton is a process of associating the logic of reason (deduction and the tracking of cause and effect) with the memory system of the brain, so that when nervous associations are made through the causality of the sensibility, the impressions provoked are now *organized* in a new way in the brain. Again, the mind does not act on the body or the brain, but in thinking rationally, the mind affects the *being* that is at one and the same time mental and physical. This is how the material systems express, in their own way, the affections of the mind. (And we can note here that this conceptualization of the relation between reason and the body's physiological systems can help explain Spinoza's effort to implement a kind of psychosocial engineering in the *Tractatus*. One will behave rationally, Spinoza explained, even if one does not possess a properly functioning "spiritual automaton," as long as one's brain is affected by the right kind of ideas [e.g., stories, morals, aphorisms] that introduce surreptitiously the *operation* of the logic of reason into the very matter of the brain.)[31]

Beyond Reason: Intuition and Unity

We can now return to the issue Spinoza raised in his early letter to Oldenburg concerning the conjectural bloodworm. Given the theory of the spiritual automaton, Spinoza's machine of reason that extricates an understanding of cause and effect from the chaos that is sensibility, how can we even know that we are truly understanding the nature of things when we cannot clearly identify the genuine boundaries of organized systems? Recall that the bloodworm has a rudimentary rationality that allowed it to map the features of its material environment, to know, that is, how things behave in a regular manner so it may predict the movements relative to its survival—this is not mere induction. Human reason is profoundly more sophisticated and with the proper technology of method can not only understand causal relations and predict the future but also understand *why* things behave the way they do—at least as material entities. However, what reason *cannot* do is chart what I would call the topology of parts and wholes; the identification of what we call "bodies" is deficient because it is dependent on a vastly limited perspective. If all parts obey the mechanical laws of nature, and can thus be tracked and mapped by reason without any error, the identification of a genuine *essence* is another thing altogether. The mere existence of interrelated bodies does not prove any divine "harmony" after all. Spinoza is denying, then, that reason could ever be the path that will lead the mind outside of its own contingency toward an understanding of *unity* in nature, let alone the unity *of* nature itself.

Spinoza says in the *Ethics*: "The order and connection of ideas is the same as the order and connection of things" (2p7). If we seek the essence of something in ideas *only*, we will get at best the "idea" of the *universal*, in the sense Hobbes gave the term. For Spinoza, these universal notions (man, horse, dog) arise from the multitude of images that are formed in the human body, which blur together to erase the slight differences of singular images; however, this is mainly because the imagination has a limited capacity and cannot keep in mind "a determinate amount of singulars." The universal is just a forgetting of difference, the consolidation of a set of common affections of the body. As Hobbes emphasized in his own argument, these universal notions are by definition derived from the affection of singular experiences of particular bodies. As Spinoza writes, universals "are not formed by all in the same way but vary one to another" (2p40s1). For Spinoza, this is why "it is not surprising that so many controversies have arisen among the philosophers, who have wished to explain natural things by mere images of things" (2p40s1).

What Spinoza does in his philosophy is turn completely away from the effort to discover the essence of things through abstraction (finding the "common" properties) and by interrogating with reason the particular experiences of things—the effort to wrest the universal from the singular.[32] That process, however systematic and governed by a flawlessly automatic spiritual machinery of reason, will always be contaminated by the contingency of the particular and its essential lack—the lack of any presence or trace of the comprehensive unity of an essence, or further, a trace of the essential unity of nature itself. Spinoza's alternative is, in fact, to return to the *singular* but now in order to follow a wholly different path.

Spinoza will introduce a radically novel capacity of the understanding, one that is not simply an extension of reason (the "adequate ideas of the properties of things"). This new capacity he calls *intuition*:[33] "And this kind of knowing proceeds from an adequate idea of the formal essence of certain attributes of God to the adequate knowledge of the [formal] essence of things" (2p40s2). The adequate idea of such a formal essence cannot, of course, be derived from the ideas that represent the sensible immediacy of the body. However, it seemed to be the case that any insight into the formal essences of things would require a comprehensive knowledge of the universe to even begin cognizing in the first place. How can we reconcile this with Spinoza's claim to intuitive knowledge?

If we recall Spinoza's disdain for those who pretend to see in nature the "harmony" of parts that come together in, say, living bodies or cosmic systems, it is clear that Spinoza is not arguing that we can perceive directly the necessity of any *organization* that exceeds the mechanical laws of cause and effect. As he noted, the living body is just a machine that can be understood as a matrix of causes and effects. However, as Spinoza went on to claim in the *Ethics*, there is something about the living body that we *can* perceive and that is not at all in conflict with our understanding of the causal machinery that is physiology. This insight is into the very *union* of the system itself.

Spinoza denies we need to know the genuine organization of the body in all its physiological complexity in order to comprehend the essential unity that is the body. The mere fact that parts of the living body can be replaced by other parts without a change in the functioning of the organism *demonstrates* the very existence of a nature (something that "is what it is" despite its substantial or modal transformation—think of Cartesian wax) even if the essential "topological" order of that nature is still unavailable to human intellectual perception. The only spiritual lever needed here is the adequate idea—intuitively recognized—that there *exists* some such essence, which is to say, there is some form of *bounded unity* that maintains the body's integrity despite the metabolic construction and destruction that takes place throughout a life history.

The insight into the existence of a bounded unity, a singular unity, that is, opens up the mind then to an even more radical intuition: the unity of the body (which exists certainly even if the precise specification of its individual organization escapes us) is immaterial and constitutes an individualization of an attribute of the divine—the comprehensive unity that encompasses all substance and all modification or representation of substance. There is no "reason" a living body can replace its materiality. But the *fact* that it does so discloses the existence of some essential nature that remains continuous in this transformation. It is also the case that the human mind, which operates according to a logic *entirely distinct* from the causality of the body, cannot avoid being conscious of the necessary unity that mediates the body as physiological system, and the mind as an artificial machine of reason. The insight is not *produced by reason* but is prepared by the awareness of the constant tension between the corporeal logic of causality and the psychic logic of reason.

In both cases, then, the mind has an intuitive knowledge of a unity that exceeds corporeal materiality (the autopoietic system) *and* a unity that is the necessary ground for the interaction of psychic and bodily activity. Both forms of activity are defined by Spinoza as resolutely particular and therefore essentially limited in their contingency. There is not (as we encountered in earlier thinkers) any hint of a mystical or occult "connection" between the mind and some higher power or reality. This is to all to say that for Spinoza the intuition of the "attribute of God" that is nothing other than unity itself is predicated on the direct confrontation with a resolutely singular and therefore *particular form* of unity—one's own experience *as* an experience unified with one body must be only ever just that experience and not any other, and that experience is only ever dependent on the actual affections of that body. As an individuation of the being that is the unity of mind and body as system, the mind has access, intuitively, to the unity that must *precede* the individualization of unity that is the human being—the final unity that defines the absolute integration of everything. What Spinoza calls God.

4
Spiritual Automata Revisited

Leibniz and Automatic Harmony

> Since the efficient containeth all ends in it self, as it were the instructions of things to be done by it self, therefore the finall external cause of the Schooles, which onely hath place in artificial things, is altogether vain in Nature.
>
> Jean-Baptiste van Helmont

The early modern debates concerning the nature of the autopoietic body recapitulated the issues raised by Spinoza concerning the relations between parts and wholes and the boundaries of the living being. The question of the living being as *system* could at this moment hardly be bracketed from the broader question of the relationship between systems and the comprehensive structure of nature as a totality. To even frame the problem of human reason and other intellectual faculties required deep immersion into both the physiology of the body and especially the nervous system and the brain *and* the relationship of the body and mind to its own organismic unity. The question was never whether or not the human mind was explicable in terms of the body conceived as an automaton. The question was whether or not the organization of the automaton could explain the intelligence and rationality exhibited by the mind (and, perhaps, the "wisdom of the body" in its own terms).

In 1699, Gottfried Wilhelm Leibniz received from Joseph Hoffmann a dissertation titled "Mechanical Nature of Medical Diseases." Leibniz was impressed, not least because Hoffmann affirmed many of his own ideas concerning organic living bodies. Hoffmann's main argument was that the body, understood as an organization of systems (blood, sense, humors, muscles), is explicable purely in terms of "physical corporeal causes," without introducing any spontaneous activity, even when considering the organic "states of exception"—that is, disease and injury. The key to the argument was the idea that the machine had been constructed with such intelligence "it possesses sufficient forces for preserving the machine and defending it against putrefaction and the destruction of its combinations of mixtures, but also for repelling and avoiding all internal lesion that would tend to ruin the machine."[1]

The body did not need independent intelligent control because it was designed to function automatically; its "reason" was entirely internal to the body's organization. The body was able to sense interference, provoking a "violent and extraordinary [*extraordinarium*] flow of spirits . . . for correcting or eliminating the pathogenic matters" (104).[2] For Hoffmann, and by extension Leibniz, it was crucial to recognize that the mechanical "nature" of the body acts not only in the internal administration of the parts but also through the "whole mechanism or organism" [*totum mechanismum organismum*] (102)—but without any "soul" acting as a representation or incarnation of that unity.

And so, a few years later, Leibniz read with great interest Stahl's "Disquisition on Various Mechanisms and Organisms" (1706) and the *Theoria media vera* (1708), in which the "Disquisition" was reprinted. For Stahl, the living being was not simply a coordination of various mechanisms, but rather a "functional integration" with specific ends of its own.[3] The artificial machine, for Stahl, perfectly exemplifies this functional integration, with the caveat that the artificial machine's "agency" is *external* to the machine, whereas the body's is internal to its organization (106). Stahl's "organism" exceeded what was meant by "mechanism" because he believed that it was not possible to explain the guided activity of the whole being without some kind of conduit for that unity to act (105). This is why Stahl insisted on the concept of soul, something that could operate through the "organicus nexus" to preserve, defend, and restore the material body's functions (107). Stahl's views were in line with other thinkers (most notably Locke, who was of course trained as a physician), those who postulated a "material spirit" of some kind, one that acted wholly *within* the body to form and maintain its order—unlike Cartesian or Neoplatonic immaterial souls that carried out higher intellectual functions that mirrored or channeled divine rationality. What exactly was an automatic natural body? Where could one locate the *force* of unity and integration in moments of disruption and pathological breakdown?

Organismic Unity and Cosmic Order

When Leibniz decided to write a critique of Stahl's views on the physiological organization of the body, the metaphysical stakes were rather high. As Locke had argued, in the *Essay concerning Human Understanding*, the identity of any living thing was to be found in the "organization" that allowed "fleeting" materiality to be unified in a single functioning body.[4] The status of organization was the site of key controversies concerning mechanism and vitalism (or organicism). Leibniz's position was in the end highly idiosyncratic: he refused to accept that there was any fundamental difference between the usually opposed concepts of "organism" and "mechanism."[5] Even more cryptically, Leibniz affirmed both the existence of the soul and the independent "machine" of the body, arguing along the lines of Spinoza that "when the soul desires some outcome, its machine is spontaneously inclined and prepared by its inherent motions toward accomplishing it."[6]

Essentially, Leibniz was claiming that the two spheres, spiritual and mechanical, were entirely *independent* of one another, yet also perfectly coordinated. Adopting Spinoza's language directly, Leibniz wrote, "The operation of spiritual automata [*automates spirituels*], that is to say, souls, is not at all mechanical; but it contains in the highest degree all that is beautiful in mechanism."[7] How to make sense of the interaction of these two domains? Given the terminology, Leibniz's attempt to bring together intellectual spirits and physiological mechanisms through the infamous doctrine of "preestablished harmony"—what I will call here *automatic harmony*—might seem like an evasion of the problem that early modern thinkers struggled with rather than a satisfying solution. And so to understand Leibniz we have to see how he was redefining and reconfiguring the foundational concepts of seventeenth-century thought.

The first thing to note is that Leibniz was an early adherent to the mechanical philosophy and never wavered in his view that in the world of bodies, *everything* can be explained in terms of the size, shape, and motion of matter. What is not so obvious is that Leibniz early in his career recognized that any theorization of mechanism had to confront the fact that the fundamental elements of this philosophy were necessarily defined by their *form*, which is to say the atomic element possessed a kind of "soul."[8] Leibniz's argument is not too far from Locke's: the laws of physics, for example, do not explain themselves and cannot be deduced from the hypothetical starting point of simple matter. There must be something that *determines* the interaction of material bodies. As Leibniz explains, it was necessary to reject Aristotelian forms in favor of a minimalist conception of the cosmos. However, if the universe is a collection of aggregated matter, the essential atomic starting point could not be a mere mathematical point, an abstraction. A material thing can, according to the mechanical philosophy, always be divided. But it could not therefore be *indivisible*, a "true unity," if it was only matter—something Leibniz recognized early in his career.[9] "Hence it was necessary to restore, and, as it were, rehabilitate the substantial forms that are in such disrepute today," explained Leibniz in 1695.[10]

The philosophical argument is that the essential nature of the atomic element is *force*, which means that there exists at this fundamental starting point of all natural things something akin to sensation and appetite, "so that we must conceive of them on the model of the notion we have of souls."[11] It is easy to misunderstand this claim. What Leibniz is saying is that there must be something internal to the organization of the atom that would specify its "direction" (which would then be the origin of the basic laws of reaction) and at the same time something internal to its organization that would allow it to receive from outside the configurations of force inherent in another atomic body. Sensation and appetite are not anthropomorphisms at all but structural necessities if we want to have a pure mechanistic philosophy. Otherwise, we would need to introduce *external* forces and organization that would coerce the dead material elements to behave in certain ways. This would mark a return to the arbitrary (and for Leibniz, incoherent) Aristotelian genre of matter and form—what we now call hylomorphism—or recent updates, such as the Cambridge Platonists' concept of plastic nature.[12]

So when Leibniz goes on to assert that the analogy between the atom and the human soul only goes so far, because some souls have not been "thrust into matter," he can justify the distinction. For the most part, he explains, God "disposes of other substances like an engineer handles his machines." But with rational souls—minds—God "governs like a prince governs his subjects, and even like a father cares for his children"; this is because minds are "like little gods," "made in the image of God, and having in them some ray of the light of divinity."[13] These are not mere assertions. In other words, Leibniz will try to demonstrate the necessary distinction between what he calls "spiritual automata" and corporeal machines,[14] in order to explain how minds are "above the upheavals in matter" *and* resolutely synchronized with living bodies, that is, living *machines*. The crucial link is the concept of soul, or form, a problematic but essential dimension in all previous early modern theories of the rational mind and the organized body.

Divine Machines, Human Minds

Leibniz lays the groundwork of his new approach with a thorough clarification of the repeated error of "confusing natural things with artificial things." The mistake is that in comparing technical objects with living, natural bodies, philosophers have inevitably come to the conclusion that the difference between the two is simply one of scale. Of course, as we have seen, sophisticated thinkers have noted that the distinction between automata and organisms lay in the "externality" of organization in the former. But Leibniz's argument here is more radical. How can we actually understand the *nature* of this "integration" of organization and materiality in the organism-machine? Even if it is "naturally" present in the living body, conceptually speaking, the form of organization is still exterior to its materiality. This is why Leibniz proposes a new way to distinguish natural and artificial—and not by the presence of organs and organization. The "immense distance" between masterpieces of the mind and the mechanisms of divine wisdom lies in the fact that the machines of nature have a truly "infinite number of organs, and are so well supplied and so resistant to all accidents that it is not possible to destroy them."[15] For Leibniz, the key is that natural machines are still *machines* (i.e., organized entities) in their "least parts" and maintain their unitary form despite the "different enfolding it undergoes" as it is extended, compressed, or concentrated.[16] The artificial machine simulates only one aspect, namely, the comprehensive unity of organization. Because the organs and parts are finite, they cannot maintain themselves and therefore cannot help preserve the enclosing organization of the body operating as a whole.

Addressing the mind in this context, Leibniz argues that the soul (like "any other real unity") is entirely self-enclosed as a structure: it has a "perfect spontaneity relative to itself." This is what he means by "spiritual automaton" (which is not at all what Spinoza meant by this phrase, as we have seen). Yet at the same time, despite its internal autonomy, the mind also has a "perfect conformity to external things." Leibniz explains that our internal sensations (the experience of the mind, that is, not the material activity of the brain and sensory system) are "merely phenomena which follow upon external beings"; they are, he infamously states, just "well-ordered dreams."[17]

The question is why do these sensations arise in the soul? There can be no *causal* relation between the ordered body and the soul, since they constitute two radically different modalities. Therefore, the logic of sensory appearance must be purely spiritual and derived from the nature of the spiritual automaton itself. The crucial link between the events of external beings and the experiences of the internal mind is the concept of representation: the soul as "form" has a representational nature, capable then of representing external bodies "as they relate to its organs."[18] This is not an ad hoc solution, for Leibniz has already demonstrated that the world of material beings is already, at the very foundational level of reality, organized by form. The form of a material entity (even at the atomic level) is what allows it to "represent" what is external to it through a kind of *transduction*, and then it is the internal logic of the form that produces—like a machine—a new effect in response.

So if every substance (as formal organization of materiality) represents the "whole universe" from its particular perspective (since anything it "senses" is a result of the total configuration of the machinery of the cosmic system), then the soul can be understood as an *immaterial* representational organ. That simply means that its experiential order represents the universe in the same way, and furthermore, that the transformation of its experiential content is a function of the fundamental organization as spiritual automaton, that is, according to the laws of form alone.

The perfect automatic synchronicity of the body and the soul (despite the complete lack of any causal relation or other form of connectivity between the two realms) is the challenging question that Leibniz sets himself, and is often a stumbling block for readers.[19] Leibniz's own examples (the infamous dual clocks that keep time together) are not always helpful, since he implies that there are, literally, two fully independent machines (body and mind) that just so happen to have been "organized" constitutionally that they mirror one another with precise exactness: perceptions occur in the soul simultaneously even though it has its own unique laws, "as if in a world apart."

The critical issue, however, is the relationship between the formal order of the *body* as one kind of representational sensing and the formal essence of the *mind* as an altogether different mode of representation. To understand Leibniz on preestablished harmony we must tease out the specificity of the mind-soul relation (and note this does not at all entail any notion of *causal* connection). The crux of the argument, I propose, will lie within a claim Leibniz has already made, namely, that every substance represents the whole universe from its own particular perspective.

What Leibniz suggests is that the relation of mind and body is not a relation between two substances so much as it is the perfect superpositioning of two different perspectives (two "forms" of organizational unity) at *exactly the same point*. As a unitary body, the living organism "senses" the external bodies of substance by re-presenting them as transformations of its substantial being—we could say, as the deformation of the topological organization that relates organs (and organs of organs) in the natural machine. The "point" of the body's perspective is not a physical point, since there is no such thing, for Leibniz, as a purely "physical" space—we are always in the realm of force and relation and hence organization. The point of perspective is the location of the *unity* that is the body's organizational form, which is an aggregation of subordinate forms and organs.[20] The next step of the argument, then, is to show that the formal unity of the soul must occupy that *exact* same point, so that they coincide precisely—something that is impossible for two *material* substances, given the solidity and impenetrability of matter itself. However, there is no obstacle to a purely spiritual form occupying precisely the same perspectival position, if we remember not to think of this position as existing in some neutral Newtonian or Euclidean "space"; it is located rather in something like a complex topological domain, an infinite system of relations.

For Leibniz, the order of the organism is self-enclosed (which is why it is a unitary body), as is the mind—which is to say it is therefore absolutely *impossible* that they could ever "disturb" each others' internal laws. Any "communication" between soul and body must really be a parallel synchronicity, guaranteed by their common perspective. The body will act by itself given its representation of the external reality that affects it; the soul is affected by the same representation of the same external reality and responds with its own reaction. This means we see the machine act *as if* the soul "willed" it, because each responds in its own way to the representations *common* to both—"the spirits and blood then having exactly the motion that they need to respond to the passions and perceptions of the soul." The soul is "in" the body, it is an "immediate presence,"[21] Leibniz explains, just as the organismic unity emerges from a multitude of other unities. The internal bodies (tissues, organs, and their components) are all unities of organization, but they are organized in a layering of superpositioning that identifies them as *one* body (again, the superpositions are not strictly speaking "spatial" ones). The soul is just one more level of superposition—a unity like the unity of the form of the body, a unity that effectively *repeats* the very perspective of the body. The real challenge for Leibniz (as it was for Descartes) is to explain what value the soul adds to the body if it never truly (actively) participates in its functioning.

The first point to make is that Leibniz positions the soul "in" the body so that we can see how it functions as a *representation* of the body itself.[22] The soul is *not* the entelechy of the body. But what happens in the body as an organized system is always "expressed" in the soul, just like any other entity in the universe, although to be clear, the laws of the spiritual automaton's own organization are what govern that expression.

What makes this all much more legible is Leibniz's discussion of the body. Every organ of the body (up and down the infinite levels of the divine machine) acts according to its own independent laws, derived from the formal essence of its being. To have a coordination or organization of organs does not ever entail the violation of these laws. Everything going on in the "private" entelechies of individual bodies must *agree* (lit., "conspire") with the formal order of the body's own entelechy, but they are not necessarily *derived* from it. The illuminating example is the perspectival projection of a two-dimensional architectural plan. The two forms express each other, but each has its own organizational logic. The soul therefore expresses the body whose perspective it occupies, more like a strange palimpsest written over the body that changes in response to the transformations (representations) of the body as it responds to the impact of external bodies on it.

The function of the soul might still seem superfluous in this context. What purpose would the soul serve for the body? And what function does the body serve for the soul? We can start with Leibniz's rethinking of causality in this context. Remember that any body responds to another only in the sense that the external is sensed (represented) by a deformation of the topology, and then the body acts according to its natural transformation of this original deformation. So the action is not "caused" by the external entity; rather the body "expresses" in its own logic the relationship of the external to itself. Now imagine the body as it forms a singular perspective on reality—think of being in one part of a building or in a large city.[23] If the soul is draped, so to speak, over this body so that it is in the exact same topological "space" defined as a particular relation in the networked order of the whole universe, then it must also represent that same *exact perspective*. The difference between the soul and the body is that the organization of the soul *cannot* be affected by material substances. So what is the soul representing?

As is well known, Leibniz will eventually adopt the term "monad" to refer to any organizational unity insofar as it is a pure unity that cannot at all be divided into parts. He allows for the interaction of monads but also what we can describe as the superpositioning of monads. The mark of unity of any substance, for Leibniz, is not a totalizing imposition of a complex organization on some material substrate but instead the way that a number of monads can come under the "domination" of one monad. The natural machine is a unity of monads, and that unity is itself a monad. The Aristotelian idea of a single form for any unified body gives way here to a more complex notion: a unified body is an organization of organized bodies, which helps explain the extraordinary capacity of the body for self-repair, reproduction, growth, and so on. Organic bodies are "machines of divine invention, and "divine machines have this noblest feature beyond what is had by those machines that we are able to invent, that they can preserve themselves and produce some copy of themselves, by which the operation for which they are defined is further obtained."[24] The highest monad defines a *new function* that does not violate the subordinate functions of the other monads—but does take advantage of them. A living body is such a unity, not a mere assemblage of independent entities—the difference between a living fish and a pond of fish, as Leibniz will explain. One is what we will call autopoietic, the other a merely contingent, or apparent, unity.

Each monadological unity of animate life is therefore "purposefully designed for a definite kind of function," Leibniz writes, and the function of the divine machine of the human is clearly stated by Leibniz: it is to display *reason*. However, the reason of the body is resolutely unconscious—like a keyboard player performing unconsciously by habit.[25] Leibniz's explanation of this unconscious reasoning is subtle. He notes, in an evocative example, that a human being with limited skills, a manual worker for example, can easily make use of a semiautomatic calculating machine. (Leibniz had of course designed one such machine himself, as had Pascal.)[26] No actual knowledge of the reasoning process (or the machine) is necessary to generate the result. Similarly, Leibniz argues, the "plastic nature" of the divine machine as embodied layers of organization is also an *expression* of rationality—even if it is completely unconscious. As he explained in the letter to Masham, the organization of the individual bodies and organs that make up the organic being constitutes a rationalization of matter itself: the monadological entity is now the vehicle of a means-end relation as it senses and reacts to what it is *not*. The difficult question concerns the relationship between the unconscious body as a rational system of individually rational subsystems and the self-conscious soul that co-occupies the singular perspective of the dominant monad that gives some direction to the whole integrated assemblage.

The fact that the soul appears to be "the primary source of action" and the body a merely passive instrument does not at all mean that the soul is somehow affecting the body and causing it to deviate from its own innate principles, Leibniz emphasizes. The soul must be operating in *agreement* with these intrinsic principles. What the soul does is *perceive changes* in the body but without the body "disturbing" the laws of the soul, or vice versa. However, the soul in perceiving the changes of the body is perceiving (or better, representing, consciously or not) the very *rationality* of these changes at the same time. Therefore, the soul never responds to the body but rather to the *representation of the rationality of the body's own operations*—which is to say, the perception of a change cannot be separated (logically speaking) from an understanding of both the monadological continuity (i.e., structure of unity) and the laws of its transformation or "deformation" (a process analogous, it seems to me, to a functional transformation in the mathematical sense of the term).

The memory that is maintained in the soul is therefore not a storehouse of "perceptions" akin to a collection of single experiences. The memory of the soul is a collectivity of insights into the body's functionality. So when the soul responds to a specific perception of change in some moment, it is perceiving a change and at the same time draws into its conscious field new perceptions from its own prior (albeit sometimes "confused") perceptions, intensifying the insight into the logic of the body as a monadological system of monads. As Leibniz described it, in a reconceptualization of Locke's "dark closet" of the understanding, we have to imagine a "screen" in the closet, a kind of sensitive membrane, one that continually receives the impressions of the senses that are represented by its movement. However, this screen is not merely a passive transducer, according to Leibniz. We need to think of it as under constant tension, imbued with a "kind of elasticity" that makes it dynamic and interactive. Therefore, when the screen vibrates in response to one new impression from without (the "images and traces in the brain") it *produces* new tones and forms because the screen has been continually "folded" (organized) in response to *past* impressions.[27] Each new impression (and for Leibniz, we should remember, the soul represents *all* states of the body, even if it is not conscious of them, or only dimly aware) is mediated by this complex topology of prior experiences. Moreover, the mind can itself *think* new ideas, which means that the elasticity of the screen will allow for internally produced vibrations.

Here we get closer to seeing how the human soul can function as a spiritual automaton without seeming to be a useless representational mirroring of the body. The soul mirrors the *logic* of the body, whatever the specific impression might be, and however conscious or unconscious that impression might be. Whatever the logic of any one impression (which is always an impression of a transformation), what is registered is the specific rationality of a response. The soul's collection of representations, even if they do not display perfect organization, are all internally consistent in the sense that all of the responses of the body—from the atomic to the molecular to the level of tissue and organ—conspire in agreement with one another under the domination of the primary monad. The clearest and strongest impressions are those of the senses, for those are the organic systems designed to react to the external world in a peculiarly focused manner. So the soul perceives most clearly through the senses—and gains even more insights via the representation of the common sense in the brain, where features of sensation are mapped together, overlain one with another, revealing commonalities of structure and differentiations.[28] The common sense becomes, for Leibniz, a space that generates the first concepts—in their material form—which are then represented in the soul.[29] The soul acts, so to speak, as the amplification of the harmonies revealed first in the body's own sensible operations. The first kind of reasoning—inductive reasoning—is a logic based on these conceptual repetitions and can be expressed both in the body as a learning machine and in the soul as an induction machine. The extent of the soul's reasoning is opened up by the experience of the body but also limited by its contingency.

Still, as Leibniz will emphasize, the soul as a thinking entity might require external senses, but sensory information is not in fact its *essence*. The soul is capable of active reasoning, even if reason always has its medium sensory impressions.[30] What is this "inborn light," and how does it *escape* the body while still being subject to the doctrine of automatic harmony? As we know from Descartes and Spinoza, inductive reasoning on the senses can never lead to insight into a universal truth. To understand absolute necessity, one must go beyond particular examples. This will be the distinction between human and animal reason. Leibniz's position is tricky here. He wants to avoid the implication that the "light" of the soul is somehow absolutely distinct from the origin of experience in the sensible response of the living body, because it is then impossible to understand how the two domains interact or maintain any relation. This was Descartes's trap.

Leibniz insists, then, on the fact that the rationality of the body and its sensory apparatus in particular are already constituted by the integration of the immaterial and matter, the monadological mode of being. The representational status of the body's immaterial form is in fact the key to the soul's own representational activity; it perceives how the body is *organized*, even if those impressions are only glimpses. We can see, then, that the field of representation in the soul, for Leibniz, is also a revelation of the laws governing *external* bodies as well, since their presence in the body's own system of representation ("sense") remains traced in the corporeal transformations of sensation—although of course again these laws might only be glimpsed.³¹

Here Leibniz can identify the special nature of the human soul: our capacity for knowledge of order and not just *participation* in order. "This makes us resemble God in a small way," because with the conscious knowledge of order the human is capable of *imitating* that order. We can, he says, organize "things within our grasp in imitation of the order God gives the universe." The incorporeal automaton is one whose natural end is to *create*, to produce the artificial in the very midst of the natural—that is, to *reorganize* the natural in imitation of creation. The artificial machine is, in a real sense, a kind of artificial monad, a materialization of the activity of the mind, the spiritual automaton.³²

This radical exteriorization of the human spirit into nature, the invention of the artificial machine, introduces interesting and also challenging complexities into Leibniz's thought, most notably how to understand preestablished harmony in the context of a new zone of materiality—artificial embodiment—that establishes a problematic new relationship between the human living body and its spiritual palimpsest.

We can begin to unravel this thread by looking at Leibniz's thought earlier in his career, when he attempted to invent a new comprehensive symbolic logic. There he emphasized, as many others in the period, the importance of artificial techniques in the development and progress of human knowledge. However, what distinguished his approach from Descartes, Hobbes, Locke, or Spinoza—all of whom advocated a certain form of methodical training of the mind to prepare the way for true knowledge—is Leibniz's focus on *technical instruments*. Spinoza's spiritual automaton must, Leibniz suggests, be formed through a machinic intermediary (and not just on an analogy with technology, as with Hobbes's idea of "registration," for example.)

When Leibniz wrote to Oldenburg in 1675, a time when he was developing his "universal characteristic" based on an ideographic and diagrammatic theory of meaning drawn from both philosophical sources and theorizations of the Chinese language, he believed that the errancy of the mind could only be regulated by a method that was quite literally mechanical. Belief in concepts should be withheld "until they have been tested by that criterion I seem to recognize, and which renders truth stable, visible, and irresistible, so to speak, on a mechanical basis." As is well known, Leibniz had in mind here a kind of algebra of thinking, but note the specific practice of algebra was an image of a universal technique that was a kind of *technology*:

algebra will show that "we cannot err even if we wish and that truth can be grasped as if pictured on paper with the aid of a machine."[33] The goal was the "perfecting of the human mind," and the result would be a future where we could know God and mind as easily as numbers and invent machines as easily as creating geometry problems. This "mechanism" would lead to the revealing of nature's secrets. As he put it in another text, on the method of universality, the "characteristic" akin to the algebraic fusing of function and variable is what "gives words to languages, letters to words, numbers to arithmetic, notes to music. It teaches us how to fix our reasoning, and to require it to leave, as it were, visible traces on the paper of a notebook for inspection at leisure." The method "enables us to reason with less effort" by relieving the mind.[34] This method is, Leibniz says, a *tool*: once symbols of concepts are established, "the human race will have a new instrument which will increase the power of the mind much more that optical lenses strengthen the eyes and which will be as superior to microscopes or telescopes as reason is superior to sight." With this form of *artificial intelligence* humanity would have a "lodestar" to "navigate the sea of experiments."[35]

Enforcing Automatic Harmony: Regimes of the Mind

Leibniz's dialogue on the connection of words and things is illuminating on this score. Positioning himself between the more radical nominalism of a Hobbes and the more mystical Neoplatonic theories, Leibniz shows that while a system of substitute characters is always arbitrary, nonetheless, the *relations* of the networks connecting the characters are not. And more radically, Leibniz proposes that without any signs, rational thinking itself would not be possible. Basic arithmetic requires numerical signs, for example, and geometry requires diagrammatic figures. Characters themselves can be vehicles of ratiocination. Truth resides not in the arbitrary element of the sign but in its *permanent* element—what connects it to the relational order that gives all of the signs their specific position.[36]

Perception functions in a similar way for Leibniz, in that the material specificity of a body's transformation is also a manifestation of a monadological principle ("shapes and motions")[37] that is not at all exhausted in the moment but in fact relates the monad to every other dimension of nature as a *system*. The complexity of an animal, with its "heightened perceptions" (§25), allows the central monad to have more avenues of response to its environment. It mirrors the universe in a more detailed and intensified manner, which is to say, more *directly* because it receives ("senses") more external bodies and also then has access to the *relations* between impressions and transformations taking place in all the individual "incorporeal automata" (§25).

Genuine reasoning, for Leibniz, is an escape from the complex interrelated sequences of perceptions flowing from the body's dynamic organs *and* an escape from our intellectual automaton with its own sequences of stored memories. As Leibniz says, the sequences of memory are an imitation of reason, in that any striking perception will evoke a memory of that perception as well as any other perception that was attached to it. This in turn evokes the original sensations and allows for a kind of anticipation or "reasoned" action. For example, a dog will be shown a stick, remember the pain it caused them, and then flee before it can be beaten again (§26). Leaving aside the problem of explaining how the soul and the body are coordinated in this example, the important point is that reason is what gives us access to "eternal and necessary truths" and not just the empirical inductions that stem from accumulation of experiences in memory. The rational soul disengages from the relations of contingent occurrences in order to perceive the operations of the automata themselves—the sources of all action—and then the necessary *relations* that organize the monads in their mutual coordination.

The mind's capacity for reason is not a supplemental feature but derived from the nature of the monad itself. If the concrete materiality is the "arbitrary" element of a body and the entelechy its "permanent" aspect, we can also trace another feature of the monad that reveals the nature of the system of universality itself. The monad is defined by its fundamental unity of organization, however simple and limited that unity might be. "God alone is the primitive unity or the first simple substance," Leibniz writes, which is to say that distinction within the universe of nature is the result of the creation of monads within that first substance. The key idea here for Leibniz is that monads are generated by what he calls "continual fulgurations of the divinity" (§47). The monad is not simply a fragment of the divine totality. Yes, a monad is a limited form of organization with a limited form of "receptivity." And yet every monad also contains a *perfect* representation of God— namely, in the *presence of unity* itself.

The rational soul of the human body, draped over the central monad that organizes all of its subordinate organs, is capable of representing all the "sensations" and appetitions of the body, but at the same time it can represent *consciously*, in a way precluded from the monad that is the body's defining modes of order, the unity of the monad.[38] The soul perceives the order of the body in its activity, yet can also within that order intuitively comprehend the defining unity that is the law of the living organism. As we saw with Spinoza, every genuine unity of organization is an individualization of the total unity that is the comprehensive integration of all nature. The unity of the monad has no material substantial form and therefore cannot be accessed through the sensible representations.

The soul is therefore a conscious articulation of the unity of the laws of the organic body and "rules" that body (if that is the right word) as a "rector."[39] "Rector" is a term from Roman law that refers to an appointed representative of the state in a local region—what we can call the governor. The rector, significantly, did not possess "imperium" and therefore could not change or suspend the law. The rector only *makes visible* the law that is already in operation; it articulates its presence and thereby gives clarity to the law that already defines the order of the locality.

What Leibniz describes here is a process wherein the automatic operations of the monadological organization of the body are first represented in the soul and thereby made "visible" to the human being, and then the order relating the operations is also represented and made visible, before finally the unity of the order of the body is also expressed in the soul as a representation. We can see how this is a process in which the body *writes* itself onto the soul; it is a kind of exteriorization of the body's automatic system into a representational *diagram*. The value of this representation is its clarity; the soul knows more clearly what the ends of the organic body are and is able to see more clearly what would possibly attain those ends most efficiently. Nothing the soul does is in contradiction to the system it represents in its own characters. What we call volition, Leibniz says, is an action that is clearly visible in the operations of the soul; however, there is nothing actually interfering with the body's functions. Similarly, there are times when the body seems to act on its own, yet here it is just that the soul has not clearly represented that particular event. The value of the visibility of the soul's logic of operations is not that it reveals genuine volition; rather, the soul expresses the complex logic of the body, an automatic system that is often absolutely opaque to observation. The writing of the soul is produced by all operations, remember: all perceptions of the monads are represented. The conscious, and then consciously systematized, representations are condensations of these originary representations. The soul's cognition is a symbolization of the living being, a window into its divine structure.

What escapes the logic of the body is the soul's own capacity to read the order of nature *through* its own representations—and to produce for itself what are essentially artificial monads, these imperfect human imitations of the unitary systems that are organized forms of materiality: machines.

> Souls, in general, are living mirrors or images of the universe of creatures, but that minds are also images of the divinity itself, or of the author of nature, capable of knowing the system of the universe, and imitating something of it through their schematic patterns [*échantillons architectoniques*] representing it, each mind being like a little divinity in its own realm. (§83)

At this point, we look ahead to the importance of technicity for thinking the human as both in and out of nature. The structure of monological order will have to be rethought under the sign of the organismic body—how a natural human will bring itself out of nature, artificially, precisely through the use of artifice, the tool as mind, the mind as tool.

5
Hume's Enlightened Nervous System

Habits of Spiritual Automata

The early modern idea of "spiritual automata" figured the mind in terms of a productive automaticity that nonetheless preserved the intellect's capacity to see beyond the body, and even beyond reason itself. The spiritual automaton was predicated on the recognition—even idealization—that human cognition was an *artificial* automatization of the mind. The figure of the automaton in early modern thought, so often understood to be a model for the mechanistic reduction of life and the life of the mind, is perhaps more importantly a way of understanding the artificial *construction* of the mind.

As Pascal wrote in the *Pensées*, "We must not misunderstand ourselves: we are as much automaton as mind [*esprit*]. And therefore the way we are persuaded is not simply by demonstration. How few things can be demonstrated! Proofs only convince the mind; custom provides the strongest and most firmly held proofs: it inclines the automaton, which drags the mind unconsciously with it."[1] The automaton here is not at all the body-machine of Descartes; it is the automatism of culture impressed on malleable nervous systems. The automaton of habit could be transformed into an *automatic reasoning machine*, although of course the risk of wayward habits was extraordinarily high. Pascal puts it this way:

> We must acquire an easier belief, one of habit, which without violence, art, or argument makes us believe something and inclines our faculties to this belief so that our soul falls naturally into it. When we believe only through

the strength of our convictions and the automaton is inclined to believe the opposite, that is not enough. We must therefore make both sides of us believe: the mind by reasons which only have to be seen one in a lifetime, and the automaton by custom, and by not allowing it to be disposed to the contrary.[2]

The important point to be made here is that Pascal is fully aware that the automaton that is unconscious thought is not a *natural* machine but instead very much an artifact. It is only because this automaton is an artificial construct that it is possible to *reconstruct* it through new formations.

For Nicolas Malebranche, this forming of the mind is a matter of forming the very structure of the brain. Thinking in particular is dependent on the capacities of the brain, as we see so starkly when, in old age, people lose their mental acuity and even their reason, all because of the drying and hardening of the material of the brain. Where the child's fibers "are soft, flexible, delicate," in older people they are "inflexible, gross, and intermix'd with superfluous humors."[3] There is always a potential for material interference on the part of the brain, as it deteriorates, as the "solidity" of the fibers also solidifies error and resists "the attempts and force of reason." Malebranche argues that these facts show the urgency of properly forming the youthful mind, so that its habits are well established in the brain. For habit is nothing more than memory consolidated in the traces "imprinted on the brain."[4] And like a new machine that runs best after some practice, the brain consolidates its impressions the more there is *repetition*, as the animal spirits work the fibers in such a way so as to make them respond promptly in the same fashion in the future. We cannot overemphasize the physicality of this process as described by Malebranche—like a sculpture molded by instruments, or the engraving of a printing plate, the depth of the impression is a direct result of the *force* of impression. The variation of impression—we might say, *inscription*—in individual brains constitutes the very diversity of minds we see in human society.

Much in line with the speculations of Malebranche and others, Locke admits that we are often subject to "strong combinations of ideas" that are "not allied by nature." Hence the wide diversity of human beings, with their "different inclinations, education, interests." The power of habit reigns in all spheres: "Custom settles habits of thinking in the understanding, as well as of determining in the will, and of motions in the body." Surprisingly perhaps, Locke believes that it is likely habit can be reduced to the "trains of motion in the animal spirits, which once set a-going, continue in the same steps they have been used to." Repetition wears the path smooth so that "motion in it becomes easy, and as it were natural." The musician only has to start a melody and the rest of it flows automatically "without care or attention." The point is not that these thoughts are reducible to the operations of the brain, even if they are the "natural cause." What is crucial is that the natural automaticity of the animal spirits can help us understand how, in the psychic sphere, ideas seem to be produced and connected so "naturally" once they are made habitual.[5]

The mind must be able to direct its own train of thought, which means rejecting ideas that are not relevant to inquiry. However, unlike the body, with its automatic processes of self-repair, the unity of the psyche, for Locke, is not self-organizing. The foreign thought is not so easily excised from rational discourse of the mind: "This may be, if not the chief, yet one of the great differences that carry some men in their reasoning so far beyond others, where they seem to be naturally of equal parts."[6] Hence Locke's proposal that we inure children to "attention" rather than distract them with admonitions when they stray from the rational course—this will, he says, "introduce a contrary habit."[7] And so even if reason is a *natural* faculty, attention is not. Wandering of the mind, forgetfulness and unsteadiness, these are all natural forms of errancy at work in the youthful mind. But is there such a thing as a "habit" of attention? What kind of automaticity would be at work in such a habit? If the mind does not already naturally have the ability to control its own psychic organization, how can it be transformed into a machine of indifference? Because to "wander" in thought is to be forced away from the weak—because difficult—lure that is the work of reason.

If early modern thought had consistently positioned human autonomy as something parallel to creation itself, in the most cosmic sense of that term, with the advent of the Enlightenment the challenge was to understand how the human mind could be at once a product of nature *and* nature's own exception. This would prepare the way for theories of intelligence and freedom that could not be fully captured by any strict distinction between the technical and the natural, or a collapsing of these categories. The human only emerges on the threshold of that distinction, that is, as an expression of what could possibly comprehend this duality.

In the eighteenth century, the idea of an independent realm of spiritual thought was dismissed. The mind had to be understood as a product of some kind of biophysical *self-organization*.[8] The movement from the natural condition of the self-regulating body to the artificial nature of the rational mind demanded a new form of philosophical inquiry, which points us to David Hume's revolutionary and, we can say, ruthless dismantling of early modern epistemology.

The Errant Automaticity of Reason in Hume's "Treatise"

> Some small touches given to Caligula's brain in his infancy, might have converted him into a Trajan.
>
> Hume, *Dialogues on Natural Religion*

Hume's critical philosophy takes aim at one of the central conceits of the line of early modern thought I have been tracing here—that the comparison of organized machine technologies with natural systems of order will reveal something important about the relationship between human cognition and the inner secrets of the universe. His avatar for this position is Cleanthes, who, in the Dialogues, compares the universe to houses, ships, furniture, and machines. But as Hume's Philo points out—and here we can locate one of the key markers of a new naturalism characteristic of what will be called *Enlightenment*—"thought, design, intelligence, such as we discover it in men and other animals, is no more than one of the springs and principles of the universe, as well as heat or cold, attraction or repulsion, and a hundred others, which fall under daily observation."[9]

Hume's philosophy is much more than a critique of the analogy between mind and nature. Hume will in fact leverage his interrogation of "natural" intelligence into a dismantling of all claims concerning the nature of things. Even if we admit that aspects of nature affect others, why, he asks, do we privilege one principle, especially one "so minute, so weak, so bounded a principle, as the reason and design of animals is found to be on this planet? What peculiar privilege has this little agitation of the brain which we call thought, that we must thus make it the model of the whole universe?"[10] Hume's critique draws on a novel way of conceptualizing human thought: the essential ground of intellectual cognition, the laws of Leibniz's incorporeal automaton, can constitute the foundation of a human nature but only if we acknowledge their fundamentally *artificial* character. Hume's skepticism, on this reading, is a meditation on the weakness and fragility of an artificial form of intelligence as it confronts the overwhelming natural forces of the mind. But even these forces have already been disrupted and have reorganized themselves, by the no less artificial invasion of alien thoughts. Here Hume will point to a new way of conceptualizing the relationship of mind and brain—and underline the importance of material forms of intelligent thought.

As Gilles Deleuze argued in his penetrating analysis of the *Treatise*, Hume's account of the mind begins without a beginning, with a radical questioning, that is, of the status of the spiritual automaton governed by a cognitive logic independent of physiological organization. As Deleuze puts it, for Hume the mind is *not* nature; it has "no nature" because at its origin it is just "identical with the ideas in the mind." An aggregate collection of ideas is not a *system*.[11] To identify a "mind" in the midst of the givenness of *experience*, we have to first *separate* the mind from its experiences, and this requires the mind to be constituted as a new organization of the givenness of experience. But it is crucial to remind ourselves that the constitution of the mind as organization is predicated on the initial *separation* of experience from the mind. This is what Hume means by the almost insensible "reflection" that accompanies all experience.[12] So the mind is neither given (as a faculty of organization) nor constituted *from* experience—it is at once inside and outside of itself from the very beginning. Hume's name for this liminal space is "the imagination." There is no logic of the imagination; rather it is the name for the possibility of freeing the idea from its givenness. It is the space where ideas enter a new zone of possible connection.

We can easily misread Hume as if he were continuing the tradition of analyzing the *functions* of the mind (whether corporeal or intellectual) and through his analysis reveals for the first time their deep epistemological failing. However, taking our cue from Deleuze, the skeptical epistemology is much better understood if we acknowledge that the starting point of Hume's philosophy of mind is the *absence* of any mental function altogether. The mind begins as an empty space of liberation where new things happen. These events are what constitute the mind as a new system, a new logic of organization. Reason, intelligence, memory, all of these "faculties" are reconceptualized by Hume as not necessary within the foundational economy of experience. These functions are all derived, then, from new behavior on the part of ideas, behavior made possible only by their *separation* from the very origin of these ideas.

One way to think about the origin of the mind, then, is to emphasize with Hume that the mind is not some kind of faculty; it does not impose structures on given ideas. The ideas that come from the "impression" of the senses are transformed by *no longer being given*; they appear, now, only as the possibility of an association.[13] The challenge of reading Hume is this: How do we give a critical account of the mind when we begin with the recognition that the mind is not the origin of association, only its space of activity, so to speak? If cognition cannot be modeled according to the "principles" of order governing its operations, then what is the function of human thought? How can we critically evaluate reason and unreason, productive and unproductive thinking, without *norms*? The advantage of Hume's radical starting point is that we no longer must explain at one and the same time the nature of the mind and the possibility of its *violation*. The difficulty, however, will now be establishing any critical perspective whatsoever. My suggestion is that the slippery boundaries between nature and artifice in Hume's thought can point to a new way of addressing these questions. As Deleuze suggested, the space of imagination *becomes* a system when the subject appears as the vehicle of a new synthesis of the mind. Yet this synthesis is never given; it is not, we might say, automatic.

The opening of the mind for Hume is the constitution of the imagination as a pure possibility of new relations between reflected ideas. It would seem that Hume believes that there must be an intrinsic logic to the mind, "some universal principles, which render it, in some measure, uniform with itself in all times and places." Without such principles, ideas would be "entirely loose and unconnected," and they would be joined only by chance. How else to explain the regularity of "complex ideas" (not to mention the correspondence of these ideas across different languages) if there wasn't some kind of "associating quality, by which one idea naturally introduces another" (1.1.4.1)? At the same time, we know that the pivot of Hume's critique of induction will be his claim that the imagination can *undo* any connection between ideas. The resolution is found in Hume's careful distinction between the association of ideas and the mental perception of association. The mind therefore does not *discover* resemblance between two ideas, say, through comparison or some other method; "resemblance will at first strike the eye, or rather the mind," and the same goes for any other relation—there is no "enquiry or reasoning" involved (1.3.1.2). So what is the activity of the mind as such?

Crucial for Hume is not so much the immediate affirmation of a relation between ideas but instead the suggestion that the mind is where associations are transformed by repeated experiences into what Hume calls "habit." That is to say, what the mind introduces to the economy of sensory association is the consolidation and *artificial intensification* of always singular experiences, which makes possible an experience that will not have yet taken place: the experience of an "anticipation."[14] Once experiences of associations are layered in this way within the mind, with the new experience of a single idea "our imagination passes from the first to the second, by a natural transition, which preceded reflection, and which cannot be prevented by it" (1.3.8.2). This is the first articulation of "human nature" for Hume— not the association of ideas (which we will see is a logic external to the mind in an important way) but the integration of associations and the production of automatic *anticipations* of a future that does not in fact exist.

This logic of anticipation (what Deleuze called a kind of "fiction") is stronger than the "reality" of even our own bodies and nervous systems. Hume notes, for example, that someone who has lost an arm or leg to amputation "endeavours for a long time afterwards to serve himself with them" (1.3.9.18). Both the initial association of ideas in experience and the production of a "habit" that naturally produces artificial anticipations are "independent of, and antecedent to the operations of the understanding" (1.3.14.28). For Hume, it is the cognitive construction of habit that underlies our particular reality, a system that does not simply store ideas and their associations in memory, but actually constructs out of past associations and present impressions a comprehension of what is, a comprehension grounded on a field of interconnected expectations of how this reality will behave. The fact that expectation is the *basis* of our cognized reality means that the system of stored associations can be artificially supplemented by "habits" that have not in fact been personally experienced by the subject— what Hume sometimes calls "custom," although the terms are often used interchangeably.

The important point is that *anticipation* is equally natural in habit and custom, whatever their origin. For Hume, the "easy transition" is the result of a "curious and almost artificial preparation" (1.3.9.16). So the artifice of education (the repeated installation of prepared ideas in the young mind) is thoroughly consistent with the logic of habit precisely because there is no difference for the mind. The artificial idea of education can even overwhelm the ideas that are "naturally" produced by sense, memory, or reason when they are "strongly imprinted" on the mind (1.3.9.19).

Of course, Hume's philosophical reputation rests to a large degree on the way he dismantles the presumption that the anticipations of repetition that emerge from mental operations are in no way implied by the original experiences themselves, and any belief in some underlying "identity" that would guarantee the anticipation is absolutely unfounded. But what is the function of reason in this system? If belief and even judgment itself are an effect of some "secret operation" that is completely *unthought*, how can the mind's habitual tendencies (if that is the right word) be internally critiqued?

It is very clear that for Hume reason is not an independent critical faculty of some kind that could intervene and interrupt the natural logic that is habit and custom. Hume describes reason as another form of *automatism*: "Reason is nothing but a wonderful and unintelligible instinct in our souls, which carries us along a certain train of ideas, and endows them with particular qualities, according to their particular situations and relations"

(1.3.16.9). This instinct, to be clear, is not physiologically grounded. Like habit, reason is a function of the system of ideas, associations, and experiences. Indeed, reason is folded into habit: "Nature may certainly produce whatever can arise from habit: nay, habit is nothing but one of the principles of nature, and derives all its force from that origin" (1.3.16.9). That force is *felt* by the mind but remains absolutely impenetrable. The system of association and the logic of anticipation that produces new ideas appropriate to whatever is the state of things—this system is quite literally for Hume formed by a "magical faculty in the soul" (1.1.7.15). This magical power is at the heart of reasoning. The specific kind of expectations that are inferences will automatically recall counterexamples from the past if the new ideas are contrary to experience as organized in the mind. The overlooked possibilities "immediately crowd in upon us, and make us perceive the falsehood of this proposition." The failure of automaticity marks the "imperfection" of the faculties, "often the source of false reasoning and sophistry." Still, Hume notes that the failure is usually in cases where ideas are unusually abstruse and compounded. "On other occasions custom is more entire, and it is seldom we run into such errors" (1.1.7.8). Reason is the automatic detection, then, of a "mismatch" between the expectation implied by some proposition and the *system* of expectations forged from natural and artificial memories of association.

Hume is of course intensely interested in the ways in which habitual thinking consolidates untenable beliefs. In particular, he admits that the admixture of opinions and notions introduced by learning, which takes "deep root" from infancy, is impossible to eradicate "by all the powers of reason and experience" (1.3.9.17). To be clear, Hume is arguing that we could not explain the power of automatic belief and expectation if ideas were associated only by "a reasoning and comparison of ideas" (1.3.9.17). But the consequence of this view is that reason will not be able to interrupt that automaticity of belief.

In this context, Hume's somewhat marginal remark in a section discussing animal reason is quite significant. One way to allow the mind to constrain its own operations is to circumvent automaticity through the use of a technological prosthesis. In order to see how we can maintain a complex line of thinking (i.e., "reasoning") to produce an inference *without* relying on the steadiness of mind, we can look to the example of the merchant: "In accounts of any length or importance, merchants seldom trust to the infallible certainty of numbers for their security; but by the artificial structure of the accounts, produce a probability beyond what is derived from the skill and experience of the accountant" (1.4.1.3). This externalization of cognitive relations into the artificial structure is a disembodiment that short-circuits the weakness of rational thought in the face of intensive, affective experience.

For Hume, this must be understood in relation to the physiology of the nervous system. The problem with reason is that it tracks the implications of ideas that have not always been experientially associated; rather the links are made through "general terms" and propositions. In this kind of thinking, "the action of the mind becomes forced and unnatural, and the ideas faint and obscure" (1.4.1.10). Here the train of thought is always subject to competition from the "more natural conception of ideas," and attention can falter. This is because "the posture of the mind is uneasy; and the spirits being diverted from their natural course, are not governed in their movements by the same laws, at least not to the same degree, as when they flow in their usual channel" (1.4.1.10). The materialization of habit as second "nature" cannot be countered by the artificial and weak thought that is rational inference—unless that weak thought can be materialized elsewhere, safe from the interference of powerful neuronal flows. To "strain" the imagination is to disturb that flow, and the stronger the agitation of the brain, the more difficult it will be to resist that flow. As Hume explained earlier in the *Treatise*, "It would have been easy to have made an imaginary dissection of the brain, and have shewn, why, upon our conception of any idea, the animal spirits run into all the contiguous traces, and rouze up the other ideas that are related to it" (1.2.5.20). This process is physiologically imprecise because the spirits seldom maintain their direction and "naturally" deviate from their track, falling into nearby traces and thereby rousing the wrong (or at least alternatively associated) ideas.

Philosophy is the name for the activity that is not quite purely artificial but that does operate in the absence of "nature," whether of the habitual sort or not. It is thinking with the *minimum of passion*, the effort to escape the natural—if sometimes deviant—course of nervous communication, to think in the absence of the animal spirits, we can say, in other words, to think through the *relationship* between ideas and not through the always material, because physiological, *association* of ideas. However, as we might expect, thought cannot so easily disembody itself. "When the soul applies itself to the performance of any action, or the conception of any object, to which it is not accustomed, there is a certain unpliableness in the faculties, and a difficulty of the spirit's moving in their new direction" (2.3.5.2). And in the process of thinking passions may well be awakened. Novelty itself (as Descartes also argued) can enliven the spirits with the passion of *wonder*. As Hume explains, this passion *agitates* the spirits. Hume's point is that philosophy as attention to the course of ideas is always threatened by the economy of the brain and the automatisms of habit.

> There is a great difference betwixt such opinions as we form after a calm and profound reflection, and such as we embrace by a kind of instinct or natural impulse, on account of their suitableness and conformity to the mind. If these opinions become contrary, it is not difficult to foresee which of them will have the advantage. As long as our attention is bent upon the subject, the philosophical and studied principle may prevail; but the moment we relax our thoughts, nature will display herself, and draw us back to our former opinion. (1.4.2.51)

Nature and reason are therefore figured as opposed and unreconcilable enemies; the mind must feign a kind of "dual existence" in order to satisfy the demands of these two automatic systems.

Interestingly enough, it is the very weakness of radical reason that is its saving grace for Hume. The relentless and destructive skeptical conclusions he follows to the bitter end would make the life of a sociable human being intolerable. So, as Hume admits, it is nature, that is, the neurological *force of habit*, that "cures me of this philosophical melancholy and delirium," relaxing—literally—the philosophically focused mind, or alternatively attracting it with "lively impressions" (1.4.7.9). As Hume wrote elsewhere:

> To what purpose should I pretend to regulate, refine, or invigorate any of those springs or principles, which nature has implanted in me? Is this the road by which I must reach happiness? But happiness implies ease, contentment, repose, and pleasure; not watchfulness, care, and fatigue. The health of my body consists in the facility, with which all its operations are performed. The stomach digests the aliments: The heart circulates the blood: The brain separates and refines the spirits: And all this without my concerning myself in the matter.[15]

Still, if the "mild" sentiments of philosophical thought cannot "interrupt the course of our natural propensities," the *trace* of philosophical doubt remains, that "cold and general speculation"(1.4.7.13) lingers, perhaps, even in the "heated" brain.

Reason and Plasticity: Enlightened Habits

Hume's thought makes visible a threshold that marks the transition away from an early modern preoccupation with the relationship between the order of nature (and especially the organization of nature's bodies) and the order of cognition—not, as we usually think, to establish the connection between the mind and the body but in fact to isolate the privileged state of the intellectual sphere in comprehending the order of the natural world. The predominance of machinic metaphors and analogies in this period does not at all reveal an inclination to what is assumed to be a mechanical metaphysical position. The key concept of *organization* was interrogated with respect to both artificial machines and what Leibniz called "divine machines." The mind, then, was in turn conceptualized as both an independent form of organization and a capacity to identify and *reproduce*—or even create—forms of order. What Hume's method reveals is the possibility that the organization of the mind is inherently contingent, dependent first on the array of impressions given to it by the sensible machinery of the body but also dependent on the secret forces that make and consolidate the associations between ideas that produce the anticipations that order our experience of the world. Moreover, the mind is susceptible to "alien" associations in the form of imported customs—that is, education. Ultimately, for Hume, the two logics (of mind, of body) are steered by the dynamics of the nervous system—the formation of habit and custom through the literal *writing of the brain*. The possibility of a pure logic of thought (itself a form of automatism) depends on the contingency of the state of the brain and the passions that excite it. (Figure 5.1.) Only in the absence of lively impressions and heated passions will the brain be willing to succumb to the pursuit of reason—the tracing of a cognitive logic of ideas, often against customary association, often forging a path through the brain that has never been followed. Fleetingly, Hume will suggest that the technologies of disembodiment—writing, in particular—might liberate reason or at least offer it a space of refuge.

In the wake of this turn we can see that if reason and intelligence in the Enlightenment will be played out on the terrain of the nervous system, it is not in the way we might suspect. The question will not be the how the mind is connected with the brain, and whether or not this implies a reduction of the mind to the body. With Hume, we can now ask: How does the contingency of an *organized* mental system relate to the organizational capacities of the sentient body? Against early modern philosophy, Enlightenment thought recognized that the challenge of actually explaining the development and reproduction of organic bodies precluded any easy method for

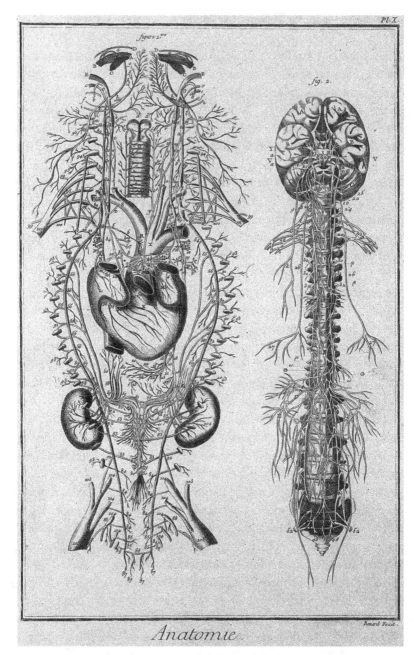

Figure 5.1. From Denis Diderot and Jean le Rond D'Alembert, *L'Encyclopédie* (1751–66). Source: Wellcome Collection. CC BY 4.0 / https://creativecommons.org/licenses/by/4.0/legalcode.

linking these living systems with the systems of thought generated within minds. For these systems were not strictly speaking analogs of one another: they had independent logics of development. How could the contingency of belief, reason, passion, anticipation, memory, categorization, language be mapped on to the complex physiological systems that maintained their unity and reproduced it without any obvious vehicle of organization?

The sterile opposition of mechanism and vitalism fails to grasp the crucial significance of contingency in Enlightenment concepts of mind. Yes, there were many schools of thought competing to understand the nature of the organized body and the nature of the human mind. However, what links these efforts is a radically new naturalism that takes as central the origin of organization, or, to put it another way, the possibility of *self-organization*.[16] If in the seventeenth century the technological object could serve as a model of order as active organization, in the eighteenth century the automaton demonstrated a crucial absence: how the living being could *construct* itself. Or at least: how the living system could produce and reproduce its own organization, without the assumption of some immaterial presence of purely formal order, the Leibnizian "monad."

In the Enlightenment, then, the contingency of the mind's formal development could be linked with the constitutive openness of the dynamic body rather than the body's abstracted (and theologically framed) formal unity. Enlightenment psychology, unlike its predecessors, offered models that would account for the recognized regulation of our mental life *and* the singular human capacity for creative insight and invention. Habit was the liminal zone that laced together organismic order and distinctly human minds, via the neurophysiology of automaticity.[17] As Charles Bonnet put it, cultural education marks a "second birth," since the brain is literally rewritten.[18] Denis Diderot argued that the life of the mind was a constant writing and reading of the brain *by itself*.[19] But the original natural order of the nervous system never disappears entirely. As David Hartley would explain, the human is defined by twin automaticities: the automatic physiological organization of the body and what he called the "secondarily automatic" functioning of cultural habit, the deformation and re-formation of the brain from the *outside*.[20] It will fall to Kant to systematize these two automaticities, the a priori logic of the mind and the empirical formations—and often errant deformations—of concrete experience.

Threshold

KANT'S CRITIQUE OF AUTOMATIC REASON

6
The Machinery of Cognition in the First Critique

"The role of a human being is perhaps the most artificial [*künstlichste*], the most arduous, but also the most exquisite in all of this planetary system."[1] Kant offered this dense but illuminating reflection in notes written in the 1780s, on the topic of metaphysics. The relationship between these terms is not entirely obvious. Given the legacy of Enlightenment thinking on the nature of the human mind, or at least the organization of the mind, Kant might well be echoing Rousseau or Herder, suggesting that it is the special difficulty of human existence, its radical *lack*, that is, that necessitates the turn to artifice, to the nonbiological prosthetics of technology and language—thereby becoming the most magnificent of all creatures. Perhaps. It is also possible that Kant is defining the human here as the *most* artificial of all creatures, and hence it is human life that is the most challenging.

From the perspective of Kant's critical philosophy, at least, the essence of the artifice peculiar to human beings is worth exploring carefully, since the term is entangled with the crucial problem of judgment, linked as it is to novelty, spontaneity, and the independence of the organization of the mind from any external causality. Still, the slightly incongruous reference to the "planetary system" is puzzling and does not seem to point to any particular issue raised in the description and analysis of the architectonic system of cognition. The astronomical frame does recall the more cosmic speculations of the early modern thinkers navigating the thresholds separating and connecting the order of nature, the capacities of the mind, and the divine unity but now in a new, Enlightened context.

Decades earlier, Kant (in his first, anonymously published work on universal natural history, which appeared in 1755) had in fact already addressed the grand question of the position of the human in nature, combining leading edge neurophysiological theory, natural history of the organic body, and new psychological accounts of cognition in order to revisit the fundamentally theological tradition of metaphysical thought. For Kant, in this early work, the distinction of the human would be found, as Rousseau argued in the near-contemporaneous *Essay on the Origin of Inequality*, in some kind of lack—or at least deficiency. The limitation of human thought and action, Kant said, was due to "the coarseness of the matter into which his spiritual part is sunk, in the rigidity of the fibers and the sluggishness and immobility of the fluids that ought to be obedient to its stirrings." The cognitive psychic system was trapped, so to speak, in the nervous system, and overpowered by the sheer intensity of organismic demands on its attention.

> The nerves and fluids of his brain supply him with only coarse and unclear concepts, and because he is unable to balance the stimulation of sensory sensations in the interior of his faculty of thinking with sufficiently powerful ideas, he is carried away by his passions, dulled and disturbed by the tumult of the elements that maintain his machinery.[2]

This might serve as a relatively faithful précis of Hume's *Treatise*. What distinguishes Kant in this work, however, is the belief (or maybe faith) that the mind is somehow capable of *increasing its strength*. The logic of cognition is not expediency but reason, and Kant understood reason to be in a constant struggle with the body's own existential orientation. Kant laments, "The efforts of reason to rise against this and to expel this confusion by the light of the power of judgment are like the flashes of sunshine when thick clouds constantly interrupt and darken its brightness."[3]

Why this struggle in a creature of nature? As Kant explains, the mind has a function in the economy of the organism: the "invisible spirit" is what conceptualizes the external world, and repeats and combines these concepts with its specific form of *internal action*.[4] The problem is that the coarseness of the fibers makes this internal reflective activity especially difficult; the soul is, he says, in a state of "constant exhaustion" because the body resists the reorganization of its sensorial flow. So the mind succumbs, and "soon falls back into the passive state by a natural tendency of the bodily machine since the sensory stimulations determine and govern all its actions." But this is for Kant *not natural*, in that the spirit of rational cognition should be governing the "lower" functions of the physical body. Our "excellent abilities" are being wasted, and with age, the situation is even more dire, as weak circulation and increasingly inflexible fibers and ossified spirits make thought even more sluggish and prone to error. It is this unnatural, or at least highly problematic, condition that stages Kant's analysis of the human mind.[5]

However, it is important to note that for Kant the question is not to what extent the mind can *liberate* itself from its materiality, to aim, that is, for a certain kind of disembodiment. What is at stake is the *coordination* or harmony between two related but also radically independent systems that together constitute a living, sentient being—namely, the psychic and the physiological.

In contrast to Locke's speculations on the cognitive capacities of angels, Kant's conjectures concern the relative rationality of extraterrestrials in the solar system.[6] Given that the other planets are most likely inhabited, Kant argues, the structure of the sensory and nervous systems of these alien creatures would vary according to their distance from the sun: the fibers of a resident on Jupiter or Saturn, say, would have to be extraordinarily fine in order to be able to respond to the slightest impression from the weak heat and light, whereas those on Venus would be even coarser than our own. Our middle position on Earth puts us in the middle rank of intelligence, where the power of the mind struggles against the coarseness and inelasticity of our sensory and nervous organs. This is what makes the life of the human so arduous. Unlike virtually all other living creatures, who live, reproduce, and die in a more or less seamless existence, the human being struggles because its nature is not so fixed. As Kant puts it, the human can both debase itself and strive beyond itself, seeking "gratifications" for which "he is not organized and which conflict with the arrangements that nature has made for him." This is to say that the human is the being capable of *disturbing nature*, leaving his "post" in creation. Kant's argument here is that the human must rediscover its place in nature, its original sphere—because "outside the sphere of a human being he is nothing, and the hole that he has made spreads its own damage to the neighboring members."[7]

I recognize the danger of overreading these passages in light of Kant's future philosophical development. Still, it does seem appropriate to frame his critical project with these questions, for one thing Kant is trying to accomplish in the three Critiques and related writings is a systematic survey of the *exact* capacities of the human mind, so that we can understand the tendency to stray from the very strict limits of reason and cognition and only then formulate a certain intellectual regime that will guard against straying and guide us in the pursuit of knowledge. Unlike all previous thinkers, however, Kant will refuse to *separate* the logic of corporeal cognition from the intellectual faculties or capacities, and he will resolutely avoid the metaphors of liberation they entail. Kant will emphasize, then, the unitary *integration* of all forms of cognition, drawing on the early modern philosophical and medical concept of organismic unity that we can trace back to at least the seventeenth century.

The key issue motivating my reading of Kant here is reconciling Kant's relentless systematicity (i.e., the essential *automaticity* of the cognitive functions and their interrelationships) with the profoundly important role that disruption (in the form of error, interruption, and pathology) plays in the very constitution of the psychic space *and* the topological organization of the organic body. This reading suggests that there is something cognitive that Kant will locate outside of all understanding, a presence in thought of what is necessarily radically *outside* of all thought. To explain this, I focus attention on the status of the human as artificial. Kant will directly (and often not so directly) explore the ways in which human artifice, and especially human *technical* artifice, can interrupt the automatic integrity of the human mind. In so doing, he will open up new possibilities for the reintegration of the materiality of the body and the organization of the mind.

Automaticities in Conflict

We can start with a look at Kant's various lecture series on logic. Here the cognitive faculties are staged, as psychic phenomena, independent of any determining causality; however, that did not mean they were not natural and hence not subject to their own natural laws. This claim was important for Kant because it allowed him to evade some philosophical traps. For now, I want to focus on the implications of Kant's invocation of law in his lecture notes on logic: "We can only become aware of error through our understanding, and thus we can err only when the understanding acts contrary to its own laws. This, however, is impossible. No force of nature can act contrary to its own laws if it acts alone." Error, then, is a deviation resulting from the interference of other natural laws—just as the motion of a body subject to gravity can be deflected by the force of air resistance, Kant remarks. The mind is, therefore, a kind of spiritual automaton in Leibniz's sense of the term. "The understanding by itself does not err because it cannot conflict with its own laws." "Pure reason does not err." And so on.[8]

But what is the function of this automatic system? If, as Kant says, we can conceive the continuity between the form of an external object and the way that object is represented in the nervous system and brain, and perhaps even a continuity between those brain representations and the psychic representations that constitute the mind's sensible experience, it is not possible to assert any continuity between the *active* functions of the mind (combining ideas, assigning predicates, isolating properties) and the very nature of the world itself. To argue (as Spinoza and Leibniz seem to) that there is a relationship between the two spheres established by God is to abandon philosophy, according to Kant. Hume offered the radically skeptical alternative: human cognition is just a kind of grand fiction. For Kant, as is well known, the third alternative is a systematic grasping of the *nature* of cognition in and of itself. The laws of cognition are what determine freely acting humans. But not as if humans were mere machines with no alternatives. What Kant says in the lectures is that different laws can interfere with one another; to be free is to be determined by the independent and autonomous system of the understanding and reason. These are the laws *of* cognition and not a displaced, externalized origin of action. We must recognize that we are not machines driven by the nature of our embodied experience—the logic, that is, of embodiment. To be free is to follow rational laws.

The function of cognition is to see and process relations that are *not* in the "appearances of nature" that constitute the sensory field of experience. That is the point of the understanding for Kant: to draw from cognition's own logic of organization an orientation in nature, a knowledge that is not itself given by nature.[9] It is also the source of the struggle that fractures the integrity of the human. A natural automaton would be driven by the logic of instinct, the laws of automatic behavior that flow from the organization of the living body. As Kant says, the understanding "makes everything unruly when it fills in for the lack of instinct"—when it tries to substitute itself, in other words, for that other nature, to imitate the laws of the organismic and therefore material automaticity governed by physical laws rather than operating properly as the purely incorporeal (or "spiritual") automaton that it is. The freedom of the understanding is precisely the fact that it "enjoys an exception" from dependency on any sensible ordering.[10]

This challenge can serve to prepare the transition to Kant's critical philosophy. Kant will, beginning with the First Critique, at once determine the "external" limits and internal boundaries of the forms of human cognition *and* the nature and power of its systematic integration. Ultimately, the deviations, disruption, and pathologies of the mind will point to an even more radical interruption of the laws of cognition, namely, the *spirit of technicity*. A rethinking of the place of unity in Kant's work will be necessary to make sense of the emergence and development of *techne*.

The Automaticity of Pure Reason

> For a first rough orientation the structure of the *Critique of Pure Reason* should be thought of in terms of various materials falling into a machine where they are then processed; and that what then emerges as the result of this processing is my knowledge.
>
> Theodor Adorno, *Kant's "Critique of Pure Reason"* (1959)[11]

The *Critique of Pure Reason* cannot be reduced to the architectonic typology that it articulates with such precision and systematicity. By that I mean Kant begins with the fact of mental experience and proceeds to identify and demonstrate the necessarily independent logic and economy of each faculty. That independent logic is what Kant refers to as the "pure" instantiation of that faculty, even if that particular subsystem cannot at all be entirely separated from the others in reality. The mind must, then, be approached as a system with rules that bring together these individual capacities: "By an architectonic I understand the art of systems. Since systematic unity is that which first makes ordinary cognition into science, i.e., makes a system out of a mere aggregate of it, architectonic is the doctrine of that which is scientific in our cognition in general, and therefore necessarily belongs to the doctrine of method" (A832/B860).[12] The beginning point of the system must not be thought of, then, as a foundation for subsequent modes of organization and reorganization. For as Kant shows, the nature of *appearance* for an experiencing being implies the necessity of a law of appearance, and it is only by pinpointing the different forms of appearance that we can see the interaction of different faculties and their role in cognition more generally. If we assumed that the "raw material" of cognition was the givenness of sensory information, which is "worked upon" by the faculties, we would be unable to locate the origin of thought because it would always be conditioned by what was *not thought itself*.

The beginning of thought, then, is the first structuring of the experiential wholeness of the field of thought *as* thought, whatever may be the connection between the body's sensory organs and our experience of the world. For Kant, the first internal division of experience is the appearance of what he calls intuitions. Literally a form of "seeing," the intuitions are what delineate in experience the givenness of structure, the structure first of objects and their properties. Without this primordial differentiation of the unity of thinking, there could be no possibility of any perception or identification of relations *within* thought.

Having established the necessity of the division and distinction that intuition introduces, Kant demonstrates that in order to maintain the internal unity that has now been fractured into individual fields of organization, the mind must reunite the intuitions. Or to put it another way, the internal differentiation cannot lead to a multiplicity of experiences (since that is impossible in one mind), but if the mind *is* one system of experience, then the unity must reappear in some form: that is the law of the imagination, which must, simultaneously with the appearance of difference, reunite the intuitions by forging a real unity out of this plurality—what Kant calls the manifold. At this stage, all that is required of the imagination is that it exist in order to synthesize back into unity what was initially divided. The objects of experience now *appear* to us as aspects of one genuine singular world, necessarily—that is, not as simply juxtaposed in an ad hoc and contingent manner.

Now, if we can see how this double move of intuition and imagination resolves the continuity of the cognitive unity of experience, it also raises a new question, a new challenge, that was not present to the mind in its original undifferentiated state—namely, what laws govern the manner in which this new unity of experience is produced and maintained. What marks the synthesis of the manifold plurality? In the synthesis of the manifold, the imagination must now *relate*, using principles of organization, what was not fundamentally capable of any relationality before the initial cuts of articulation. This is why Kant can demonstrate the absolute necessity of a new faculty, the faculty that can provide the law for the appearance of regular organization within the synthetic unity of experience.

This is the understanding: the faculty that produces the "forms of thought" (B288) of relational structure that could give a substantial form to what would only be a "pure" but empty synthesis. The categories are the basic, foundational forms of relational order that are intrinsic to that synthesis, applied to the manifold so that we cognize it as organized in a seamless way. The categories (spatial position, number, succession, etc.) all depend on the prior articulation of the field of experience and the synthesis of that field: the need for relations stems from the *fact* of creating independent sites of organization that nonetheless demand to appear under an embracing unity that relates them. That unity is essentially "immaterial" and must be actively reengaged with the sensibility.

> We can, however, trace all actions of the understanding back to judgments, so that the understanding in general can be represented as a faculty for judging. For according to what has been said above it is a faculty for thinking. Thinking is cognition through concepts. Concepts, however, as predi-

cates of possible judgments, are related to some representation of a still undetermined object. (A69/B94)

Kant's point here is that the fact of articulation and the need for a new synthesis does not imply any particular form of relational order. However, it is not possible for us to actually cognize an alternative order, since the order that appears to us appears as necessary, a priori, and hence absolutely universal.

Kant's method protects him from any obligation to account for the relationship between the laws of cognition and the "reality" of any world that is not just external to the mind, but radically *outside of experience itself*. Whatever the connection might be between our thought and the nervous system, or our thought and the physical world, or even our thought and God, thought itself cannot conceive the relation of what *is* mental experience to something that is *not* experience. This is not to say that the mind is not part of nature. Rather, metaphysics in this new key will consist of the consideration of "everything insofar as it is," which is to say: insofar as it is thinkable in terms of the operation of this incorporeal automaton that is mind. Kant occupies an uneasy middle ground between Leibniz and Locke: the first philosopher "intellectualized" (*intellectuirte*) the appearances in order to link them with the order of things in themselves, the second "sensitivized" (*sensificirt*) them to bind them to an empirical reality (A271/B327). As Kant says, a cognition that could cognize what it is *not* is one that would have to belong to nonhuman beings, "beings we cannot even say are possible, let alone how they are constituted" (A 278/B 334).

That said, the initial portrait of the mind seems excessively limited, in that the automaticity of the interplay of sensible intuition, imagination, and the categories that understanding continually applies via judgment, this all leaves no room for what we might call active and productive cognition that is spontaneous and not, we might say, automatically necessary. We have seen that the division of the faculties can be read as an unfolding of an internal and consistent organization from within a foundational unity that is always preserved in this internal differentiation. The interruption of the purely automatic determinism of thought (the origin of individual cognition) is not, for Kant, an intervention from a lawless "outside" of thought but rather a consequence of the fundamental contingency of the content and even formal order of the intuitions—appearing as they do in the dynamic flow of experience that cannot, as sensible, ever be derived from the laws of thought, even as their appearance and possibilities of organization are utterly dependent on those laws. It is not hard to see in this development an exact analogy with contemporary theorizations of biological genesis, in particular morphogenesis.

As Jennifer Mensch has brilliantly shown, Kant's unfolding of the faculties in the First Critique parallels exactly the kind of speculative natural philosophy of organismic development, in that the structural form of organization is given in the initial undifferentiated unity (the egg, that is), but in the very process of development, the contingencies of interaction with what that entity is *not* (however that is understood) will greatly affect what can be called the individuation of the living being from its generic universality of organization (the species of unity).[13]

This perspective is important because for Kant, the contingency of sensible properties in the synthetic form that is an intuition sparks the appearance of a new form of cognition and a new faculty that will, of necessity, emerge to maintain the foundational unity of the mind—this will be, eventually, what he calls the faculty of *reason*. Once intuitions within a structured synthetic unity appear to the mind, it is possible to mark out relations *between* aspects of the manifold, relations that may be governed by a priori principles (e.g., the way colors or shapes appear to us) but are not in and of themselves absolutely necessary. In other words they could be "unthought," or cognized in completely other ways. This identification of relations and order *within* the manifold is for Kant the formation of *empirical concepts*, which can only be made contingently on the basis of contingent appearances, though the concepts take the form of—maybe "borrow" from would be a better phrase—the normative universality of the categories, as Hannah Ginsborg persuasively suggests.[14]

Kant's theory of empirical concept formation is notoriously obscure and the topic of much scholarly debate (and speculation). But it seems to me that we can at least agree that their origin can be understood as epigenetic (that is, an internal contingent development) and not the result of an automatic function that is an a priori "preformation." The formation of an *artificial* concept would have no positive function if it did not lead to some new possibility of thought, and this is where Kant must introduce again the faculty of judgment—because without judgment, the empirical concept would never be capable of producing active thinking that is not determined by past cognitions. The general principle of the empirical concept (whether it is a formal recognition of the relational topology of intuitions or itself a kind of "meta-intuition" of structural homology across discrete intuitions) must be applied to *new* intuitions in order to generate some kind of novel insight. The function of judgment is just that: to ascertain whether or not an intuition can be subsumed under the empirical concept—a priori concepts, the categories, are of course *automatically applied*. "Synthesis in general is . . . mere effect of the imagination, of a blind though indispensable function of the soul, without which we could have no cognition at all, but of which we are seldom conscious" (A78/B103).

It is the contingency of the empirical concept that demands an active, and singular, exercise of judgment, appearing now within the contingent flow of sensible intuitions, which is to say, *not automatically*. The intuition will be recognized, that is, in light of its having been judged, via the mediation of a concept, to be an instantiation of the general principle articulated in that concept. However, intuitions are susceptible to being cognized under many different concepts. If the difficulty of explaining how, why, and when empirical judgments are made remains, it seems still that the distinction between pure a priori and empirical concepts lies in the perpetual automaticity of the first kind of judgments and the necessary contingency (i.e., always epigenetic origin) of the latter.

The crucial implication drawn in the First Critique is that the judgments mediated by the imagination are a *structural* necessity within the system of cognition (since concepts cannot apply themselves). Yet judgment is essentially spontaneous and creative, otherwise it would never exceed pure automaticity. The problem of identifying the principle of judgment arises only with the recognition of the very contingency of its empirical activities. There, its appearance is not at all predictable or even necessary, at least from the perspective of the other faculties. According to the developmental model of the mind, it is clear that the need to introduce the new, very specific faculty of reason arises in just this moment. For the contingency of the interruption that is empirical judgment cannot help but interfere with the internal organization and unity of the mental sphere. Empirical concepts, exercised through judgment, serve to *reorganize* the very appearance of things—though never violating, of course, the categories, or escaping the synthetic unities. Nonetheless, judgment is a novel and not necessary (again, an epigenetic) complication of the mind that always threatens cognitive integrity from within. This is because empirical concepts are *not necessarily congruent with one another*, as they might be triggered by different sensible experiences in different contexts, producing inconsistent expectations, anticipations, and therefore incoherent actions.

The emergence of reason, explicitly (and infamously) described by Kant as the "epigenesis of pure reason" (B167), is the emergence of a new faculty whose task is to track the *logical forms of concepts* in order to compare and make visible these threatening conceptual contradictions.[15] At the same time, reason can in the process identify conceptual relations across concepts themselves, creating new incarnations of the complex wholeness of the mind. Reason is necessarily protected, so to speak, from sensibility because it does not arise within that sphere and could not do its work if its own activity was governed at all by the contingency of that experience. Reason sees only conceptual order. And this is what makes reason so powerful, once it appears as a contingent development that nonetheless becomes necessary because of the inevitable existence of the contingent yet functionally "general" empirical concept. For reason can see past the content of any one concept and understand the logical implications of its configuration—whether the concept is a priori or not.

If an empirical concept is formed, say, the concept of "sphericity," from the layering and integration of common features in a multitude of intuitions, the leap to the claim that there is such a "thing" in nature is an obvious error, an example of what Kant would call a transcendental illusion. However, once the concept of a sphere is established, reason is capable of developing the implications of that particular organization of relationships—and discovering, for example, certain geometric truths as a result. Reason forms "ideas" that are much more abstract (but also less concrete) than empirical concepts, and these are useful precisely for organizing cognition at a structural level beyond the production, reproduction, and synthesis of sensible intuitions. The contingency of the empirical concept (its intrinsic artifice and its "accidental" sensory content) does not threaten the *necessity of the logical form of the concept*. Reason is involved in a continual and always incomplete mission to recognize, correct, and develop the *internal formal unity* of the whole cognitive system.[16] This does not mean that reason can somehow ever synthesize conceptual knowledge into a genuine unity. Reason, I suggest, is more a machine of reconciliation that aims to give presence to a unity that itself never appears as such. Reason serves as the representative of the concrete whole forged by the mutual activity of the cognitive faculties, in its endless (because specific and sequential) clarification of the internal relations of cognized and a priori concepts.

The unity that pure reason brings to the mind is therefore the unity of the understanding, because it is there that the fracture of two "universalities" (a priori and empirical) takes place. Kant's explanation is complex, to say the least. But crucial to his account is the denial that reason somehow provides a logical synthesis on the basis of some a priori principle of unity. Reason does not practice synthesis of this kind. Pure reason tracks the relations, the sequences, that condition the movement of judgments, and, as Kant says, it relentlessly seeks the *unconditioned* that must initiate any logical series of connected judgments. These judgments (applications of concepts to sense) are aimed at identifying possible *experiences*. The understanding, says Kant, "has to do only with the objects of a possible experience, whose cognition and synthesis are always conditioned" (B365). "All principles of the pure understanding are nothing further than *a priori* principles of the possibility of experience" (B294). The very condition of that "universality" is itself unconditioned, in the sense that the cognitive underpinning of the judgment is taken for granted (i.e., "the subsumption of its condition under a general rule") (A307). Reason is the necessary supplement to the understanding because in tracking the conditions and exposing the necessity of the *unconditioned*, reason is capable of identifying the field of possible cognitions.

To put this more concretely, the pure understanding is blind—must be blind—to the unconditioned ground of the conditions of its own cognitions. For example, the structural artifice of the categories makes possible the continuity of experience that experience itself cannot generate: continuity cannot be experienced in and of itself but is the condition for a unified experience (A103).

Or take judgment. In the application of an a priori or contingent principle, judgment *conditions itself* by obeying the intrinsic demand of the "universal rule" (as norm) so that it can generate a singular and necessary conclusion. Reason, in contrast to the judgments of the understanding, identifies these occluded unconditioned principles in order to critically organize the possible cognitions that could be produced from the interaction of a priori and a posteriori concepts, all of which are conditioned by their inherent *normativity*. Reason does not judge these unconditioned principles on the basis of some specified condition (which would entail assuming yet another norm). Rather, it just recognizes their status as unconditioned, the contingency of their normativity, that is.[17] Only then can reason evaluate the internal consistency of the comprehensive system of the understanding—the knowledge that organizes all experiences on the primordial condition that concepts are *normative*. "If the understanding may be a faculty of unity of appearances by means of rules, then reason is the faculty of the unity of the rules of understanding under principles" (B359).

And I would stress that the epigenetic appearance of reason is predicated on the emergence of empirical concepts that exceed the categories and the logic of intuition and therefore always threaten to introduce a disturbing element into cognition: the presence of radical disorder (i.e., conflicting norms that nonetheless share a logic of normativity) and hence the threat of a malfunctioning *system*. Which is another way of saying the death of the system qua system.

But what exactly is the principle of this system? That is probably the wrong question, because a principle must always be applied and this is exactly not what reason does. Reason must perceive the unitary wholeness of the mind as a system while at the same time taking on the responsibility for maintaining its integrity from within.[18] For Kant, reason is most definitely what must "govern" and thereby constitute a cognitive system if the mind is going to be understood as something that is not a "mere rhapsody [*keine Rhapsodie*]." A true system involves mutually supportive efforts that all strive together to advance the shared essential ends. The mind as system, as Kant will carefully explain, lies not in some independent synthetic action but rather in the *rational* unity of differentiated forms of cognition. The necessary unity of the mental sphere that Kant demonstrates is in fact an empty origin and a formless limit. The transcendental unity of apperception deduced from the continuity of thinking has by definition no actual *presence* in thought itself. "There must be a condition that precedes all experience and makes the latter itself possible" (A107). (This will become crucially important when we turn to the Third Critique.)

What gives cognizability to the system of its own systematicity is just the *idea* of wholeness—and ideas are, according to Kant, what pure reason alone produces—because the idea gives actual "content" (of a formal kind) to the integration of the differentiation of mind into absolutely separate spheres of activity. "This is the rational concept of the form of a whole, insofar as through this the domain of the manifold as well as the position of the parts with respect to each other is determined *a priori*. The scientific rational concept thus contains the end and the form of the whole that is congruent with it" (A832/B860).

The transcendental apperception of unity cannot *act cognitively* since it is *radically* inaccessible to thought, absolutely noncognizable; and since it is incapable of intervening within thought *as* thought, the mind generates an idea of the whole that will allow its unity to appear to itself, in the form of what we might call a "concrete unity." Reason produces and enforces the foundational articulation of the mind: the perfect structural coordination of the differentiated parts and their wholeness. Kant is very precise here: the parts, he states, are in fact related to the unity of the end only through the *idea* of the unity of the end, not the genuine unity of thought itself.

This is a crucial issue. As Kant wrote in the *Critique of Practical Reason*, if our actions were conditioned by our concrete determination in time, as if we were a kind of thing, then "freedom could not be saved."[19] We would be mere automata.

> Here one looks only to the necessity of the connection of events in a time series, as it develops in accordance with natural law, whether the subject in which this development takes place is called *automaton materiale*, when the machinery is driven by matter, or with Leibniz, *spirituale*, when it is driven by representations; and if the freedom of our will were nothing other than the latter (say, psychological and comparative, but not also transcendental, i.e., absolute), then it would at bottom be nothing better than the freedom of a turnspit, which, when once it is wound up, also accomplishes its movements by itself.[20]

By this he means that in either case the conditions of the determination would be *external* to our own being—even if we were still very much self-aware, even if these determinations were divinely established. As Kant explains, a human being would be just a marionette, or, with an explicit reference to the famous eighteenth-century automata of Jacques Vaucanson, an "automaton, like Vaucanson's."[21] And in a footnote to this passage, Kant would add: "Such an automaton could mistake his spontaneity for freedom, but which would be an illusion, for everything would be determined by his original makeup conveyed by God."[22]

As I am suggesting, the epigenetic arrival of reason marks the transition of an automaton seamlessly synchronized with life, reproduction, and death, within the whole that is nature, into an autonomous system capable of cognizing its own "ends" and steering itself in the face of obstacles. The organization of the living animal is what determines its action via natural needs (instincts). What turns this automaton into an autonomous[23] human is the grasping of its own organization through itself, through its own idea.

This is all to say that the unity that is the system of the mind—and this is of course the core subject of transcendental philosophy, the work of pure reason—can now act, Kant suggests, because the rational idea of wholeness "allows the absence of any part to be noticed in our knowledge of the rest, and there can be no contingent addition or undetermined magnitude of perfection that does not have its boundaries determined *a priori*" (A832/B860). What does it mean to say that the epigenetic arrival of reason itself is determined a priori? If the wholeness of the mind, like the integrative unity of the "animal body," as Kant notes in a famous passage, is an articulated whole and can therefore grow only "internally," and "without any alteration of proportion," where exactly does reason fit?

In light of a strong epigenetic reading of the First Critique, the answer might come down to this: reason is not an additional "part" of the mind, analogous to a new limb or organ. Reason is the coming to concrete awareness of the fact of the wholeness that is the "organism" of mind developing according to its internal organizational potential.

Two issues are raised here. First, does the appearance of reason entail that the mind is governing itself, which is not the same thing as saying that the unity of the mind is automatically maintained and preserved by its "parts"? The whole (whether mind or organism) must be actively governed. Or more subtly, as Leibniz put it, reason appears as a necessary supplement to the order, a "rector" of the province and not its sovereign authority. Second, Kant leaves aside the problem of conceptualizing the link between this *idea* of the rational whole and that which is represented *in* that idea. Does the mind ever make an appearance as a unity within the proliferation of difference that is the mind, in effect bypassing the representation?

Tracking the biological concepts present in the First Critique, we can turn to the status of the organism and its pathologies in Kant's work—especially the Third Critique—to elucidate these possibilities. Both mind and body will turn out to be more radically open than they first seem, and it is here that technicity will intervene as a crucial moment in the constitution of the human being qua human.

7
The Pathology of Spontaneity

The *Critique of Judgment* and Beyond

Autonomous Judgments

As explicated in the First Critique, judgment conditions itself through an adoption of the concept as principle; as Kant writes in the Third Critique, judgment "is not autonomous," "it has no principles of its own." At least, determinate judgment has no principles. However, in its reflective mode, Kant remarks, it has to do its work but without the presence of any law or principle.[1] It is clear, then, that reflective judgment must subsume objects under a law that is in fact radically artificial, in the sense that it has no objectivity. We can say perhaps that it must produce a *substitute* for the law that is not in fact present.

This would seem to go against the very nature of cognition as Kant has laid it out. Judgment proceeds from the universal to the particular because the particular cannot *reveal* anything that would be binding on the understanding. Of course, the construction of the empirical concept resolves the problematic issue because with this concept the judgment has a principle that will determine particulars in their relation to that more general or universal concept. The challenge Kant presents here in the Third Critique is this: What does judgment do when it is confronted by a "particular" that has not yet been conceptualized and hence "automatically" cognized as such? This is why our understanding works from what Kant calls the "analytically universal" to the particular, since it is only in the concept that the particular can become subsumed.

Now, Kant holds that we *can* conceive of another, alien form of understanding that proceeds from the "synthetically universal"—which is to say, an understanding that would intuit, without any mediation, the existence of a "whole" in the plurality of individuated and differentiated parts. The fact that we must "begin" with the parts means that any *construction* of a concept involves an essential contingency in the identification of a substantial unity. It is not enough, Kant argues, to simply identify unity through the appearance of purpose (as Spinoza did), because we cannot understand how the actual substance of that unity (we could say, its particular concreteness both as form and as matter) is related to its purpose. Reason simply cannot answer that question.

The only option, for Kant, is to rethink the problem. This initial substance of a body, say, *must* be understood to have an "intelligence" that enables the specific "natural forms" to develop in the "unity of a purpose"—and by "enable," Kant means that we understand the relation as causality.[2] The aesthetic judgment as reflective judgment is therefore never the application of a concept to the work of art. And it is also *not*, as we might conjecture, the process of forming a concept from the particular. The reflective aesthetic judgment is precisely singular. It forms a whole from a specific set of contingent relationships. Yet how could a singular judgment ever have any value as a judgment, or to put it more bluntly, what status would this judgment have from the perspective of the understanding and its orientation to objectivity?

For Kant, the work of art (natural or human) arrests the understanding (since it cannot provide concepts) and, at the same time, refuses to stabilize despite the efforts of the imagination to cognize it. The famous "free play" of the imagination that is involved is maybe more serious than it appears. I want to suggest that the very artifice of the work demands that it be experienced in two conflicting ways—hence the perpetual lack of stabilization. The painting or sculpture, for example, or a natural wonder, is at one and the same time organized according to a certain concrete materiality of marks or characteristics (shape, color, depth, etc.) *and* organized in the sense that the same marks can symbolize or indicate what they are *not*. The judgment in this situation seizes the moment, freed from the obligation of borrowing any principle automatically, or even of creating a general principle to then apply.

The aesthetic judgment interprets the order of the work as a unity of organization even while recognizing the radical subjectivity and contingency of that particular order. This judgment is, in effect, an imitation of an intuition of a synthetic universal. The value, for Kant, of this strange and exceptional situation, is that it reveals so clearly what we might call the "pure" form of judgment that is usually hidden by the power of the concept—namely, the power of judgment to *impose* normativity, as Ginsburg argues. The aesthetic judgment does not "intuit" objective norms in the singular experience of a unity and its parts, but that is not a failure: the judgment demonstrates the radical capacity to establish norms as normative, whatever their origin (a priori, empirical, or even singular).[3]

On the Unitary Judgment of Organismic Order

So when Kant turns to discuss the teleological judgment, again, the key assumption here is that the mind does *not* in fact have any conceptual apparatus for cognizing the way that living beings are organized and how they behave. That is, in order to understand empirically the "ends" of a living being, we would have to examine and observe that being before trying to grasp what is common or "universal" in those actions. To understand this intuitively is impossible, for two reasons. First, as in the case of the aesthetic judgment, the mind faced with an organized body cannot proceed from the recognition of a synthetic universal to comprehend how all the differentiated organs and parts of the body contribute to that "whole." Second, even with that insight into the structure of this unitary being, we could not perceive intuitively the *end* of that unity.

So what exactly is going on with the teleological judgment? If the aesthetic judgment is reflective and hence free of any concept, it is judging purely according to rules of organization that have no a priori status, no objectivity. To follow closely Ginsburg's thorough analysis of this problematic in Kant, what is objective in the aesthetic judgment is the structure of normativity itself, its appearance as a judgment that could potentially command universal acceptance—even though it may in fact *not*.

In contrast, the teleological judgment is a reflective judgment (freed, again, from the condition of the concept), but it is not judging the living being according to rules. Rather, the teleological judgment is a strange form that proceeds according to concepts but in the reflective mode—that is, not *determining* experience through the concept.[4] In both aesthetic and teleological judgments, there is a postulation that the "synthetic universal" (an unmediated access to the intuition of the "unity" of an object) exists—even if it doesn't—but not to elide the epistemological problem; the postulation frees the mind to exemplify its *own* normative creativity.

Kant's argument concerning the teleological judgment is that we have no "category" for what we call "purpose" in nature, and no a priori knowledge that nature produces beings in relation to purposes; nor do we have the cognitive ability to intuit directly the integrated unity of a systematic whole. And we cannot, of course, rely on mere empirical principles, since there is no ground for the identification of what would even be a "purpose" in nature (we would have to assume, again, some a priori concept of "intelligence" operating in the natural world).

So the solution proposed is that when we judge natural beings to be purposeful, to have ends in other words, we are, Kant says, introducing the concept that *would* be necessary for determining teleology in the natural world—but without using it for the judgment. Rather, with that substitute concept standing in for the absence, the mind can go on to judge individual natural beings as organized according to purposes—as Kant puts it, "to guide our investigation of organized objects and to meditate regarding their supreme basis."[5]

The artificial concept authorizes us to think analogically about the way organized beings function—that is, our orientation in the world of appearances allows us to see nature in only one framework, a spatial and temporal view that implies only one form of causality, namely, the sequential connection of events that undergirds modern physical science. However, organic beings evoke a kind of "wonder" (what Descartes called *l'admiration*) because they defy this kind of mechanical causal understanding. We do have access to another form of causality, Kant reminds us—but that kind of causality emerges only within the intelligent mind. Only minds (and not what appears to mind) can act according to ends. That is what freedom (spontaneity) is for Kant: the act of mind proceeds *not* from what comes before but as a consequence of what it aims *at* according to the law or principle that informs it. This kind of automaticity is *autonomous* because it is generated entirely from within the sphere of the psyche, which is to say, governed by its own rules. So it is not surprising (even if it is an error) that humans have for centuries transplanted the intrinsic purposive of cognition, its peculiar autonomy of self-generation, into the entities of nature that themselves defy mechanical explanations. However, as Kant will stress, this analogical move has no epistemological warrant whatsoever. It can have, at best, practical use.

The Pathology of Artificial Thinking: Technicity in Kant

At this point, we could be content with understanding how the teleological judgment—the idea that the organic being is self-organizing and self-generating—mirrors the aesthetic judgment. Kant's analysis of the teleological judgment shows that while the organic being is organized as purposeful, that purposiveness itself can *have no purpose*. The insight is that the human mind is organized in the same way: it is a system that strives to one final end, yet that end is in itself without purpose (at least from our own perspective), not because it cannot be objectively known, but because the purpose is in fact only understood via the unity of the system that is the mind. The unity of the mind is, like the organism, an integrated purposeful system whose purposefulness (a striving to maintain its unity) is *essentially* without intrinsic purpose. The juxtaposition and comparison of the aesthetic and the teleological serves to exemplify what is foundational to the human mind: it is a norm-producing machine that operates free of any norms. Even when it is conditioned by norms—by that I mean even the universal "norms" that are the categories—the mind must itself establish the normative power of those concepts via the judgment, which, as we saw, must always condition itself. And judgment, as revealed in its pure reflective form, can, Kant shows, operate normatively even in the absence of any concrete norm—through its capacity to *institute* norms.

It is of course difficult to conceive how judgment functions in this way, and that is perhaps Kant's point here. But there is a lingering structural issue in the Third Critique that opens up an important new line of thought in Kant's thinking, both in light of the realm of the organism and within the sphere of cognition. By paying more attention to artifice and its relationship to what we can call organismic pathology, we can excavate a preliminary—but absolutely essential—theory of technology in Kant.

Central to the analysis of both aesthetics and teleology is the idea of the "whole" as a concrete unity of integrated differentiation. Kant has us imagine someone walking along a beach in "a seemingly uninhabited country" who comes across a geometric figure, a hexagon, say, "traced in the sand." Kant explains that reflecting on this figure, "working out a concept for it," this person would, through reason, become aware (however obscurely) "of the unity of the principle required for producing this concept."[6] That *unity* is not explicable under the "natural laws" of the physical world. The figure is an effect of purpose, the purpose of externalizing a concept into a material representation. That cannot be "natural," says Kant, because it is necessarily understood to be a "product of *art*," a reorganization of the natural— the appearance, that is, of the artificial. So when confronted by a natural being, whose order and organization appears to us as "infinitely improbable" if produced only by the chance interactions of the material world, we cannot help but see the organic entity as structured and thus purposeful— exactly like the material externalization of a geometric concept traced in the sand.

Nature's causality here is then a form of *technics* (*Technik*).[7] And there are only two options to consider: either the form was produced intentionally, like the work of art, or it was the unintentional result of the blind mechanism of nature. For Kant, this dichotomy is of course undecidable, since the mind cannot conceive the necessity (or even conjectural necessity) of either option. However, what we *can* conceive (since this is the very structure of cognition itself) is a form of rationality guiding the organism *from within*. This is the only way we can conceive of nature producing organization—but as Kant argues, this is a new and illegitimate form of physical causality that radically disrupts our very understanding of the natural world. Hence the almost paradoxical vision of the organism that emerges here: a being that must be totally governed by a concept (an organizational form) so that its unity will be genuine but whose concept of organization is therefore itself *necessarily* contingent.[8]

We can now return to the important question raised in the First Critique concerning the status of the "governance" of the mind. As Kant argues in the Third Critique, if something is to be understood as a "natural purpose" it has to have two qualities. "First, the possibility of its parts (as concerns both their existence and their form) must depend on their relation to the whole." The concept, in other words, "must determine a priori everything that the thing is to contain."[9] The "whole," that is, as concept must include the exact way that parts are formed and integrated into a unitary being. Second, that concept cannot be external to the being. What this entails, for Kant, "is that the parts of the thing combine into the unity of a whole because they are reciprocally cause and effect of their form." The nature of the *whole* cannot determine the parts causally, since the whole would then be external to the products of the concept—implying, it would seem, the supplement of an internal judging and steering organ of some kind, violating the principle of total unification of concept and materiality.[10] Therefore, this "systematic unity" is cognized in the unity of the form but also concretely as the way the manifold parts are combined, its "wholeness." In later writings, Kant would say this clearly: "The definition of an organic body is that it is a body, every part of which is there *for the sake of the other* (reciprocally as end and, at the same time, means)."[11]

It is striking, then, to remark how Kant destabilizes this rigorous definition of the idea of the natural organism with his own interesting take on pathology in the living being. Given the normative requirement of part-whole reciprocity, what are we to make of Kant's observation that while "nature organizes itself," and the "pattern" is always the same, "that pattern also includes deviations useful for self-preservation as required by circumstances"?[12] And if Kant uses the ubiquitous example of the watch to point out the key difference between machine and organism (that the parts do not imply the existence of the other parts), he also goes on to say that the watch cannot *replace* its own parts by itself.[13]

But what would the organism be *as a whole* if it lost one of its essential parts? How does the unitary concept act in this situation? The term "self-organization" elides the location of causal activity. Even more problematic, Kant says that the organism, unlike the machine, "can compensate for this lack by having the other parts help out."[14] Yes, Kant introduces here the idea of a "formative" and not merely motive force of action in the organism, which as we know he borrowed from Blumenbach.[15] But what is important here is the fact that the compensation referred to here must be understood as a reorganization of the original part-whole synthesis that defined the unity of the original being. The parts are no longer what they are, once they are adapted to function *otherwise*. The formative force must deviate from the conceptual (functional) determination of the organism, unless it is assumed that both matter and organized tissue are completely receptive to what would be the automatic reproduction of the concept through this formative force—without anything like an organismic analog of the imagination.[16]

However, the possibility of this sort of automaticity is undermined by Kant's remark that even the accidental deviations of some individual organisms would still have to be judged in terms of a purpose—but now displaced into an internal process of radical transformation and reorganization to preserve a unity without concrete locality. As Kant notes, with some mutant beings, "we find that the altered character of these individuals becomes hereditary and is taken up into the generative force," an alteration that must be cognized, then, as a necessary development lurking in the original formation of the species. In other words, the accident of mutation that is taken up by the formative force transforming the concrete unity of the organism as a result is still consistent with the "undeveloped" predisposition of that same being.[17]

But there is no *reason* why the deviation could not be entirely "accidental" or purposeless—contingent, in other words. What is important here is not whether this makes any sense scientifically (whether in the eighteenth century or in light of contemporary biology). What I want to suggest instead is that Kant's analogy between mind and organism becomes more interesting and more complicated once we admit the possibility, and perhaps even the necessity, of these pathological conditions of life. For they point to a way in which the concrete organization of an integrated being (whether organic or psychic) can be internally transformed in a radical way, without losing the essential continuity of its identity. This is, for our purposes, most suggestive and interesting when we move from the organic world (where we have absolutely no real conceptual access to the way in which beings develop and function) to the very origin of the analogy, namely, the rational order of the mind, where we have intimate experience.

What is at stake in the question of the pathological is not the return or reproduction of the integrated organization of the plurality, where part and whole emerge as co-constitutive. It is precisely that organization considered as a specific concrete order which is literally violated by the injury to the organ or the collapse of local zones of order. If, as Kant says, the part can be replaced or even compensated for by a new organization, there must be an internal principle capable of effecting that change. And yet there *cannot be* such a principle, since then it would also be integral to the original given organization. What is needed is not a normative faculty of repair but rather a capacity to establish a new norm to replace or modify the original normative order of concrete unity—without violating the principle of unity itself.

Given the absence of any "super" norm *within* the entity (cognitive or organismic), that is, one that would be "steering" the concrete order in exceptional conditions in light of a specific notion of unity, the establishment of the new norm must be understood to be absolutely devoid of all necessity, utterly contingent—except for the fact that the norm must serve the continued existence of the being itself. What would this mean? In the context of the organic living being, we would have to say that the very unity of the being (and not its wholeness as a concrete form) must somehow intervene in the concrete order so as to reestablish itself anew. However we cognize this (if we can actually think this at all), Kant's theory would seem to foreclose the possibility of some pure unity ever appearing *as itself*. It cannot function as a concept, since it has no formal organization. The unity of the "life" of the body considered existentially must therefore have its representative, a stand-in, one capable of intervening somehow and yet operating completely outside all of the actual "norms of life."

If we turn now to the psychic unity, we can see that pure reason, which produces the rational idea of the "end" that integrates the mind's functions as a concrete whole, is incapable of enacting the refoundation of unity in a new organization. This is all to say that given Kant's own argument in the Third Critique we have to admit what should be impossible—that something *outside* of cognition (namely, the pure unity of the transcendental apperception that can only be deduced) makes its appearance *within* cognition, as a force of radical novelty, a radically indeterminate determination of the logic of cognition itself.

The vehicle of that pure unity is of course the judgment, whose capacity to establish norms is predicated on the impossibility of *normativizing the very act of foundation*.

The radical origin that is the judgment (hidden in the determinate form but there nonetheless, lurking behind the facade of self-conditioning) has no place in the system if it is defined as the concrete unity, organized by the faculties and its concepts. The judgment is therefore *outside* of the system as it appears within cognition, the representative (and decidedly not representation) of a pure unity. If the mind is "steered" by reason as the governor of the law that is the formal organization of psychic activity, in times of crisis or emergency, when this spiritual automaton is failing, reason is not deposed but instead offered a new order, a bifurcation within the *organization itself* that can preserve the continued existence of unity in a new, concrete order of mind.

The pathology of the spontaneous judgment is therefore the only path to a new normative existence. What Kant will imply, here and in other writings, is that the artificial form of order—namely, the technical object—is both an origin and a result of this productive aberration.

Cosmic Orders and Human Artifice

As Kant wrote in a fragment collected in the *Opus Postumum*, there is an "immaterial" principle of "indivisible unity" necessary to understand the integration of cognition. However, this unity is not given: "For the manifold, whose combination into unity depends on an idea of a purposively (artificially) acting subject, cannot emerge from moving forces of matter (which lack the *unity* of the principle)."[18]

But there is *another* form of artificial order we must confront, as Kant does in the *Anthropology*: "Natural understanding can be enriched through instruction with many concept and furnished with rules," although the power of judgment itself can only be "exercised," not instructed.[19] However, there is something important going on here, as Kant notes. The vehicle of instruction is the *technology* of memory. Citing Plato's *Phaedrus*, Kant writes:

> One of the ancients said: "The art of writing has ruined memory (to some extent made it dispensable)." There is some truth in this proposition, for the common man is more likely to have the various things entrusted to him lined up, so that he can remember them and carry them out in succession, just because memory here is mechanical and no subtle reasoning interferes with it. On the other hand, the scholar, who has many strange ideas running

through his head, lets many of his tasks or domestic affairs escape through distraction, because he has not grasped them with sufficient attention. But to be safe with a notebook in the pocket is after all a great convenience, in order to recover precisely and without effort everything that has been stored in the head. And the art of writing always remains a magnificent one, because, even when it is not used for the communication of one's knowledge to others, it still takes the place of the most extensive and reliable memory, and can compensate for its lack.[20]

But there is more to artificial memory, in fact. Kant will observe that through reading one can have what is nothing less than an artificial illumination: "clear representations" are a kind of gift, a flash of illumination on a large map. "This can inspire us with wonder over our own being, for a higher being need only call 'Let there be light!' and then, without the slightest co-operation on our part (for instance, when we take an author with all that he has in his memory), as it were set half a world before his eyes."[21] More prosaic but equally important: the blind man can overcome the lack of a sense, not just with some prosthetic aid (the cane), but also with the introduction of the *conceptual order* of the sense that is being substituted. The blind mind internalizes the organizational power of a missing sense, which is to say, it grasps something that is alien to its own sensory system through the symbolic mode of artificial memory. The new "part" of the manifold is capable of then being integrated because the unity as unity is precisely not concretely given: it can accommodate new organization just as the pathological body can in making use of its own "parts" (and perhaps mechanical ones as well) as prosthetic substitutes for original organs.[22]

The Technical Predisposition

I will end with Kant's reflections on the human mind as exception with the suggestion that what constitutes our cognition and separates it from the bestial is just the capacity for this normative interiorization. The human is defined by its more radical openness, a more radical "lack" that will forever preclude a systematic automaticity because the unity of the mind is not a relation between whole and part but between a unity and a *gap* that will never be bridged.

That gap is the origin of what Kant calls the human "technical predisposition." This is the capacity for *manipulating things*, a process that is at once "mechanical" and psychic, one of the sufficient markers of humanity (along with the pragmatic and moral disposition) that separates it from "all the other inhabitants of the earth." Key here is the idea of *manipulation* (*Handhabung*); the focus is literally on the hand. As Kant explains, with the advent of the human hand, a certain necessary spontaneity, plasticity, interruption, whatever we want to name it, is revealed concretely.[23]

The hand is the *hinge* between nature and artifice, between the automaticity of reason and the possibility of norm-generating judgment, because the hand can both enact the will of the mind and reorganize the world itself. As Kant will clearly state, "The characterization of the human being as a rational animal is already present in the form and organization of his hand, his fingers, and fingertips; partly through their structure, partly through their sensitive feeling." Why is the hand a sign (a material instantiation) of rationality? For Kant: "By this means nature has made the human being not suited for one way of manipulating things but undetermined for every way, consequently suited for the use of reason; and thereby has indicated the technical predisposition of skill as a *rational animal*."[24] *Undetermined*. The "form and organization [*Gestalt und Organisation*]" of the hand makes the hand capable of *organizing the unorganized*, of introducing novel order into the world; the hand is the origin of both the technical and rational dispositions. If the rational animal is by nature undetermined it must therefore determine itself. But this novel, original, improbable determination must always be integrated and continually reintegrated into the unity that is the mind—a unity that can never be fully present to itself because the "parts" that constitute it are always *changing*. And especially, we might add, the patterns of thought involved in the spontaneous creation and manipulation of these prosthetic implements. That is, of tools.

PART TWO

Embodied Logics of the Industrial Age

8
Babbage, Lovelace, and the Unexpected

> Were we required to characterise this age of ours by any single epithet, we should be tempted to call it, not an Heroical, Devotional, Philosophical, or Moral Age, but, above all others, the Mechanical Age. It is the Age of Machinery, in every outward and inward sense of that word; the age which, with its whole undivided might, forwards, teaches and practises the great art of adapting means to ends. Nothing is now done directly, or by hand; all is by rule and calculated contrivance.
>
> Thomas Carlyle, "Signs of the Times" (1829)

The Spirit of Technology in an Industrial Age

As social and economic life in industrial Europe became increasingly industrialized, human life would become more and more habituated to the operation of the machine. And the machine was now more than a machine. In *The German Ideology* Marx and Engels had defined the human as the special creature that created its own form of life through productive—that is, technological—labor. But as Marx would later see so clearly, the modern industrial economy was grounded in a complex system of integrated machinery. In other words, human life as a technical form of being was being reorganized in terms of the logic of networked technology.

> An organized system of machines, to which motion is communicated by the transmitting mechanism from a central automaton, is the most developed form of production by machinery. Here we have, in the place of the isolated machine, a mechanical monster whose body fills whole factories, and whose demon power, at first veiled under the slow and measured motions of his

giant limbs, at length breaks out into the fast and furious whirl of his countless working organs.[1]

In this system, according to Marx, the individual human functions as a supplement to the increasingly automatic operation of the factory and, we can also say, the interconnections of communication and energy.

No longer is the machine a technology utilized by the worker to produce something tangibly related to the hybrid of human-machine. Marx reflected on this development in his "Fragment on Machines." Now the automatic systems of machines (and not just the single automatic machines of early industrial production) require human intervention as a "conscious linkage" that enables the system to function as a dynamic interconnection of technologies. The actual machines are not intermediaries or artificial organs: they possess their own skills and their own strength, and they feed on their own forms of fuel. The worker is now regulated by the machines. Going beyond the concept of alienation of labor, Marx proposes an even more sinister condition. The alien power of the machine—this new form of artificial life—is acting through the worker, who becomes an accessory of the machine, or to put it more radically, the individual human worker is an organ of this system, which Marx does describe as a "mighty organism."[2]

A tool was once a "modified natural thing," a productive turning or deflection of nature that was used in the service of human nature, the nature that must produce its own life. With industrial automation and industrial systems of production, human technology is not a deflected nature but in some ways an independent *process* of nature, or at least an artificial simulation of such a process. What does it mean then to work and live in the midst of tools that seem to have escaped the original "organic" logic of supplement, the time when the tool was precisely what was *not* natural, an artificial organ of our natural life, of the life of the brain, the hand, the mind, and even the social body as collective being?

Mechanical Intelligence and Industrial Automation

The appearance of the first major technological advance in mechanical intelligence—namely, Babbage's Difference Engine project (and later, the more sophisticated but unfinished design of a general purpose Analytical Engine)—coincides with the intensification of industrialization in Europe and the increasing automation of large-scale industrial processes of production and distribution. These automatic calculating machines are always positioned as important forerunners of the modern computer (philosophically and technically) and hence haunt any understanding of the postwar project of *artificial intelligence* that emerged seemingly inevitably from the development of digital machines in World War II. But we can ask another question here at this historical juncture: How were these engines understood in relation to the strands of thought we have traced in early modern thought up to Kant's radical critique? Can the organization of the living body and the cognitive powers of the psyche be rethought in light of this important foray in what was not just automatic reason, but an artificial embodiment of that reason?

The project of the Difference Engine started as a dream—quite literally. Babbage's dream took place in 1812, at least according to his recollection, and was only written down decades later. At any rate, the instigation of the dream was the daunting challenge of the logarithmic table, an essential yet error-prone tool for every practical mathematician and engineer in the period. As Babbage would note, the standard published tables were (understandably) rife with errors, which were often corrected in errata sheets that of course themselves might contain errors. (Babbage would cite the Nautical Almanac of 1835, with its "Erratum of the erratum of the errata of Taylors' *Logarithms*.") As Babbage saw, only a "mechanical fabrication" would ever ensure absolute accuracy. This was what Babbage supposedly dreamed as he dozed in a stupor one evening at the Analytical Society: "I am thinking that all these Tables (pointing to the logarithms) might be calculated with machinery."[3]

The fascinating and complex story of Babbage's epic quest to design and build the first fully automatic calculator (which would also automatically print its own results, bypassing another weak human link) has received significant attention and historical analysis. The close, even constitutive relations between the processes of industrialization, human labor, social organization, and new conceptualizations of intelligence have been astutely critiqued, none better than by Simon Schaffer.[4] As William Ashworth has detailed, Babbage and other elite representatives of the new industrial era in Britain, such as Babbage's colleague William Herschel, argued that the model of the efficient factory would be usefully applied to the mind; it would function best if it was well organized to retrieve ideas ("data") without unnecessary *waste* of energy or its storage capacity. The "speed and precision" of an industrial technical process depended on a rational organization, so the aim was to increase the productivity of the mind in an analogous manner.[5] The Difference Engine project can be understood in this light, according to Ashworth, as a technical model of an intelligence modeled on industrial technology: "Babbage's work on his calculating machine was the march of the material intellect set to the rhythm of the factory."[6]

However, this analogy (which can hardly be denied) does raise questions about the "materiality" of the intellect, questions that exceed this framework altogether. What was driving the mechanization and thus materialization of human intelligence in the Engines if human intelligence was itself the inventor of this vast new industrial organization?

If we go back to one of Babbage and Herschel's early texts, namely, the preface to the *Memoirs of the Analytical Society 1813*, written precisely in the period of Babbage's "dream" of mechanical calculation, it is clear that intelligence is not strictly defined by industrial metaphors. Intelligence (like human life itself perhaps) is something that can be *perfected* through the deployment of its own technical inventions. "Modern calculators"—and here they mean by that term error-prone human mathematicians—have been able to devise a highly refined and general system of mathematical analysis that allows the mind to "trace through successive developments" the many "varied relations of necessary truth." The analytic itself is a *system*, a tool that supplements the human mind, allowing it to pursue "trains of reasoning, which, from their length and intricacy, would resist for ever the unassisted efforts of human sagacity."[7]

While it could of course be argued that all forms of mathematical inquiry function in this fashion—as artificial, instrumental prostheses of reason—the singular advantage of analytics, for the two authors, lies in the particular accuracy and conciseness of its wholly *arbitrary* and hence dispassionate mode of symbolization. This symbolization is not the substitution for intelligence at all: it is the zone that allows the necessary exercising of human *judgment* at every step in a train of reason to be much more precise and therefore less prone to error. This is because the ideas are presented so clearly and particularly in one delimited space. Unlike the texts expressed in common language, with "all its detail" and texture that often force the intellect to "suspend its decision" while it seeks related clear conceptions to compare with one specific idea it is holding in mind, the "mechanical tact" of the new symbolic language allows the mind's eye to see at "one glance" the most intricate relations, "shortening the road to discovery, and preserving the mind unfatigued by continued efforts of attention to the minor parts, that it may exert its whole vigor on those which are more important" (ii). Reasoning is anything but automatic here; it follows, in many ways, the structure outlined by Descartes.

Moving to Babbage's most important and influential text, the *Ninth Bridgewater Treatise* published in 1837, it is immediately clear that even in the wake of developing the Difference Engine, he is hardly interested in reinventing or even reconceptualizing the mind as an automatic machine analogous to the Engine. Babbage instead proposes to use the machine as a way of arbitrating a central question in science, that of the universality and automaticity of the laws governing *nature*. Is the universe a system where the "contriver" intervenes at times with a "restoring hand," or is it one whose order is maintained automatically via "the necessary laws of its action throughout the whole of its existence" (32–33)? What is interesting is that Babbage argues that the distinction between these two possibilities can only be illustrated with "recourse to some machine, the produce of human skill"—because, like the natural universe the machine is a "contrived" or artificial order. The challenge is whether the seemingly unfathomable complexities of natural systems could ever be reduced to the operation of a "few simple and general principles" (32).

Babbage's deployment of his Difference Engine in this thought experiment is ingenious and goes in a completely different direction from that of the analogical investigations of the early moderns. He uses the results of his calculator to demonstrate (and not merely conjecture) that an utterly mechanical and purely *automatic* object can produce what appears to be something like a "miracle," or at least an exception to a "natural" law. Babbage explains how he can arrange (we would say program) the machine with a function that will print a series of numbers—an exceptionally long series, say, the series 1 to 100,000,000—that clearly reflects the principle of adding one numeral to the previous one. Using inductive reasoning, a human mind would no doubt assume the existence of a fixed and determined law governing the production of the series, especially because the calculator is a mere machine. There is therefore great astonishment, Babbage says, when the law suddenly appears to change as numerals are printed that are clearly being produced on the basis of an entirely *new* principle (38–39). (Babbage supposedly delighted in performing this demonstration with guests to his home.) The point was to show that a "hidden" law was behind the manifestation of both the original law-like series *and* the sudden transformation that interrupts the continuity, at least for the mistaken human mind.

As Babbage points out, any natural phenomena with similar "catastrophic" aspects (like the historical geology of the earth revealed in strata) need not count as a violation of natural law if we understand that the abrupt changes themselves "have been throughout all time the necessary, the inevitable consequence of some more comprehensive law impressed on matter at the dawn of its existence" (48–49). The fact that the Difference Engine can—*without any intervention*—mimic these abrupt shifts proves that it is at least possible to discover these laws that would govern even what seems to be a discrepancy, a deviation, or exception, even a total transformation of the law itself. As Babbage says, the Engine that works automatically to generate its own simulated "anomalies" without interference from the operator is not itself "more intelligent" than a machine that would produce that same effect but only with an interventional adjustment from outside.

This is not about artificial intelligence per se. The Difference Engine as an automatic exception simulator is just an exemplar of greater contrivance, of more intelligent and elegant artifice, and therefore a better model for an understanding of how the laws of nature could be unified at a foundational level by relatively simple general principles (40–41). The presence of incredibly complex differentiation in nature—Babbage points in particular to the way organic forms fit so perfectly into the successive upheavals of the world—demonstrates the incredible technical skill of the original engineer who established the set of laws that could generate such complexity.

So, far from advocating a mechanistic understanding of human thought or the automation of intelligence in the *Bridgewater Treatise*, Babbage underscores the significance of technology as an expansion of natural intelligence through invention. "It is not a bad definition of *man* to describe him as a *tool-making animal*." After replacing the "skill of the human hand" in the industrial era, newer technologies "of a still higher order," like the calculating engines, are substituting for "the human intellect."[8]

The rapid advance of humanity in modern times is explained by Babbage as the consequence of the materialization of thought in the printed book. In oral cultures, human knowledge was a function of the "accidental position" of the individual. In society, the knowledgeable mind could only address a small number of other minds (52). Truths would be lost, opinion maintained through traditions. With printing, the cultivation and improvement of the mind first becomes possible. Externalized storage and distribution of thinking is like the analytic method, a prosthetic instrument that prepares the mind for self-transformation. "Until the invention of printing," Babbage writes, "the mass of mankind were in many respects almost the creatures of instinct" (51). The printed book is not for Babbage merely a substitute or extension of the traditional form of instruction; the key to the book is its material objectification. The book exists in a new community of thought that purges it of its individual origin; it is an instrument that transmits in an intensified form the distillation of thought across space and time to one single individual, the student. The production of knowledge as material, circulating objects enables a "sifting" of expertise from a vastly collective effort, and an efficient means for "communicating thought from man to man" (55). Society can now benefit from a new form of collaboration where the "reasoning" of one class can unite productively with the "observations" of another.

However, there is no question that Babbage understood his Analytical Engine to be something like a universal reasoning machine. Designed on the model of the Jacquard automatic loom, where punch cards would encode a sequence of operations to be performed, this automatic machine was far more flexible than the calculating engine that preceded it. Still, the internal openness of the Analytical Engine—the fact that the "patterns" were not limited by any one sequential logic—reveals starkly the essential role of the "artist" who creates and encodes the pattern; indeed, the encoder of a textile design for the automatic loom could be described, Babbage says, as a "peculiar artist."[9] While it is true that Babbage strove to protect the operation of the Engine from human error (by insisting, for example, that it punch its own cards and by including an automatic error detection system for occasions that required human assistance) the operating structure of the engine depended on what we would call its algorithmic programming.

Ada Lovelace recognized immediately that the brilliance of the Analytical Engine was its capacity to embody a general "science of operations."[10] That is, instead of merely calculating *numbers*, the Analytical Engine was a "material and mechanical representative" of analysis, the symbolic representation of *relationships*: the machine literally "weaves algebraic patterns" (696), Lovelace commented in a memorable line. The engine is an acceleration and intensification of the very method it embodies. "The mental and the material" are thereby "brought into more intimate and effective connection with each other" (697). She explained:

> It were much to be desired, that when mathematical processes pass through the human brain instead of through the medium of inanimate mechanism, it were equally a necessity of things that the reasonings connected with operations should hold the same just place as a clear and well-defined branch of the subject of analysis, a fundamental but yet independent ingredient in the science, which they must do in studying the engine. (692)

In the end it is Lovelace—often heralded as the first computer programmer based on her algorithm designed to calculate a particular function on Babbage's Analytical Engine—and not Babbage who crosses the line into what we now call the cognitive sciences. It is not that Lovelace wants to make analogous the mind and the automatic reasoning engine. She proposes (in a letter from 1844) to mathematize "cerebral phenomenon"—to discover, that is, "a law, or laws, for the mutual actions of the molecules of the brain; (equivalent to the law of gravitation for the planetary and sideral world)." In an unconscious reenactment of Descartes's research, Lovelace imagines herself becoming a "skillful practical manipulator in experimental tests; and that, on materials difficult to deal with; viz: the brain, blood, and nerves, of animals." As she declares, "I hope to bequeath to the generations a Calculus of the Nervous System."[11]

9
Psychophysics

On the Physio-Technology of Automatic Reason

Lovelace's speculations on a calculus of the nervous system exemplify an important current of thought in the nineteenth-century—what will come to be known broadly in the period as "psychophysics." Figures straying across the border between philosophy and the biological and medical sciences tried to identify both the inner dynamics of cognition and their connection with the activity of the nervous system—this in a period before the accelerated advance of brain science and the emergence of a new experimental psychology later in the century. Indeed, the lack of any solid consensus concerning the relations between mind and body in the period stimulated much new thinking that went beyond traditional philosophical inquiry. As Alfred Smee, author of the book *Process of Thought* (1851), noted, "I was remarkably struck by the unsatisfactory account of the functions of the brain, and I was surprised that so little appeared to have been done in connecting mental operations with that organ to which they are due."[1]

The development in this period of machines like Babbage's Engines, William Jevons's logic piano, and Alan Marquand's automatic electric reasoning machine did not, I suggest, signal a comprehensive mechanization of the human mind; their automaticity, however, is a feature worth dwelling on. Or at least, the "silent" operations of the artificial logic machines point to an important zone of inquiry in the nineteenth-century theorization of intelligence. The flash of insight that occurs "suddenly, without effort, like an inspiration,"[2] and the "inconceivable" presuppositions that fuel the creative leap of hypothesis, the cognitive challenge of the exceptional and the anomalous—this kind of thinking could now be framed as essentially unconscious cognition. The leap of intelligence was a radical gap, to be sure, in the experience of insight. However, that gap could be reconstructed by tracing the unconscious roots of conscious reasoning—and, ultimately, the role of the brain in the production of the creative thinking that was the hallmark of the human mind.

The question of the cognitive unconscious, then, helps us understand what was at stake in the debates about the brain in the nineteenth century, and beyond. How was thought—especially intelligent and *creative* thought associated with discovery, intuition, and epistemological insight—mediated or produced by these physiological information systems? If it was at least possible to conceive of a purely automatic mechanism generating mental life, as Huxley advocated, the question that still lingered from the early modern debates—the question of whether there was something in human thought that demonstrably *exceeded* the mechanisms of sensibility—took on new significance in this new era of brain research.

Smee's own effort to chart the mind results in what he takes to be the "fixed principles" of thought. It is because these principles are fixed that Smee has the inspiration to represent the complexity of the mind's interactions with a "mechanical contrivance" that would give results "which some may have considered only obtainable by the operation of the mind itself."[3] That is, artificial thought. Smee's conjectural machine would encode every word and sequence possible in the mind—thinkable in abstraction but an impossible engineering task, since such a thinking machine, he admits, "would cover an area exceeding probably all of London" and its movement of parts "would inevitably cause its own destruction" (43). However, he does detail the construction of a restricted version of this "relational machine." The important point, however, is that such a machine could give "an analogous representation of the natural process of thought." Which is not to say that the mind is itself a machine. Indeed, Smee admits that the mechanical representation only shows how superior is that knowledge "which is obtainable by the mind through the operation of the brain" (45).

If that is not satisfyingly clear, what is obvious is that Smee positions his thinking machine as more than just a modeling tool for understanding the mind. It can, just like the notational system of analysis invoked by Babbage, function as a supplement to the ephemeral nature of thought and memory. It can also act as a tool for the breaking of habitual prejudices or the eruption of passion, not because it reveals a truth, but because it displays in objective form the correspondence and nonconcurrence of particular logical statements. Ultimately, Smee will map the machine back onto the brain, suggesting that the device could reproduce a reasoning process "according to those principles which regulate the action of the brain in such circumstances," in effect claiming that the externalization and mechanization of the brain activity has the advantage of artificially isolating it from other activity in the body or mind, activity that could, as Hume well knew, "interfere with a sober and correct judgment" (49). Still, if the machine could reason "by a process imitating, as far as possible, the natural process of thought" (50), no human artifact could instantiate what Smee (echoing Kant in a way) calls the power of *spontaneity*. This reveals the infinite superiority of the "cerebral organization" compared to our most ingenious contrivances. Left unsaid is exactly how the biological "machine" is capable of such autonomy.

George Boole's much more influential work, the *Laws of Thought* (1854), explored the same terrain—"the fundamental laws of those operations of the mind by which reasoning is performed"—and was also concerned with translating those principles into a system of representation, here a "calculus" that would function as a new methodological tool for the mind in expanding its logical capacity. As Smee already asserted in his work, the logic of the mind was, for Boole, facilitated by our use of discursive symbolization: "Language is an instrument of human reason, and not merely a medium for the expression of thought."[4]

The limit of these projects is clear: the form of reasoning that could be reduced to fixed principles, even mechanized—namely, deductive logic—was already a function of a kind of internal psychic technology, which is to say, our use of a symbolic system of notation. Even more problematic was the widespread recognition that the human mind exhibited forms of thought, such as insight, analogy, and intuition, that defied rigid definitions of syllogistic reasoning inherited from the classic logical tradition. For philosophers and proto-psychologists, as well as practicing neuroscientists, these forms of thought were often understood to be the most productive, even essential to the discovery of new ideas and other forms of invention. So the scientific challenge for mid-nineteenth-century thinkers studying human reason and intelligence was how to understand both the formal procedures of logic and the more ephemeral capacities of the mind, all in the context of a "natural history" of the sensible body and nervous system.

John Stuart Mill perhaps best exemplified this trend, with his influential and controversial rethinking of logic as, at its foundation, a natural and intuitive dimension of the mind and not the result of a formal symbolic procedure. Reasoning was a fact of our *embodiment*. Mill's examples highlighted the significance of the layering of experience and the largely unconscious process of integration (especially inductive formations) that would prepare the way for insights that emerged in the moment of decision. Central to his theory was analogy as both a cognitive function of the mind underlying rationality and a basic form of operation for the nervous system and brain. His example here is the *coup d'oeil*,[5] the insightful glance of the experienced mind.

> An old warrior, on a rapid glance at the outlines of the ground is able at once to give the necessary orders for a skillful arrangement of his troops; though if he has received little theoretical instruction, and has seldom been called upon to answer to other people for his conduct, he may never have had in his mind a single general theorem respecting the relation between ground and array. But his experience of encampments, in circumstances more or less similar, has left a number of vivid, unexpressed, ungeneralized analogies in his mind, the most appropriate of which, instantly suggesting itself, determines him to a judicious arrangement.[6]

Other figures would characterize the often-maligned category "hypothesis" as a crucial dimension of human reason, indeed a core capacity that was indispensable for the discovery of truth. As William Thomson, in another work addressing the perennial topic of the "laws of thought," published in 1849, pointed out, whatever the vagaries and risks of hypothetical conjecture, without the insight that often comes in a "flash" when investigating nature the scientist would be lost. This is how he described it: "This power of divination, this sagacity, which is the mother of all science, we may call anticipation."[7] It was obviously not clear how anyone could ever formulate fixed *principles* for this kind of inference—what Thomson called "the logic of anticipation, the philosophy of the unknown."[8]

This is all to say that a nineteenth-century preoccupation with mechanical instantiations of logical processes and automatic devices in general hardly indicated a general project to frame human thought or even human reason as an essentially mechanical process—or even to say that human thinking was some kind of output of cerebral activity. William Jevons, who actually built a sophisticated reasoning machine in the 1870s, the so-called logic piano, did characterize the process of logical reason as in essence a process of "substitution,"[9] which was akin to the industrial system of assembling products with standardized replaceable parts.[10] This kind of action was the model for the logic machine, which would operate automatically to arrange and "process" trains of reasoning based on the precise substitution of terms. The formal instruments of symbolization and method could now be liberated from the mind entirely and performed accurately and repetitively. As Jevons noted, logic was often called an "organon," and in many sciences the assistance of a vast array of technical machines was commonplace (107). The logic piano was just such a machine, to be used for logical assistance. Jevons was not at all willing to set up the logical piano as a kind of *substitute for human thought itself*. As he remarked, "It cannot be asserted indeed that the machine entirely supersedes the agency of conscious thought; mental labour is required in interpreting the meaning of grammatical expressions, and in correctly impressing that meaning on the machine; it is further required in gathering the conclusion from the remaining combinations" (111 n.).

Even more important, the machine was incapable of what Jevons called *intuition*. For example, given the proposition A = B, the machine could not infer on its own that B = A; one had to supply both propositions because it could not intuit the identity. This was no mere technical observation; Jevons called it a "remarkable fact." This is because he was, in his larger project on the "principles of science," interested in showing the importance of such intuitive thought (Mill's "induction") for the advancement of knowledge: "Hypothetical anticipation of nature is an essential part of inductive inquiry, and . . . it is the Newtonian method of deductive reasoning combined with elaborate experimental verification, which has led to all the great triumphs of scientific research" (ix). As Jevons would explain, these anticipations of nature are the leaps that make possible the kind of research and observation that make possible new spheres of understanding. The foundation of this kind of thinking was not amenable in any way to the mechanisms of his logic processor. As he wrote, "The truest theories involve suppositions which are inconceivable, and no limit can really be placed to the freedom of hypothesis" (557). What was crucial to this kind of creative thinking was the cognitive attention to *novelty*, especially its often unsettling appearance (644).

Here Jevons moves us further and further away from a machinic vision of reason. Not only is the machine incapable of even the most basic forms of intuitive anticipation, the inductive reasoning natural to the mind; it was therefore also incapable of the even more creative turns that come with the failures of anticipation. We are aroused to "mystery" precisely when the "routine" of our knowledge is affirmed in unvaried "everyday observation." The confrontations with the "exceptional" need to be understood as "the points from which we start to explore new regions of knowledge" (644–45).

Peirce and Machinic Thought

Since calculation is so much of an external and therefore mechanical business, it has been possible to manufacture machines that perform arithmetical operations with complete accuracy. It is enough to know this fact alone about the nature of calculation to decide on the merit of the idea of making it the main instrument of the education of spirit, of stretching spirit on the rack in order to perfect it as a machine.

Hegel, *The Science of Logic* (1816)[11]

In one of his early essays in logic, published in 1871, the great philosopher and semiotician Charles Sanders Peirce gave a formal analysis of the different modes of rational inference, supplementing the traditional categories of deduction and induction with that of "hypothesis"—that strange and somewhat mysterious capacity for insight into a problem that is not predicated on any specific evidence.[12] True to his pragmatic orientation, Peirce redescribed logic here as different forms of inference suitable to the varying needs of the embodied mind. If deduction is the careful, attentive process of deriving conclusions from trusted principles of knowledge and induction is the at times unconscious reasoning that formed of habits from repetitions of observations in similar contexts—habits that gave the mind the security and efficiency of prediction—Peirce positions hypothesis (which he will in later works rename "abduction") as *emergency thought*. Hypothesis becomes necessary, that is, in times of crisis, when our knowledge, our trust in certain principles, say, or our inductive expectations about the world, fail us.

The essence of the hypothetical conjecture is, for Peirce, surprise, the "unexpected." Hypothesis begins with the startling result. And because of its jarring appearance, the result at first has absolutely no context for understanding, or to put it more precisely, this result *defies* what our understanding would entail in this specific situation. Hypothesis is therefore entirely conjectural.[13] As Peirce explained, the mind must find some principle associated with the anomaly and then imagine, with no evidence and not even an indication of direction, a way that the anomaly could have been generated. The value of the hypothesis is not in its speculative character but rather lies in the fact that without its fantastic invention, the mind would not know where to begin to look for clues. Discovery (especially in the sciences) was first of all a purely cognitive act. The hypothesis can then be verified through directed observation and experimentation.[14] As the French philosopher Ernest Naville would note, in his own study of hypothesis from 1880, scientific progress cannot be accounted for solely by induction and deduction reasoning. To understand "the special process of discovery" we must introduce hypothesis, for which Rougemont, echoing Peirce, had coined the neologism "conduction" in 1874, Naville remarks.[15]

But what was the source of a hypothetical conjecture? Peirce, like so many others in this century and even before, emphasizes its sudden and unpredictable nature, which is to say, its essential mystery. "The abductive suggestion comes to us like a flash. It is an act of insight, although of extremely fallible insight."[16] This was the kind of thought that had, by definition, no method or system that would guide us to the promising solution to the exceptional problem. Peirce was more than willing to admit that deduction, the classic example of rational inference, was essentially mechanical in that it relied on strict relations that could not vary. As he wrote in a review of recent technologies from 1887, "The logical machines of Jevons and Marquand are mills into which the premises are fed and which turn out the conclusions by the revolution of a crank."[17] Peirce worked closely at Princeton with Alan Marquand, who was developing an electrical variation of Jevons's logic piano. Remarkably, decades before Claude Shannon's pivotal work on logical relays, Peirce offered Marquand the brilliant suggestion (never taken up) that electrical circuits could be arranged in such a way as to perform logical operations directly.[18]

In any case, it might seem surprising to find Peirce admitting that the difference between the logical machine and the human reasoner was merely the "place" where logical inference was taking place. As he once wrote:

> All that I insist upon is, that, in like manner, a man may be regarded as a machine which turns out, let us say, a written sentence expressing a conclusion, the man-machine having been fed with a written statement of fact, as premiss. Since this performance is no more than a machine might go through, it has no essential relation to the circumstance that the machine happens to work by geared wheels, while a man happens to work by an ill-understood arrangement of brain-cells; and if there be room for less, still less to the circumstance that a man thinks.[19]

However, it is clear right from the seminal essay on logic of 1871 that Peirce felt that deductive reasoning was by far the *least* interesting form of human cognition, just a technique for "paying attention."[20] So if Hobbes's notion of logic as computation, Babbage's theory of artificial engines, and the more recent invention of logical machines left nothing to be desired in terms of computational precision, Peirce insists that all these machines "have no souls that we know of." This was not some vague mystical rejection of materialist understandings of the mind. For Peirce, the machine "does not appear to think, at all, in any psychical sense."[21] In synthetic leaps of thought like the abductive conjecture of an imagined case, machines (governed by strict procedures that conditioned their movement from one configuration to the next) would be at a complete loss.

In the end, Peirce would express doubts about the mechanical performance of even the strictest "necessary reasoning," noting that a machine like Babbage's or Jevons's was incapable of generating the "unexpected truth" that comes in the process of reasoning, where initial premises, for example, can be reintroduced in creative ways. Even in chains of deduction, then, there are moments when the reasoning path opens up "several lines" of possibility and therefore demands a decision. Genuine human cognition will always depend on the exercise of *invention*, which is present even in deduction but was utterly essential to the radical leap of conjecture. Crucial to genuinely human thinking was the use of certain *tools*, from the relatively mundane (such as pen and inkstand) to the complex (chemical instrumentation) in scientific pursuits.[22]

As Pierre Janet put it, in an 1889 thesis on automatism, and here we can note an echo of Descartes, the "judgments" that are so essential to human reason are phenomena that are radically different from the experience of sensations and images; the "idea of resemblance" between two perceptions, for example, is not itself something that could be a "sensation."[23] While detailing the many automatisms central to human existence, Janet would argue that what distinguished the human was the presence of some "other activity superadded [*surajoutée*] to this automatic activity." What I want to underline here is that the status of the brain was in question in these conceptualizations of human thought and did not stand in for mechanistic reductionism. As Janet said, "the automatic act is rigorously determined" (476) and the exercise of judgment reveals a "true independence" (477); it was a "power of synthesis capable of forming entirely new ideas which no previous science had been able to predict" (478).

The brain that emerged in the middle of the nineteenth century as an object of serious experimental and physiological study was positioned as something capable, at one and the same time, of both automaticity and the very *disruption* of that automaticity. Janet, echoing Peirce, wrote that judgment was a "new and unexpected [*inattendu*] phenomenon, like consciousness itself, appearing in the midst of phenomena of mechanical movement" (477). What was the origin of this break that was the unprecedented intervention? This was not really a purely philosophical question for nineteenth-century intellectuals. As Janet reveals so clearly, the problem of creative thought was defined in terms of what we can call the natural history of the organism, in that the function of "novel" and unprecedented cognition was resolutely linked to the survival of the organism; the synthetic capacity is likened to the political decision that must be made in "new conditions" because the conservative forces of automaticity can no longer produce "equilibrium" since they derive only from past experience (486–87).

The spirit of the intellect was not, then, a sign of some radical division within the organism. This spirit emerged in the moment of the *exception*, when the norms of automatism might fail.

> When the mind is normal, it consigns to automatism only certain inferior acts which, conditions having remained the same, can be repeated without inconvenience, but it is always active in carrying out new combinations at every moment of life, which are constantly needed to maintain equilibrium with changes in the environment. (487)

The pathology of "automatism" that is addressed in Janet's clinical work is not identified with automaticity itself but in fact the lack of this adaptive and creative capacity. He writes, "But whether the environment varies because of misfortunes, accidents, or simply changes, an effort of adaptation and new synthesis will be required, otherwise [the individual] will collapse into the most complete disorder" (487).

10
Singularities of the Thermodynamic Mind

The emergence of thermodynamics and the study of organization as a function of energy and its dissipation provoked important new conceptualizations of organismic vitality. Claude Bernard on the internal regulation of the body (the "internal milieu") would be one notable early example. However, we can also see how the thermodynamic approach to technology (the transformation and direction of energy) offered another angle on the question of the self-interrupting spiritual automaton. That is, thermodynamic models and the mathematical tools of this science opened the way to an understanding of how the lingering issue of this Cartesian spirit of judgment could operate concretely within automated and self-regulating systems (whether these were biological or artificial machines). The nervous system was now often directly analogized to the communication technologies that regulated vast networked systems like railways or colonial infrastructures. Crucial to this vision was the idea that the direction and organization of vastly powerful material forces would be dictated by the most ephemeral of physical forces—that is, electricity.

The dangers of a materialist reduction of the nervous system (not to mention the complex mega-systems of the industrial and colonial age) are obvious. So it is important to tease out of these intertwined domains the conceptualization of the points of intersection within these machine-information hybrids. What were the limits of automation? Where, in the human being in particular, would be revealed a potential space for intervention—not a space for the transcendence of the automaticity of the machinery but rather for the appearance of indeterminacy or uncertainty, conditions that demanded decision precisely because there was always something that escaped the automatisms organizing this system.

One sphere for offering new scientific resources for this challenge was thermodynamics. In his famous and influential essay, "On the Interaction of Natural Forces," which appeared in 1854, Hermann von Helmholtz takes the time to give a concise but thorough history of automata, characterizing the early modern interest in constructing self-moving machines (his examples include Jacques Vaucanson's infamous defecating duck and his flute-playing machine that mimicked the actual blowing techniques of human musicians) as an exercise in overly ambitious scientific enthusiasm—to rebuild nature, to construct a model of the functions of men and animals. However, by the nineteenth century, interest in the automatic was focused squarely on the functions of the machine in the industrial age. "We no longer seek to build machines which shall fulfill the thousand services required of *one* man, but desire, on the contrary, that a machine shall perform *one* service, but shall occupy in doing it the place of a thousand men." So how to conceive the automaton in light of the new science of energy? With the principle of the conservation of energy comes the problem of entropy, the loss of energy in any physical system due to friction and collision, and the inevitable conduction of heat from warmer to colder until homogeneity is reached. As Helmhotz put it, if the universe is left undisturbed in its physical motions, "all force will pass into the form of heat, and all heat come into a state of equilibrium."[1] At this point, there will be no change, no transformation whatsoever—what we now call the "heat death" of the universe.

The question of the organic being must be posed anew in this context. The automata that were models of living "machines" (from Descartes on) were "clockwork" systems, Helmholtz observes, but there was never any consideration of how these machines were wound up, *energized*, so to speak. But what if we compare the living body to the steam engine? The machine, the body, both must be seen as active, even *destructive* beings—constantly consuming and transforming energy in order to maintain themselves in changing environments. And, at least with organic beings, the worn out or damaged parts of the system must also be repaired and replaced from within.[2]

The automaton as something fighting against disorder—rather than simply expressing order—opened up new concepts of will and action (and "mind") within machines and bodies.[3] For Balfour Stewart, writing in 1874 on the conservation of energy, there were (as he wrote in his appendix) "vital and mental applications of the doctrine."[4] He points to the kind of physical structures that are balanced on a knife edge, in moments of radical indeterminacy, where an "unstable equilibrium" will lead to an unpredictable outcome. Or at least, the ultimate cause that pushes the structure one way or the other is so "exceedingly small" that it is beyond all measurement. Stewart's point is that there are different kinds of machines (whether in the organic or technological sphere)—namely, those that are organized to be highly *predictable* and stable so their behavior is calculable and those whose "delicacy of construction" entails an essential incalculablilty (99–100). Depending then on the goal of the machine, it either tries to reduce instability, or it takes advantage of it. If the "timepiece" is the quintessential example of the first, the other kind of machine, where "the object aimed at is not a regular, but a sudden and violent transmutation of energy" utilizes as its means the "unstable arrangements of natural forces" (158). Any machine with "trigger" points (quite literally the example of a rifle, say) would count as an instance of this category—although significantly, Stewart will emphasize that in reality the machine consists of both the rifle and the human that wields it (159).

Stewart's provocative claim here is that the organism must be understood to be this kind of incalculable machine; its internal instability is the only way to explain, physically, its profound flexibility, what Stewart calls its "freedom of motion." He will suggest that "life is always associated with machinery of a certain kind, in virtue of which an extremely delicate directive touch is ultimately magnified into a very considerable transmutation of energy" (163). There is, Stewart argues, something obviously absurd about a human being calculating its own movements, or those of any other human. The argument is drawn from chemistry and physics, not theology or philosophy: living beings are constructed of unstable substances and must operate in unpredictable and often hostile circumstances. What guides the organism—"life" is what Stewart calls it—might be likened to a "great commander" at the end of a chain of organizations in a "vast army" at war (161). Telegraphic wires lead to the "well-guarded" room of command, and the commander can send orders back through the same cables. And as Stewart says, what flows here is *information*. Life is not a sovereign presence actively coercing the structure of the body. What "causes" the army to move in a coordinated way is that there is a flow of commands along the chain. The "cause" of the whole movement, then, is one "delicate directive touch" in that room—and Stewart suggests that the brain chamber is where we might locate the directive force of life, "so well withdrawn as to be absolutely invisible to all his subordinates" (162).

Citing Stewart's work, the thermodynamic theorist and physicist James Clerk Maxwell would refine this model by noting that certain physical systems might be, in general, predictable and calculable based on their configurations, yet these systems may pass through "certain isolated and singular phases, at which an infinitesimal force may determine the course of the system to any of a finite number of equally possible paths."[5] He makes an analogy with the figure of the "pointsman" who, at a railway junction, has to decide which trains go to which tracks. Maxwell will also speculate that we could imagine the soul as the engine driver of the train, who by means of valves directs the vast and powerful machinery of the steam locomotive. But he would go on to think of the mental direction of the body primarily in terms of the pointsman, or "steersman," who intervenes precisely at the critical moment when the predictable path of the particular entity reaches a point of decision—what will be called a bifurcation in mathematics.[6] "When the bifurcation of path occurs, the system, ipso facto, invokes some determining principle which is extra physical (but not extra natural) to determine which of the two paths it is to follow."[7] Maxwell is interested in applying to cognition and mental action these kinds of examples in physics where natural laws consist in influencing direction—the process of molding and determining, for instance, as we see in the case of organisms. Not, strictly speaking, "forces" in themselves, they are unconscious in the body but become conscious in the realm of thought.[8] The laws of thought are related then to the laws of organization and not "physical" laws of force. The will is figured here as a possibility of direction but not in the usual sense of participating in the action and reaction of material entities. The will is a capacity for deflection, and must be conceptualized within the framework of Maxwell's development of the mathematics of the singularity.[9]

The organic being could not actively resist the second law of thermodynamics if it meant using force to counter the effects of physical disorder. The soul that directs the body is therefore not an active source of energy but rather what Maxwell will describe as "that of a steersman of a vessel—to not produce, but to regulate and direct the animal powers."[10] The traditional division between "free will" and determinism is replaced by Maxwell with the idea of machines that are either largely *stable* and therefore predictable in their motion or in some way essentially unstable and therefore susceptible to being drastically affected by relatively small impacts on the system. The free will is more like a Lucretian atom, swerving and deviating in an "uncertain manner from their course" (820). The implication is that at definite moments in the life of a being there are points of bifurcation, and in these moments an "imperceptible deviation" will be the determining factor for which path will be followed. The difference between lower organisms and higher animals or humans, and the distinctions within artificial machines as well, will be measured by the degree of complexity. For the more complex an organized system is, the more "singularities" will appear in the system. And it is in these singular moments that deflection is possible. "All great results produced by human endeavor depend on taking advantage of these singular states when they occur" (822). Maxwell can redefine the preeminence of the human now not so much in terms of a more sophisticated or complex organization itself, but rather as a consequence of the mathematically demonstrable instability and incalculability of these systems. "It appears then that in our own nature there are more singular points—where prediction, except from absolutely perfect data, and guided by the omniscience of contingency, becomes impossible—than there are in any lower organization" (822).

Maxwell was offering a novel way of integrating the tension within the human, between the demonstrated automatisms of both physiological and mental life, and the irruption of spontaneity and invention in the midst of regularity. The singularities are of course defined by the functions governing regularity, and they are by their very nature isolated occurrences. In any case, Maxwell's call for a science of *singularities and instabilities* is aimed at removing the dominant "prejudice" in favor of determinism that permeates contemporary research. The exception here is no longer the artificial miracle generated by Babbage's Difference Engine; it is the sign of a new metaphysics, one that allows a certain contingency or even looseness in the concept of the natural law. The "steersman" idea pointed to a concept of a human mind that had its own systematic function but that could also enter into the open space that was the organismic singularity to radically influence the normally automatic behavior of the physiological system.

In a similar vein the French mathematician A. A. Cournot would argue that the laws of the physical world and the laws of organization that held in the world of living beings were two different things. Physics could not account for the phenomena of life, yet the seed of life develops from the "soil" of physical laws. But in the physical world, an exception—the *surnaturel*—could only ever be a miracle.[11] The definition of organic life and the *evolution* of organic life depended on what was a kind of violation of the physical laws, in that it was the occurrence of small "deviations" that produced complexity. This propensity to novelty was what continued to distinguish organic life from the merely physical for Cournot. In moments of crisis, organisms were capable of "extraordinary dispensations."[12] The exceptional was an exception but not really a *violation* of the physical laws of the universe; rather, the exception was a violation of the usual biological norms of existence. In the singular case, nature abandons its usual course and is not governed here by any other law. As was once said of the great monarchs of the Old Regime, "si veut le roi, si veut la loi." In other words, the appearance of the radical exception in organismic beings just means that there is no law that can explain this appearance; the singular is not really a violation of the law but a moment where the law simply does not function.[13] Writing against the Lapalacian idea of total prediction made possible by the calculability of physical law, Cournot points to the organic sphere.

> He would have been forced to recognize in each species [type] all the characteristics of a law which the legislator repeals and replaces according to his views. At this count, for insects only, there would have been several hundred thousand laws to register in the Code of Nature, and laws which have changed several times supernaturally, that is to say extra-legally, by a sort of revolutionary measure or *coup d'état*.[14]

As we have seen, the constitution of freedom within the systems of organismic and technical machinery lay in the capacity for *interruption*, and the instability or disorganization of a system (especially a living system) was pinpointed as the localization of *indetermination*. But what exactly was this capacity, or to put it another way, where was the point of instantiation for the leap into the unprecedented that provided the opportunity for reorganization and creative invention, especially with respect to the organic, living body? In the late nineteenth century, the brain would be a key site for framing this critical question.

11
The Dynamic Brain

The history of brain science in the nineteenth century has often been framed as a struggle between the (ultimately successful) advocates of "cerebral localization" and those who saw the brain as a decentralized organ.[1] Without diminishing the importance of these debates for the history of neurology, it is crucial to recognize that the boundaries between these positions were never always clear or rigid. And, as I want to emphasize here, what links the two perspectives is the underlying if not always explicit problem of staging within the brain key thresholds, the lines between, say, animal and human, or between instinctual, habitual, and creative (or, we might say, *artificial*) cognitive behavior. If the brain was understood to be a kind of neural machine, the human brain raised very particular challenges, since it was identified as the site of the human exception—the site, that is, of a special kind of intelligence that exceeded mere physiological or environmental demands.

Indeed, the ambiguity can be seen clearly as these debates open in Europe, with the startling and seminal experiments carried out by Marie-Jean-Pierre Flourens, beginning in 1815. When Flourens surgically removed various sections from animal brains, these animals would subsequently exhibit various pathological conditions, most notably the complete lack of perception, initiative, and judgment when portions of the cerebral spheres were lesioned, for example, and then the disappearance of balance and mobility when the cerebellum was taken out. Flourens showed the complex intersection and interactions of tasks taking place in the brain by disturbing the functions. One could eliminate the perception of visual phenomena, for example, yet the automatic systems of visual sensation (such as the operations of the iris) would still operate independently.[2] While the theory of brain localization was not new, before Flourens it was all essentially pure speculation, which complicated the controversial efforts of Franz Joseph Gall and his collaborator, Johann Gaspar Spurzheim, earlier in the century, to found the new discipline of "phrenology," which infamously mapped the surface of the skull as an indication of the spatial location of certain intellectual functions.

However, Flourens's demonstration of the functional division of the brain must be juxtaposed to his fundamental and much-debated claims: first, that the nervous system was fundamentally a *unity* (ch. 12) in which the "energy" of each of the parts influences the others, creating a "unique system" that would then coordinate the parts in a hierarchical order of control and subordination (243); and, second, that *within* the distinct zones of the brain neural tissue had a general capacity to instantiate these functions. This theory would come to be known as the "equipotentiality" of the brain tissue, though Flourens did not use this term; today we would say plasticity. The theory was based on the idea that if the intelligence of an animal had a "determined location," that is, its own organ (namely, the cerebral lobes), then the "property" of intelligence (perceiving, judging, willing) was clearly unitary in its operation, and therefore could not be localized *within* the cerebral lobes (244). The key demonstration of this hypothesis was the fact that even with localized surgical destruction within the cerebral lobes, the exercise of intellectual function was usually not disturbed (as long as the destruction did not go past certain bounds) (265). As Flourens explained, the lobes must operate as an ensemble, so "it is completely natural that one part can compensate [*suppléer*] for another, that as a result intelligence could either continue—or come to an end—through any one of them" (264).

Generally speaking, France was considered the center of leading research on the brain in the wake of Flourens's work, and for the following decades there was a cautious consensus against the most radical localization theories, in line with Flourens theory of equipotentiality and the cerebral distribution of functions in separate but interacting spheres. Yet in 1861, Paul Broca's startling discovery of a "language" area in the human brain—he identified postmortem highly specific damage in the brain of a patient who had suffered from a well-studied severe language disability—opened up more concerted efforts to "map" the location of more discrete functions in the cerebral organ.[3] The introduction of electrical stimulation methods in both animal and human experimentation allowed for a much more precise and disciplined methodology, leading to a new era in neurosurgery once doctors could seek out lesions in specific zones according to symptoms and avoid damaging sensitive areas. By 1908, the different zones of the brain, carefully differentiated by distinct tissue types, had been identified, and the German neurologist Kordanian Brodmann, in collaboration with several other researchers, published a definitive map (still in use today) that provided a comprehensive topography for the subsequent localization of various functional capacities in experimental studies.[4] (Figure 11.1.)

Yet the isolation of these specific functions and their spatial coordinates only displaced the crucial problem of explaining the crucial fact of integration and coordination of the myriad operations of the brain. And Flourens's initial demonstration of neural "compensation" in the face of damage or injury could hardly be accounted for by a strict localization hypothesis. The development of the neuron doctrine in the late nineteenth century, instigated by revolutionary staining technologies that revealed for the first time the microstructure of nerve cells, raised radically new questions concerning the nature of neural connectivity and communication. In this context, the interplay of automaticity and flexibility would be reimagined. The psychic operations, which were understood to be mostly unconscious, need not be considered "fixed" just because of their automaticity and routine nature.

Revisiting the eighteenth-century theory of habit, brain researchers hypothesized an *active* brain, dynamically adjusting itself to the environment. As Henry Beaunis argued, in 1876, following on his study of somnambulism, "Cerebral organization, a necessary condition for psychic phenomena, can be continuously modified under the influence of impressions coming from either outside or from our body itself."[5] With repetition came habit, and the process could be explained as a transition from conscious attention and reaction to unconscious automaticity *within* the brain as a unitary organ—a dynamic of maintaining an "equilibrium." The capacity of the brain to change itself was fundamental, and as Beaunis notes, it is not only through repetition that reorganization takes place. With a single, intense experience (we might say an extraordinarily exceptional one) "the modification produced only one time can become permanent, and this modified nerve center thus reacts differently than it would have done before the modification."[6]

Beaunis indicates here the openness of the brain to interference. Describing the power of the hypnotist, he explains that an "unexpected shock" delivered to the nervous system can interrupt its course. The success of the hypnotic intervention in clinical cases relies on this cerebral shock because it initiates the modification and hence reorganization of the brain. The implications are obvious: To what extent, Beaunis asks, is the brain subject to

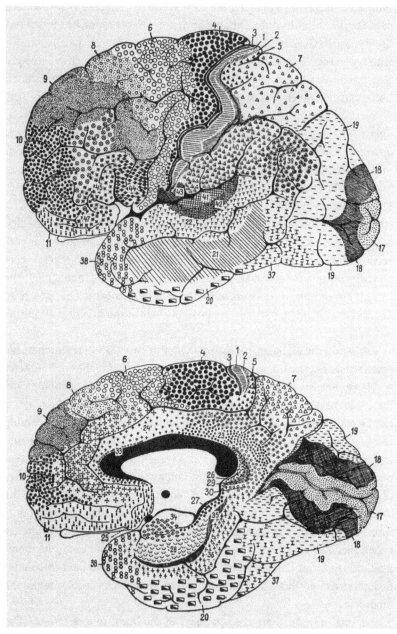

Figure 11.1. From Korbinian Brodmann, *Vergleichende Lokalisationslehre der Großhirnrinde* (1909).

such shocks in other situations, perhaps *producing* the pathological (or normal) state itself as a series of radical self-enclosed modifications? So while Beaunis affirms his commitment to localization of functions in the brain, his depiction of habit as the acquisition and superposition of new cerebral organizations that supplement any "innate" order reveals at the same time a vision of a fundamental neural plasticity.

In lectures given in 1877, the neurophysiologist Charles-Édouard Brown-Séquard put this issue of cerebral openness at the center of his doctrine. Admitting that different functions have different organs in the brain, he would argue that the organ was not a fixed *spatial* object but instead a network of connected cells "diffused in many parts of the brain," cells that form a "whole" through the union of fibers made possible by constant communication between the cells.[7] Virtually all the cells of the brain, he suggested, could take on any of the various functions. The best proof of this theory was the many cases where brains overcame radical damage, for example, children who had half their brain diseased and yet still developed normal language and intellectual abilities. To commit to localization therefore meant conceiving of a new *topology* of organization that exceeded three-dimensional mapping.[8] The famed pathologist Camillo Golgi, who developed a revolutionary staining technique, would criticize in 1883 the prevailing research that focused so much on localization, which was not, he said, "in perfect harmony with the anatomical data." He insisted on the importance of the neural *networks* formed by the extensive interconnection made possible by the multiple links between individual cells and others throughout the brain. A "rigorous" localization of function made no sense in this framework, as Golgi argued for a new concept of organization based on the "diffuse network" that conjoined cerebral "provinces."[9] With the network idea, it was plausible to see the higher intellectual functions, which were responsible for the more abstract and flexible modes of behavior, as more or less freed from spatial determinations altogether.

Although the Scottish neurologist David Ferrier is renowned for his contributions to brain mapping and functional localization, and drew criticism from Golgi as a result, the question of the higher intellectual capacities of the human mind was not for him simply a matter of locating the area of the brain associated with those capacities. Ferrier distinguished the human from the animal (and marked the gradations of all higher animals) according to their potential for *volition*. Based on extensive experimentation ablating animal brains (he would, in fact, later be the first one charged under new anticruelty laws), he saw that rabbits, for example, compared to dogs were more governed by "autonomisms" and therefore were not disturbed as much by destruction to the cortical centers. Conscious movement was *not automatic* and required what Ferrier called education. So in the human case, where "volition is predominant," extensive "and laborious" education is necessary in order even to function capably in the environment. Most birds are virtually "conscious automata," with little in the way of developmental potential, while "the human infant can scarcely lift a finger on its own behalf."[10] Linked to the nonautomatic human brain was the capacity to retrain even after catastrophic injury. So while Ferrier believed that the intelligent brain was a function of the complex operation of both hemispheric halves, nonetheless, even with destruction of an entire hemisphere, individuals have been able to acquire the full complement of functions, even those that seemed to be consistently localized.[11]

The tension revealed in the brain debates on localization and distribution of function points to the constitutive tension at the heart of human cognition, the existence in one being of both natural automatism and an equally natural capacity for exceeding in some way cerebral determinations. Ferrier, like so many others in this period, was quick to identify the cortical lobes as crucial for human intelligence, given their predominance compared to other animals—and equally quick to point to the supposed "great differences in the development of the lobes" among the races, marking the racial hierarchy of "mental powers." At the same time, his theory of the brain denied any specific "seat" of intelligence as it emphasized the significance of *association* with the activity of other "centers" and their "respective cohesions."[12] Finally, Ferrier would remark just how critical the use of symbols was for human intelligence, which could be linked directly to our ability to use abstract ideas and form conceptual relations—leaving open the question of how symbols arise in the brain in the first place.

Much of the comparative and anatomical work on the brain in the late nineteenth century was of course intertwined with the evolutionary understanding of the development of species and their respective place in the process of specialization. It was not just that the brain was understood to be progressively developed from the appearance of sensory organs in simple creatures to the highly complex nervous systems of the higher mammals. The brain itself could be interpreted as a kind of palimpsest of evolution itself.

The idea that the brain was a recapitulation of its own evolution was most explicitly conceptualized by John Hughlings Jackson, whose work, especially in epilepsy, led him to the conclusion that seizures revealed what he described as a condition of "de-evolution," where various regions of the brain were deactivated, allowing for a regression to certain automatisms generated by lower brain centers.[13] Hughlings Jackson articulates a thread implicit in nineteenth-century brain science, a thread that crosses many of the disciplinary and theoretical disputes, namely, that the brain was not a final product of evolution but in fact an organ still in the midst of a *process of evolution*. The tension between automatism and creative "freedom" in the brain was resolved in a way similar to the theory of habit: original responses are, through repetition, consolidated in the brain's structure as an automatic reflex. But this theory of habit is now played out along a long temporal axis. Hughlings Jackson argues that the most basic autonomic controls of the brain have been long fixed by evolution and transmitted through a genetic inheritance, while other topological spaces of more recent lineage have a more or less essential determination (as Ferrier and others also argued).

What is most tantalizing about Hughlings Jackson's theory is the idea that the highest functions of the human mind are located in the most *recent* cerebral acquisitions, areas that were the site of a stage of neural evolution that was still ongoing. The cortex now understood as a space of evolutionary dynamism, testing, adapting, improvising—all this entails a fundamental openness of structure. This means that the *least* organized centers are at the same time the most complex, according to Hughlings Jackson, in that they are capable of the greatest variety of behaviors. This capacity is what can explain the exceptional status of the human mind without invoking metaphysical distinctions and without rejecting the main tenets of evolutionary theory. The flexibility and creativity of our intelligence stems from the flexibility and creativity of the higher brain's own *lack of determination*. "If the highest centres were already organised, there could be no new organisations, no new acquirements."[14]

There was no longer, then, an abrupt division between "automatic" and volitional for Hughlings Jackson. If the human was an automaton, he remarked, it was never a *perfectly* automatic one, give that the brain revealed evolutionary layers arranged from "most to least automatic." "A perfect automaton is a thing that goes on by itself. . . . To say that nervous arrangements go on by themselves, means that they are well organised; and to say that nervous arrangements go on with difficulty, if at all, by themselves, is to say that they are little organised."[15] The very lack of organization opens up more room for what he calls "correct adjustments in new circumstances." The automatic automaton, with its total determination of cerebral responses would be perfectly adapted to the circumstances of its original environment, but in the context of evolutionary transformation it would be incapable of any adaptation to "new conditions."[16] Becoming automatic, in other words, is to become "more perfectly organized."

Thinking the brain in the late nineteenth century was therefore never simply a matter of replaying the metaphysical debates over dualist ontologies, nor was it a question of taking a materialist or vitalist stance on the nature of the corporeal "machine." Theorizing the brain as an organ meant taking seriously the fact that its operations *defied* all easy reduction and necessarily demanded new orientations precisely because the brain was at once a space of determination and *lack* of determination.

Charles Sherrington was the influential neurologist who coined the crucial term "synapse," used to describe the junction between nerve cells. For Sherrington, this "gap" was the site of an active *process* of contact and not, as others had theorized, a material bond across some kind of space. Sherrington eventually synthesized a massive amount of neurological research (including his own important experimental work) in his book of 1906, based on his ten Silliman lectures delivered in 1904 at Yale University.[17] There he argued for a strong theoretical interpretation of the nervous system as a whole, defined by what he called its essential integrative action. The nervous system, Sherrington demonstrated, was an exceedingly complex, highly differentiated, but also completely *unified* organ. Even the simplest automatic responses, often considered in complete isolation, were, he showed, in fact the result of a "total" reaction of the system as a whole to a specific irritation or excitation on the periphery (114). Unlike other "mechanical" systems in the body, however, such as the circulatory system, the nervous system could operate *simultaneously* across its network, over relatively long distances, through what Sherrington called "waves of physico-chemical disturbance" along multiple lines of stationary cells that were vehicles of these informational diffusions. And as a system, it was continually being "disturbed" internally by the activation of, for example, sensory nerves. These irruptions of activity, Sherrington would argue, citing numerous experimental studies, required a reorganization of the system, one that was tightly interconnected and constantly subjected to what Sherrington called internal "breaks" that upset its "relative equilibrium" (182). The mutability—plasticity—of the system was essential to its operation, because it had to radically reconfigure its "pattern" in an ongoing process of internal reaction and action. The nervous system was a site of *competing* reflexes, producing internal interference requiring new syntheses. Through a network something like a telephone exchange, the nervous system adjusts connection points in a spatial and temporal flow, changing its topological patterning with each interference or disturbance, like a kaleidoscope after being suddenly tapped (233).

Sherrington redefines consciousness in this context as a function of the nervous system's adaptability. Learning is the process of "adapting" certain brain centers for "other uses." The "organs of control" in the brain, the site of "adjustability," are "among the most plastic in the body" (391). In evolutionary terms, these higher brain centers show how the nervous system is not something *adapted to its environment*; rather, adaptation itself is the very function of the nervous system. "We thus, from the biological standpoint, see the cerebrum, and especially the cerebral cortex, as the latest and highest expression of a nervous mechanism which may be described as the organ of, and for, the adaptation of nervous reactions" (392–93). Conscious thought appears when the habitual reactions and adjustments no longer work effectively. As Herrick put it in an essay on the evolution of intelligence, consciousness is a phase when "non-stereotyped actions" are required because the reflexes are no longer adequate.[18]

This model of neural flexibility, Sherrington believed, can explain why human beings, in particular, are so successful: they possess the maximum "adjustment capacity" (392). But if human thinking and reasoning are defined as the ability to "forecast the future" by synthesizing past memories, what distinguishes this intelligence from similar (if less complex) processes of learning and anticipation in the animal mind?

As neurological science and evolutionary theory insisted, intelligence had to be explicated by physiological processes, and these processes in turn were always aimed at the survival of the living being—including human beings. As Ralph Lillie, the Canadian American neurologist who would later construct an artificial nerve cell, explained, the body had a whole array of "compensatory and regulatory devices" that allow for an improved "organic equilibrium." These devices, he observed, were not fundamentally different from those "used to control the rate of energy-consumption in artificial mechanisms (thermostats, governors, rheostats, etc.)." The internal alteration of structure allows for the restoration of "normal conditions."[19] Intelligence, as Sherrington already said, is what allows for an adjustment to anticipated *future* events. What, Lillie asked, would be the physiological mechanism to explain that capacity? In a strange, almost proto-cybernetic rereading of Kant, Lillie will say that intelligence is just the way that a system's "permanent organic predispositions or adjustments" (we might say, in psychological terms, categories and empirical concepts) allow it to grasp particular and singular experiences as objects of knowledge. The universal concept or frame is just one of the many permanent mechanisms of adjustment, the particular "event or condition in nature which calls this mechanism into action and so brings the organism into effective—i.e., self-conserving—relations with its conditions of life."[20]

The intelligent action is therefore no more "purposeful" than the behavior of the safety valve on a boiler, the "governor" that will be later be hailed as the forerunner of the cybernetic notion of negative feedback.

12
Prehistoric Humans and the Technical Evolution of Reason

Any effort to locate the exceptional status of the human was, in nineteenth-century scientific discourse, necessarily framed by the question of the *brain* as itself an exceptional organ, capable at once of the most automatic and dependable routines *and* the "new movement that has never occurred before."[1] This cerebral version of the human as *lack* offers a critical angle on the position of the human in evolutionary theory, entangled as it was in the progress of neuroscience and all biological study in this period. The human could be seen as both in and out of the logic of evolutionary processes. If by nature humans possessed some cognitive capacity to *invent*, then one of the key factors in human evolution (neural, social, and otherwise) will have to be our propensity for artificially organizing nature—even "human nature" itself.

The empirical development of the concept of plasticity and its role in the organization and reorganization of behavior in the disciplines of neurological science and medicine were productively reappropriated in the evolutionary debates spurred by the Darwinian thesis of "natural selection," first introduced in *The Origin of Species* in 1859. Understood as an analogue of the artificial breeding of animals and plants that has long been undertaken in human civilization, the idea of natural selection in the context of human origins and development posed a challenge for evolutionary science, not just because of the thorny theological implications. From the start the radical problem of studying the *natural* evolution of a human being defined by the capacity for artifice was entangled with the neurological and psychological blurring in this period between "natural" forms of cognition and what Hartley had called the "secondarily automatic" operations, conventional *habit*, that "second nature" (Hegel) impressed on the brain not just by "experience," but often through educational institutions and cultural practices.

As George Lewes noted in 1873, the brain is highly plastic and thus at the center of human cultural difference: "The brain of a cultivated Englishman of our day, compared with the brain of a Greek of the age of Pericles, would not present any appreciable differences, yet the differences between the moral and intellectual activities of the two would be many and vast."[2] Unlike the world of animals, human life was defined by its *cultural organization*—which is to say, different individual brains resulted from the varied historical conditions of human societies. Even our very perception of the world itself could be modified by "experience, training and habit [*Erfahrung, Einübung, und Gewöhnung*]," as Helmholtz famously argued in his influential treatise on optics.[3] As Lewes would remark, the Kantian a priori would have to be reinterpreted in light of "concrete conditions" as well as in relation to our "ancestral influences" (175–76)—not to mention the *prosthetic extensions* of industry, which mark, for Lewes, the appearance of "the equivalent of new senses" (156). Crucial to any full understanding of mind, then, was the essential assumption that human brains could be influenced by individual psychic "feeling" as well as the "residua of ancestral stimulation" (177), as Lewes put it. However, what is absolutely crucial here is the recognition that the individual brain can also be determined, quite literally, by the *experience of other minds*, as manifested in the cultural accumulation of knowledge, which is again what separates us from the animal mind, limited as it was to inherited capacities and individual learning.

The *transmission of thought*—Lewes here highlights "the great human privilege to assimilate the experiences of others." As he observed, "Our knowledge is the product of our own experiences, and of the stored-up experiences of our fellows" (177). The action of the human individual is literally "guided by the thoughts of others." Lewes wrote, "I have never seen the Ganges, nor measured the earth's diameter; but these enter my world of experience, and regulate my conduct, with the same certainty as my direct experience" (166). Indeed, for Lewes, the individual mind cannot be easily separated into personal and "shared" knowledge. The mind is constitutively formed by the collective experience of the community: "What I have directly experienced by sensible contact forms but a small part of my mental wealth; and even that part has been largely determined by the experience of others" (166).

As Lewes will go on to explain, the medium of this transmission and assimilation of mental experience is the *symbol*, the "condensation" of thought into material form. Language—at once individual and social—was the "great instrument" for minds that were constitutively defined by both participation and assimilation (160). So human societies were fundamentally cultural organizations, directly analogous to organisms, for Lewes, because individuals cooperated in light of a common goal. The strength of the mind *as human* derives from its development in the medium of the nation or the tribe. Lewes cites August Comte here on the force of history determining the present.[4]

And so the evolutionary status of the human would have to contend with the fact that the distinction between animals and humans was the *collective transmission* of knowledge. Lewes makes the analogy with technologies here: our thinking depends on its external supports developed over centuries; we literally cannot think without them, just as our social and economic orders depend on inventions like the railway to sustain life (169). More radically, technology is itself the space where the mind *as human* reveals itself. As Lewes observes, we used to dig up ancient cities and tombs searching for golden objects or sophisticated works of art. However, what is more important today is the study of the origin and history of the *mind*—an archaeology of cognition, where "every little detail which tells of the mental condition of ancestral races is now of priceless value." The archaeologists now "dig with greater eagerness for flints and the rude implements of prehistoric races, because these throw light on the evolution of Mind" (170).

Darwin on the Evolution of Consciousness

Darwin's own efforts to explain the evolutionary conditions of human development were hardly satisfactory. In his book on the expression of emotions he could not evade the problem of how to distinguish between "innate" expressions and those that are "conventional" in the human sphere.[5] Noting that he consulted with Henry Maudsley on the psychology and neurophysiology of expression, Darwin deploys an updated version of the long-standing theory of habit as the product of repetition within a nervous system that was capable of transformation (28). The challenge is that the system determined by natural (i.e., inherited) instinct is also generating *unnatural* behaviors that will come to function just as automatically and "naturally" as innate ones. Darwin's examples include dogs and horses trained by humans to walk or perform "naturally" in completely artificial ways, and he will point to human behavior as well, where it is always easy to "confound conventional or artificial gestures and expressions with those which are innate and universal" (50). Darwin will make the tantalizing suggestion that seemingly *natural* human gestures have really already been learned, just "like the words of a language" (353). Inevitably, the question of the distinction between human artifice (the power to *produce new organizations*) and the "second nature" of habit that even animals can acquire will still have to be addressed.

In *The Descent of Man* (1871; rev. ed. 1874) Darwin will indeed admit the exceptional nature of the human but will locate that exception in the moral domain rather than the cognitive more generally. If the gradual emergence of the human form from animal ancestors was taken as given, then the intelligence demonstrated by other higher animals was necessarily "capable of advancement" and differed only in degree from human intelligence. However, as Darwin states, "the development of the moral qualities is a more interesting problem."[6] The appearance of the moral sphere as a break in evolutionary continuity is conditional on the specific group practices of human social organizations, Darwin will say—but in doing so, he is displacing the problem of the human mind altogether by identifying yet *another* origin, the radical origin, that is, of a genuine distinction that can clearly (for Darwin at least) mark humans from animals, namely, our specific communal form of life.

Darwin mixes—maybe confuses is a better word—many of the strands of thought I have been tracing in nineteenth-century European and American intellectual discourse when he claims that the appearance of higher mental faculties among human groups entails that "images of all their past actions and motives would be incessantly passing through the brain of each individual" (95). This would lead, Darwin believes, to the formation of an enduring sense of the social commitment of the individual, since it is always present as a motive, unlike the other desires and instincts that disappear once satisfied. The new form of cognition makes *conscious* what is already the natural necessity of group life. Then, according to Darwin's rather casual explanation, "after the power of language had been acquired, and the wishes of the community could be expressed," *society itself* becomes conscious. At this stage, the newly aware "society" can actively impress itself on the individual, indicating "how each member ought to act for the public good" (96). What is interesting is that this new articulation of the logic of social form will be literally imprinted on the nervous system of the individual in the form of *habit*. As Darwin concludes, the moral imperatives of the group will now "naturally become in a paramount degree the guide to action" on the part of the individual (96). At the same time, Darwin is very much aware that the human has the capacity to lose instincts (105).

The evolution of human society is therefore inextricably entangled with the question of the human mind and its capacity to remake itself into something that is, if not purely artificial, at least *not entirely natural*. And significantly, Darwin will point to the tool-making capacity of human beings as one of the key indicators of the "immense" gap between the least of human minds and the highest form of animal cognition. Animals have cunning and "art" but would never have "the thought of fashioning a stone into a tool," Darwin writes, before tying the origin of technology to the human ability to "follow out a train of metaphysical reasoning, or solve a mathematical problem" (122). Almost as if he senses the difficulty of covering this gap in the origin of the human, he claims that the difference between animal and human mind is one of "degree and not of kind," noting that human infants are hardly born with the "power of abstraction" or reflection. The "highly-advanced intellectual faculties" are most likely "the result of the continued use of a perfect language" (122). Again, this is just a displacement of the problem that is more radical: How did humans become capable of *acquiring* the tools of technology and language, the implements that transform infants into civilized beings capable of advanced sequential thought?

With a nod to Alfred Russel Wallace, Darwin brings together the evolution of the human with the evolution of human culture, understood now as an artifact of human invention exemplified by the production of the artificial. Unlike other animals, human beings become less amenable to "bodily modification" via natural selection once they can use *mental* faculties to solve the problems of adaptation. The human has "the great power of adapting his habits to new conditions of life." Crucial here is the invention of tools (clothes, sheds, fires) alongside both weapons and "stratagems." So the evolution of the human is now conditional on the selection of the most "inventive" individuals whose superior technology allows society to survive in "new" conditions—which might be another way of saying *unnatural* conditions (125).[7] Animals cannot adapt so flexibly, since they must actually acquire new *bodily* tools and weapons through the vast, eons-long process of modification and selection. We can see, then, that the evolutionary perspective did not so much transform the question of the human mind as exceptional as it put renewed emphasis on that distinction. The nature of a brain that can challenge nature with technology and social organization that is *not* natural would have to accommodate both the instinctual inheritance of our corporeal evolution and the emergence of a peculiar capacity to *dis-automate* that same inheritance.

Primeval Prosthetics: On the Evolutionary Exception of the Human

This evolutionary conundrum structures George Campbell, Duke of Argyll's innovative work on "primeval man," exactly contemporaneous with Darwin's texts concerning human development.[8] The issue raised by Argyll's early effort in paleoanthropology is that any effort to assimilate humanity to a *natural* history of biological evolution will come up against the challenge of explaining how the "chasm" between humans and all other animals could ever be crossed by some kind of gradual development (21). Given the exceptional importance of human cognition, that has to be the place to start—as Argyll states clearly in his introduction: "There is an inseparable connection between the phenomena of Mind and the phenomena of Organization" (17). Every creature, he explains, has functions that correlate with its essential unity, and this means that instincts and the "mental character" must be "strictly correlated." Argyll will pose the question: What kind of organism is the human in nature? Unlike all other creatures, it seems that the physiological constitution of the human body has in fact been *de*-evolving: "it diverges in the direction of greater physical helplessness and weakness" (21). This tendency could never, Argyll concludes, be understood as the result of natural selection; it is exactly the opposite of what would be expected in that framework. And this leads to the key insight: without prosthetic, that is, *artificial*, protections, our lack of speed and strength, "the absence of teeth adapted for prehension or for defense," and our comparatively weak senses, all would mitigate against survival in challenging and competitive natural environments.

Argyll's strategy here is to show that the weakening of the human frame would only have been possible in the evolutionary context if our organs had *already* been substituted with alternative behaviors enabled through supplementation. Bipedal humans would never have been safe with their "useless arms," for example, "until the brain was ready to direct a hand" (22). The brain is the site for what he calls the "gift of reason," which was the prerequisite for this deviation of the human from the usual forces of natural selection, which always tend to the improvement and perfection of bodily forms and functions with respect to the environment. This is all to say that the physical evolution of the human was not the precursor of our eminent position in nature; rather, our physical state itself is a direct result of our "capacities of thought" (22). What is essential to humans, then, is their rationality, which is not akin to other animals' ability to adapt to the environment; humans are able to *gain knowledge and inherit it*. Distinctive to the human is the "ability to acquire, accumulate, and to transmit knowledge"

(43). For Argyll, this means that we have to understand how even the insignificant actions that might seem "the most purely instinctive" are in fact highly specific to the human mind. Throwing a stone, wielding a stick as a weapon—"both these simple acts involve the great principle of artificial tools" (48).

This capacity for artifice, for *invention*, is the attribute that is "absolutely peculiar to Man."[9] As Argyll notes, the natural animal uses implements but only those that are given as bodily organs, as guided by "implanted instincts." While it is true that some animals might make "artificial use of natural forces" (when birds drop shellfish on the rocks, for example) there is an "immeasurable distance" between these animal behaviors and the human capacity to fashion an artificial implement. And, Argyll adds, humans are constantly engaged in this activity because of their peculiar evolutionary status. Our goals require the "intermediate" organs that are tools. This is why our physical weakness is accompanied by a mind that is, so to speak, *naturally* "capable of Invention" (48). Or at least, with the human mind comes an "intuitive" understanding of causation in nature, due to a natural inductive reasoning ability. In any case, the diversity and diffusion of human cultures across the globe points to a fundamental and essential universality among humans, namely, their intelligent adaptation to conditions via the fashioning of various artificial technologies, these prosthetic intermediaries. There is no "primitive" culture, for Argyll, in that even the "rudest" nations exhibit the same remarkable and astonishing capacity for the technical organization of social life. There was no "Eskimo" Adam, for example. According to Argyll, the ancestors of these people must have come from somewhere else, driven to migrate, perhaps by war, and they would have originally had "wholly different habits" (54). And of course, if they had been, say, farmers or shepherds in their old territory, such a life would have been absolutely impossible in the icy north. The old arts would therefore have to be forgotten and new modes of life created from scratch. For Argyll, then, the origin of the tool is identified with the invention of substitutes for the "missing" natural implements of an animal body.

This idea of the prosthetic nature of the artificial tool animates the work of the German philosopher (and philologist) Lazarus Geiger, who explored the origin of human language and culture in a series of important lectures in the 1860s.[10] A key argument was developed in a talk from 1868, whose title alone is evocative: "The Earliest History of the Human Race in the Light of Language, with Special Reference to the Origin of Tools." There he proposed a theory that would distinguish between "primary" and "secondary" tools. The original tool was—as Argyll had argued—the direct substitute of an original, natural organ (41). Geiger imagines these tools emerging in a liminal period of evolutionary time, when humans will start to "recognize" somehow the similarity between certain found objects and their own natural organs, leading them to use these external objects to modify and amplify the natural behaviors enabled by the body. This was a process whereby creativity comes about through imitation (44).

However, as Geiger will go on to explain, once the tool appears in this early, imitative form, it can *itself* evolve in new directions and generate entirely new usages through a series of transformations, exceeding in the end its originary prosthetic origin. One example Geiger gives is the radical reuse of the bowstring (a weapon) as a musical instrument, which would have perhaps been occasioned by the accidental observation of the vibrating string's "beautiful sound"—to quote Homer (42). The point is that the use of tools in any one historical moment gives no indication of their *origin* since technologies, along with their social contexts, are constantly undergoing significant transformations. As Geiger emphasizes:

> The use of implements shaped by himself is more decidedly than aught else an evident distinctive characteristic of man's mode of life. For this reason the question as to the origin of the tool is a subject of the greatest moment in our early history. (43)

At the conclusion of his argument, Geiger leaves the reader with humanity on the cusp of a great epochal transformation: the *emancipation of the tool itself* in the modern machine. Gaining in power and continually being perfected, the contemporary technology had become an implement "emancipated from the hand of man, and inspiring its own maker with a peculiar admiration" (47). What Geiger demonstrates is that our historical consciousness—beginning with our "primeval origins" and looking to our unknown automatic future—will always be defined by our technical condition.

Projecting Organization: Kapp on the Evolution of Technology

Technology operated in this period as a hinge linking (while separating) the dominant evolutionary paradigm in biology with the philosophical question concerning human cognitive power and humanity's uniquely historical-cultural form of being. As Eduard Reich states, "It is certain that the hand of civilized man must be regarded as an expression of the highest organization of the gripping tool, just like the brain of civilized man is an expression of the highest organization of the organ of thought."[11] However, in making ourselves through our artificial organs, we open up the possibility of *technical evolution*, where the implements are more than mere appendages, more than just prosthetic extensions, and therefore expressions of some other kind of organization.

The proposal was taken up and pushed to the limit by the somewhat neglected theorist of technology Ernst Kapp, a German "Free Thinker" forced to emigrate after being charged with sedition who eventually settled in Texas, where he became a farmer. Kapp's idiosyncratic and illuminating book on technology, published in German in 1877, proposed that the tools of human beings are not so much "prosthetic" substitutes of organs but are more like transmissions, what he called *projections*, of our own "spirit" of organization. This process of *Organprojektion* is according to Kapp an unconscious transmission that creates a materially doubled world—our technical culture—which is animated by the same forms of order given to us in our living bodies.[12] The machine world is one in which our own organs, not defined literally but more expansively as our own possibility of organization, are abstracted, dematerialized, and sent out beyond our bodies and minds to take new forms in artifacts. Kapp's interest in this book is, in part, to study human technology as if it were a visible field that can reveal the hidden logic of our own, necessarily unconscious (because physiological) forms of life. Reading technology becomes a means of reading the human, a project that is so often impossible or at least profoundly difficult.

Kapp's history (or better, historicized typology) of technology tracks the different "planes" of organizational projection, rewriting Geiger's simplistic two-stage account. The earliest tools are quite literally projections of the basic identifiable organs of our body, as Argyll and others had already argued. These more primitive tools were at first substitutes for the hand, say, or the tooth, which are the implements of our natural condition. The early tool is the material substitution and improvement of the "inborn tool" (35). And as Kapp will proclaim, this initial substitution marks the radical origin of human culture: the human becomes "human" in this act of exteriorization. As Kapp puts it, "In the hand—the external brain [*auswendigen Gehirn*]—tools establish culture" (151).

The next phase of technical (and hence cultural) development is marked by the unconscious projection of our body's kinematic structure, Kapp will explain. No longer a concrete projection of individual organs with concrete and self-contained properties, these new tools are an exteriorization of what we can call the functional *organization* of the skeleton, that is, a material remaking of the articulation and interaction of the moving armature of

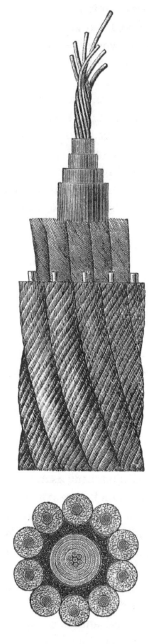

Figure 12.1. From Ernst Kapp, *Grundlinien einer Philosophie der Technik: Zur Entstehungsgeschichte der Kultur aus neuen Gesichtspunkten* (1877).

the body, the way forces pass through machinic levers, springs, and other transductive elements. The machines of technology are projections, then, of the "natural" machinery of a complex physical organization system.

The most advanced stage of projection, for Kapp, is the introduction into these complex machines of certain systems of *communication*; the examples given here are the more recent technical advances of Kapp's day, such as the electromagnetic telegraph. In this stage, we can come to know our own nervous system in a newly intimate way, as it is writ large on the vast projections of communication technologies in industrialized settings and, increasingly, at a global scale in the context of colonial cable information networks. What Kapp argues here is that the function of the human nerve is an exact "miniaturization" of the submarine telegraph cable; the cable networks, too, are the "nerves of mankind" traceable on the maps of telegraph companies. (Figure 12.1.) Neurology and communication technologies are mirrors of one another.[13] The point is that once we no longer understand tools as mere implements, however complex and sophisticated they become, we can begin to confront the nature of technical *systems* (95), which would then be the path toward the revelation of the networked nature of human life itself—a theory of social and even global organization.

Yet Kapp hints at an even more fantastic opportunity—a future when the functions of the mind itself, the *content* of the messages being transmitted by the telegraphic infrastructure of the global system, these meanings, will be unconsciously projected into new technological forms, what could be called *spiritual* machines. We can speculate here with (and, it must be said, against) Kapp: for if, as he says, language is the ultimate "self-production" of human minds, then what kind of technical apparatus or system will constitute *its* externalization? In other words, can there be "unconscious" projections of *conscious* thought? It would be interesting to speculate on what Kapp might have said about the coming era of wireless communication technologies.

At least we can say that in the context of late nineteenth-century concepts of unconscious cerebration, or theorization of intelligence as technical practice, the evolution of what Kapp termed *Organprojektion* was going to be unpredictable, to be sure. But we can also say that the possibility of a technical projection of the organ of thinking (namely, the brain) is hardly contradictory, as we well know in the twenty-first century. Less easy to conceptualize will be the projection, or exteriorization, of the *technical process of exteriorization* itself, the very constitution of the mind as a tool-making and symbol-using creature, in and beyond nature.

13
Creative Life and the Emergence of Technical Intelligence

Life, in this period of evolutionary thinking, evaded any strict mechanistic explanatory model: it would have to be conceptualized as fundamentally creative, in the sense that the organism was perpetually adjusting and regulating its own "internal milieu," to use Claude Bernard's phrase, just as it was actively responding in flexible ways to its unpredictable environment. "Life," it was said, "is perpetual auto-regulation, an adaptation to changing exterior conditions." The organism "always renews itself, and it is always the same." Life is revealed as the existential force of defense against "the enemies that attack it."[1] As so many neurophysiologists and psychologists could demonstrate, the plasticity of the nervous system and especially the open fluidity of the associational centers in the cerebral lobes were likely the locus of this creative action.

One thinker immersed in these ideas was the influential French philosopher of science, the spiritualist Émile Boutroux. In his words, "brute" matter was organized in the living being, marking a "true creation" of order, but this organization was accompanied by an "elasticity" of behavior during "breaks in equilibrium," a creative potential inherent in even the most simple organisms. This was due to a fundamental *indetermination*.[2] Boutroux's counter to any overly physiological reduction of thought was to highlight its essential activity. It was not a "function" or a material entity. Conscious thought was the active transformation of external data into internal information, a "living mold" capable of a radical metamorphosis, crossing the abrupt threshold between the self-enclosed nervous system and what was given beyond the body. The "idea" or concept was the *form* through which thought transmuted the substance of external material reality into something assimilable to the nervous system as a mode of organization (115).

So consciousness does not develop out of physiology, by definition, because it is the activity that transcends physiology—it is a "new element, a creation" (117). What is crucial to the argument here (and it is important not to dismiss this kind of work as a mere exercise in "vitalism" or spiritualism) is the independence of the logic of the mind, which enables the articulation in concepts of relations and orders that are simply not capable of being formed by a physiological system, however open and flexible and creative it might be in its own terms. For Boutroux, the appearance of the "ideal" in the organism opens up the possibility of purpose—exactly what was missing in Lillie's reductive account. What is possible in the conscious intelligent being is a new synthesis of the goal of life (self-conservation) and the goals inspired by the idealization of natural order itself, which is to say, the coming to consciousness of natural order. The organism, as complex system, is a hierarchy of heterogeneous organs and processes. But that hierarchy is only made present and active within the conceptualization of the conscious mind.

The turning point here is Boutroux's redefinition of the exception that is the human—not merely the translator between body and nature, or the mere consciousness of natural order, but in fact the vehicle of wholly new orders of existence. What distinguishes the humans (already evidenced in the evolutionary literature) is their capacity to *transform*, to make "obstacles" to life into new instruments of life. Technology is the artifact of a reorganization of the forces and order of nature, the creation of a new form of beauty (188). Through the production of organization outside the body, human beings share a world that is in nature but not *of nature*. The human is thus capable of a social existence that has an independence analogous to the independence of the human mind itself. Individual consciousnesses translate the order of the external world into the "language" of the organismic being via the nervous system, but they now also translate the artificial and exterior orders of the "ideal" that constitute the essence of a creative and technical human life.

Intelligence, then, is what is particular to animals that are not just more capable, in the evolutionary sense of their nervous systems being well adapted to their environments. As James Mark Baldwin argued in his important reconsideration of Darwinian models of natural selection, it would be impossible to imagine consciousness—as a new adaptation—ever arising in the course of normal evolutionary development. If consciousness offers a creature "choice" in the direction of its actions, how would it choose the right path and thereby become a useful adaptation? It would be more likely that the organism with its own possible modes of action would evolve according to its own testing of the environment. What Baldwin recognized is that it is possible that the "individual life history" of an organism—its specific form of life (not its internal structure)—could be what is selected for, not a *transformation* in the organism itself. He contrasts this "organic selection" with Darwinian "natural selection," which would not be able to explain any survival of organisms in conditions that are changing, since inherited "reactions" would not be adequate to the task of life.[3]

How would the elements of a life history operate in evolution? It cannot simply be the "emergence" and inheritance of some special "adaptive capacity" nor the "special creation" of every organism for its "peculiar environment." As Baldwin notes, the only option is to imagine that *new stimulations* that are provoked by the change in the environment actively modify the inherited reactions of the organism, producing new actions that are conducive to life in these new conditions. Of course, this can happen through the long processes of evolution. Baldwin is however pointing to cases where adaption occurs in the specific lifetime of an *individual* organism.[4] What he suggests is that the internal pleasure and pain system guides the organism in the acquisition of useful habits—including the habit of inhibition in the encounter with what has been experienced even one time as painful. Habit is the securing of "vital stimulations," and this is true across the organic life-world. Or, as Baldwin puts it counterintuitively, the organism is endowed with the habit before it even forms it, in that the organism uses individual concrete experiences to confirm or deny its own inner norms of evaluation. The point is that variations are conserved that lend themselves to intelligent (i.e., flexible and adaptive) modifications of the organism's own specific organization in the formation of an individual habit regime.[5]

So Baldwin will agree, with respect to intelligence, that the evolution of the nervous system can be understood as part of a process where (and here he quotes Lloyd Morgan) "the most plastic individuals will be preserved to do the advantageous things" (447). With such radical plasticity, it would be possible that an *individual* would create a new action (entirely unprecedented) "under stress of circumstances" (449) and that that action was necessary to survival. Those others that did not follow the individual in performing this action would not survive. This is the core of the "new principle" of evolution then: "The ontogenetic adaptations are really new, not preformed; and they are really reproduced in succeeding generations, although not physically inherited" (451). In the course of evolution, many creatures retain instinctive behaviors because they are well adapted to their environments. With others, *plasticity* comes to the fore, through the capacity to learn not by testing reactions but through consciousness. This way

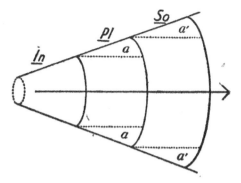

In-Instinctive. *Pl*-Plastic. *So*-social. The spaces *a*, *a'*, etc., show the increased area of facts and principles peculiar to each mode beyond those of the preceding.

Figure 13.1. From James Mark Baldwin, *The Individual and Society; or, Psychology and Sociology* (1911).

the learning that takes place in an individual is not bound to the individual.[6] (Figure 13.1.) It is a form of Lamarckian evolution because the conscious learning can be transmitted to others. Here Baldwin cites the well-known examples of animal learning and sociality and the importance of play and imitation in the animal world. As he puts it, "By the exercise of such gregarious or quasi-social impulses as these, the young are trained in the habits of life of their kind" (41). Social heredity is predicated on sufficient "'plasticity' to absorb lessons of the family and group tradition. Each must be plastic in the presence of the group life and its agencies" (42).

The emphasis in this period on plasticity and the active capacity of reorganization in the face of repeated breaks, disturbances, and interferences created a framework for understanding novelty and the appearance of the exceptional in organic nature and the human in particular. However, that very paradigm of adaptability could occlude the important and still unformed question (raised by Peirce and others): How does the mind *escape* the logic of the body—however plastic it might be—altogether and know itself and others in a way that animals simply cannot? A related question, which would become perhaps more important, was already being asked in the first forays into a theory of cultural evolution: How was the human mind capable of internalizing the experiences of foreign brains, of sharing thought itself? Baldwin's "organic selection" in the case of the human displaces the origin of this special capacity for imitation and the *sharing* of conscious thought across the boundary of the individual nervous system.

This is what Peirce described as the quality of "thirdness," the relational meaning of a symbol that is incarnated in a specific sign but is entirely *inde-*

pendent of that concrete instantiation because meaning is a function of the *network of shared intepretation* (unlike iconic signs—"firstness"—that relate to perceptual features, or indexical ones—"secondness"—that "point" to an absent cause).[7] Given the truth that intelligent thought (conscious and otherwise) was obviously embedded in the body's complex neural machinery, and given that the brain was the most plastic organs in the higher animals, the *exception* of the human could not be isolated in terms of the individual mind and body. Understood in evolutionary terms, the human revealed a chasm, namely, our separation from our natural adaptive capacities and learned habits, which we shared with virtually all animals that possessed nervous systems, and the creation of a genuine "second nature" in that natural body, a second nature formed by forces *external* to the nervous system, forces that in fact *deployed the nervous system for its own purposes.*

"Man is preeminently plastic, educable in a supreme degree."[8] This may seem a vague affirmation, and we have encountered versions of it across different lines of thought, stretching back to Descartes. However, in the context of the early twentieth century this statement is deceptively simple. Edward Bradford Titchener was *not* in fact suggesting that the openness of human minds was just the most extreme concluding point of an organismic plasticity at the center of evolutionary development. As he understood, human education seemed to constitute something altogether new: the education of the child, he said, is really the artificial *extension* of plasticity beyond its normal limits—temporally and spatially, so to speak. Yes, most organisms tended to become more and more automatic over time, as they acquired reflex habits from life experience, supplementing inherited instinctual behaviors, and these new automatisms enabled them to survive in their particular environments. Human intelligence, however, was not simply a suite of automatisms, a system of responses that came from the repetition of experiences shaping and organizing the plastic higher brain. The implication of this analysis, in line with the kind of thinking characteristic of the period, is that human thinking as intelligence involved more than just neurological plasticity. It emerged from what might be described as the creative deflection or reappropriation of *plasticity itself* as a habit.

The American philosopher Charles Judd clearly articulated the complexity of plasticity and habit in his own analysis of human evolution. Judd had trained (like so many other early enthusiasts of the new discipline of psychology) with the important experimentalist Wilhelm Wundt in Germany, and eventually translated Wundt's works into English. To attack the core of the anthropological problem, he explained, one had to recognize the fact that one set of animals somehow broke free of their biological determination. Established physiological organizations (even relatively flexible ones) at some point gave way to these unusual and exceptional creatures, who are "characterized by intelligence, meeting the emergencies of their lives by a mental adaptation of themselves rather than by a purely physical adaptation."[9] That odd locution, "mental adaptation of themselves," suggests the novelty of the human as the species that internalizes the evolutionary process itself. As Judd will propose, consciousness will serve as the threshold concept; this was what brought the external environment *into* the individual organism, not simply as sensory information (which required no thought), but as an *organization* foreign to the nervous system's natural operation. The form of consciousness "remolds" that environmental organization in relation to the individual's physiological needs that are also represented *as formally ordered* in consciousness—there is an echo here of Boutroux's theory. Judd realizes that the inner world of the higher animals, especially that of humans, is not merely concerned with physiological needs: the mind becomes capable of ever more complex organizations of its own experience.

But what is perhaps more important is that consciousness in the human case can bring together in one psychic "space" the inner order and the structures of the environment. This is why the human mind can, according to Judd, synthesize its model of both the inner world of experience and the outer world; the result is that the mind can then "attack" the environment to force it to serve the inner life of the individual. The "outer world itself is remodeled to conform to the inner pattern."[10] This foundational creative capacity is best revealed in the act of *reorganization* that is in the end essentially a technological intervention. The human being, Judd comments, "molded materials, following his ideas, into forms and combinations that never under any chance could have come into the world through the mere operation of physical forces." The human produces, in other words, not new "reactions" to the environment but rather *new forms* that become templates for the concrete "reworking of external reality."[11]

What this means in terms of evolutionary anthropology, then, is a break in the natural order, a split between creatures of adaptation and the human, who now concentrates on the "indirect methods of attacking nature"— which is to say, builds technologies and artificial infrastructures to ensure survival. Instead of, say, "cultivating strength" to oppose some natural obstacle, human beings develop "science and engineering and indirect mechanical devices."[12] Crucial to Judd's theory here is the existence of another break: humans are no longer limited to their individual consciousness.

The remaking of the world is a shared participation in a world of ideas and forms that exceeds the limits of the natural cognitive faculties. And so, Judd tells us, it is through language and symbolic life, the vehicles of this creative reorganization of the world, that we are able to catch a "glimpse into the conscious life of a fellow being." For Judd, "art carries over from man to man the inner possibilities of rearranging the physical environment"[13]—a process we might now call niche construction.

Bergson's Human Technician

This is the threshold that Henri Bergson will position as the very distinction between human and animal. The human is not defined by its flexible reactions, its habits, or even its freedom from instinctual automatisms. Like many others, most notably Willam James in this period, Bergson will (especially in *Matter and Memory*) highlight the essential "indetermination" of organic life and the open, plastic form of the nervous system. However, what makes the human the appearance of an intelligent being is the point when freedom—something that defines the very process of life itself as a creative activity occupying "zones of indetermination"—exceeds its own structuring space of activity. As Bergson put it, freedom, when it appears as intelligence, "proceeds to construct constructive machinery."[14] Human cognition, for Bergson, is not, then, a capacity for decision in the confrontation with the exceptional moment, the radical indeterminacy that defies all memory and instinct. Human action as the response to challenge now concerns itself with the *reorganization of nature itself*. The evolutionary process is now split, between the construction of new organisms with new structure and capacities and the creation of an organism that will itself, by "fashioning inorganic matter," resolve threats to its existence (142).

This is why, for Bergson, we are not so much *Homo sapiens* as *Homo faber*; our intelligence resides in the "faculty of manufacturing artificial objects, especially tools to make tools, and of indefinitely varying the manufacture" (139). As Bergson observes, all animals use tools or machines; it is just that the implement is a *natural* one, forming part of its own body, operated through inherited instincts—the faculty of using *organized* instruments. Human intelligence "as it leaves the hands of nature" is revealed in the process of *organizing* itself: "constructing unorganized—that is to say artificial—instruments" (150). For Bergson, this leap is a function of the exteriorization of thought itself into the materiality of the unorganized object. Just as the rhythm of speech is the reproduction of the rhythm of thought, the exteriorization of consciousness is the penetration of the spirit into matter, the relation of soul and body, and it is because human thought finds its way into matter that the meaning of things (language, tools, symbols) is inherently *mobile*, unlike the natural signs of the animal or insect world, which are always bound to the "thing signified" (158).

What links Bergson's theory of mind, spirit, evolution, and technology is the concept of *form*, the organization that is incarnated—but never exhausted—by its concrete instantiation. Intelligence bears on the nature of "relations" between concrete things; instinct refers to the concrete only. Intelligence is therefore the striving to make external things into *organs* of its own formal understanding. With human technology, life is *making itself visible* in external forms that are themselves radically contingent from the perspective of nature. And in exteriorizing thought itself, human minds open themselves up to the other: our own thought passes "into the soul of another."[15] Of course, this is where the question of the social intervenes. How are these artificial organizations of matter, technologies, related to one another? How are they coordinated? As Bergson will observe, in *Creative Evolution*, new technologies cannot help but disorder social relations. We are, he says, "only just beginning to feel the depths of the shock" that is the invention of the steam engine. The revolution of industrial society has "upset human relations altogether" (138). The industrial age will mark, Bergson predicts, a whole new epoch in human civilization, akin to the Bronze Age or the era of stone implements.

This new world of technology is like an organism; a new orientation will be required to understand the world of the machines. As Werner Sombart observed, in his 1910 "Technology and Culture," the question of the impact of technology demands acknowledgment of the independent logic of the technical sphere itself. We find, then, that technology is not just one dimension of a culture, or even an instrument of cultural processes and activities. Sombart's radical thesis is that the primary culture of *technology* is the basis of all human culture: "There is no cultural manifestation that is not based in some way on dependence on primary technology."[16] Even human institutions—the church, the state—are manifestations of technological inventions that make possible new techniques of control and organization—communication technologies, weapons, and so on. The cultural influence wielded by the newspaper in the modern age is itself fully dependent on technology: "It could not exist without an infinite supply of paper, the rotational printing press, or the telephone and telegraph."[17] For Sombart, we must learn to isolate the specifically technological effects in human culture, to establish, we can say, a *concept of technology*.

Prophecy

THE FUTURE OF EXTENDED MINDS

14

Technology Is Not the Liberation of the Human but Its Transformation . . .

In 1910, the same year as Sombart's letter, the Italian futurist Filippo Marinetti offered a radical vision of this idea of an independent technical sphere. Echoing Kapp's notion of organ projection, Marinetti writes that human beings are on the way to "the creation of a nonhuman species" where morality and passion are erased, where vital energy will be freed. As he wrote, "The day when it will be possible for man to externalize his will so that, like a huge invisible arm, it can extend beyond him, then his Dream and his Desire, which today are merely idle words will rule supreme over conquered Time and Space." According to Marinetti, this technical prosthesis of the human itself will hardly be restrained. It will be evolved for control; it will be evolved for power. And therefore evolved for violence. "This nonhuman, mechanical species, built for constant speed, will quite naturally be cruel, omniscient, and warlike." Looking ahead to the paroxysms of violence in the twentieth century, Marinetti imagines a technosphere shorn of its "humanity," which is another way of saying that humanity will be mirrored and even overwhelmed by its inhuman other, modern technology, this artificial being with "the most unusual organs; organs adapted to the needs of an environment in which there are continual clashes."[1]

PART THREE

Crises of Order
Thinking Biology and Technology between the Wars

15
Techniques of Insight

In early nineteenth-century thinking, the interventional character of *insight*, its structure as a gift from outside the operations of the normal mind, was internalized, assigned to the murky, and irrational, domain of the unconscious. The "genius in the man of genius" is in fact the *unconscious*, as Coleridge once put it.[1] Romantic ideas of a powerful unconscious force aligned it with the instinctive powers of the body, as, for example, in Schopenhauer's influential theory, where freedom was equated with a separation of conscious thought from the unconscious realm. Indeed, for Schopenhauer, genius was defined as the self-isolation of the intellect from the world. The brain was a "parasite" on the vital body and could lead its life separate from the physiological economy.[2] These ideas would lead, as Henri Ellenberger showed in his iconic history of the unconscious, to the theories of psychoanalysis.[3] Yet there was another trajectory of the unconscious in the nineteenth century, one that centered on its critical *cognitive* role.

This new approach to the mind was produced from the intersection of new neurophysiological research and the associationist tradition of psychological theory kept alive by thinkers such as Thomas Reid and James Mill. By midcentury, thinkers were developing "psychophysics," which, as we saw, developed theories that would explain mental activity and experience as a direct consequence of nerve action. The discovery of reflex actions, and the distinction between afferent (sensory) and efferent (motor) nerves, revealed many systematic automatisms within the nervous system, and as research progressed, it was known that even within the brain itself automatic responses could be generated. As Herbert Spencer put it, in his influential treatise on psychology, the nervous system was a space of integration. The conscious mind was in a sense a complicated fiction constructed out of a limited set of sensory experiences: "Out of a great number of psychical action going on in the organism, only a part is woven into the thread of consciousness."[4] The unconscious activity of the nervous system was continually generating responses, integrating memories, and producing automatisms.

And as evolutionary theories in the nineteenth century took hold, the nervous system could also be understood as the inheritor of a whole evolutionary history of learned responses. Reason, for Spencer, was explained as a gap in the series of automatic functions, a moment of interruption, that is, where these acquired ideas and memories, the evolutionary inheritance, this whole storehouse of automatisms, would be newly organized to help adjust the organism to its challenging environmental circumstances. Reason existed to bridge the difference between the "perfectly automatic" and the "imperfectly automatic" (566).

In this historical moment, then, the conscious thought was not reduced to its material base, nor was it wholly extricated from its vital bodily home. Rather, the conscious mind became a particular (perhaps even peculiar) function, just one aspect of a complex nervous system. This was the argument of the vastly popular (if relentlessly critiqued) best seller, Eduard von Hartmann's *Philosophie des Unbewussten* (1869).[5] Consciousness had very limited access to the vast set of operations taking place in the brain and the nerves. As such, the mind's intellectual capacity could not be mapped onto its nervous substrate. Thinking was therefore something that took place both unconsciously and consciously. The relationship was not antagonistic necessarily, but neither was it wholly harmonious. Intellectual insight came from both conscious attention and the interruptions from the automatic nervous system. William Carpenter called these processes "unconscious cerebration."[6]

The theory of unconscious cognition therefore implied a whole new way of understanding the inspirations of creative minds, or the productions of the genius. "But what is genius other than a reunion of cerebral conditions under the sole excitation of life, organic functions, and perceptions?"[7] The novelty of the insight and its subsequent assimilation into memory as knowledge were both located in the nervous system, often understood in the later nineteenth century as an electrified communication system.[8] If genius was "the power of forming novel adjustments to circumstances," these novel mental connections, noted one thinker, are produced by the nervous current flowing into "virgin soil" within the brain. Once this current "rearranges some of the molecules" of the brain, future current flows more easily: the path has been laid down, and the thinking of this association becomes largely automatic, an updating of Descartes's earlier theory of animal spirits flowing in the open structure of the brain.[9]

The distinction between conscious and unconscious was more or less a distinction between different kinds of nervous activity. As Joseph J. Murphy explained, the organs of thinking were located in the cerebral hemispheres. Such thought became conscious through a sympathetic awakening of the separate "nerves of consciousness," which were located in the sensory ganglia. This was in essence an incidental, if not sometimes accidental, relationship. At the very least, the logic of cognition was not at all located in the functions of consciousness itself. This is why the moments of insight appeared despite conscious effort and will: "Men of inventive minds say that their happiest thoughts have often come to them involuntarily, almost unconsciously, unsought, they know not how."[10] Indeed, conscious thought was even believed to interfere at times with the automatic cognition of the brain: "The rapidity and success of conception, and the reaction of one conception upon another, are much affected by the state of this active but unconscious cerebral life: the poet is compelled to wait for the moment of inspiration; and the thinker, after great but fruitless pains, must often tarry until a more favourable disposition of mind."[11] For Maudsley, insight and creativity flowed from the novelty of association within a system of nervous organization. Genius, he says, is a result of some "unconscious development" that arrives in the conscious mind like a "grateful surprise" (30).

The conscious mind could even be disparaged for its uselessness for intellectual understanding. "The more I have examined the workings of my own mind," Francis Galton wrote, "the less respect I feel for the part played by consciousness. Sudden inspirations . . . are the natural outcome of what is known as genius, are undoubted products of unconscious cerebration. [Consciousness] appears to be that of helpless spectator of but a minute fraction of a huge amount of automatic brain work."[12] In 1874, Thomas Huxley would describe human beings as merely "conscious automata," carried along by automatic processes (physiological and cognitive) without any interference at all from the subjective mind. For Huxley, Descartes was on the right track in his theories of the animal nervous system as an automatic information processor; he just did not go far enough and explain all the operations of human intelligence with the exact same model.[13]

At any rate, there was no functional role ascribed to the conscious mind. These theories of unconscious (and automatic) cognition can be juxtaposed to the contemporary development of automatic thinking machines. The possibility of artificial cognition had already been raised as early as the seventeenth century, when both Pascal and Leibniz independently invented their own mechanical calculators. In a way, the subsequent Age of Enlightenment could be called the Age of Automata, as engineers and philosophers (and political figures) speculated about the intricate mechanisms of life and of thought.[14] At the threshold of the nineteenth century, the infamous Chess-playing Turk automaton, which was exhibited across Europe and then America, staged a powerful illusion of mechanical intelligence. Not long after, Charles Babbage demonstrated perhaps the very first completely automatic thinking technology—namely, the Difference Engine, built in the 1840s.

The psychological theory of unconscious cerebration was adopted by scientists and mathematicians in this period as a way of understanding the critical moment of insight that led to the discovery of new ideas and the formation of novel scientific hypotheses. The German chemist August Kekulé, for example, famously discovered the structure of the benzene molecule while dozing in front of the fire; in a "flash of lightning" the circular solution appeared to him in the image of a snake devouring its own tail.[15] Helmholtz, who had developed an influential theory of the perceptual system as a series of "unconscious inferences," explained that the "sudden inspirations and insights" (*Zufälle und Einfälle*) of understanding were always unconsciously produced and never the result of conscious intellectual labor.[16] The great mathematician Henri Poincaré famously recounted how the solution to an intractable problem might appear rather unexpectedly in some mundane moment, while waking up in the morning or even exiting from a bus. The implication was that the serious mathematical work leading to the solution had been performed by the unconscious mind. Poincaré would therefore ascribe high intelligence to this unconscious mind, noting, in fact, that it "is not purely automatic; it is capable of discernment; it has tact, delicacy; it knows how to choose, to divine. . . . It knows better how to divine than the conscious self, since it succeeds where that has failed. In a word, is not the subliminal self superior to the conscious self?"[17]

These largely anecdotal narratives of sudden insight and psychological conceptualizations of unconscious thinking were eventually synthesized by Graham Wallas in his book *The Art of Thought*, published first in 1926. Wallas explained how creative thought followed four distinct stages of development. First was "preparation," that is, the setting up of the problem through conscious attention. Second was "incubation," a reference to the unconscious phase of cognition. Third came "illumination," when the new idea enters into consciousness. And fourth, the new insight must go through a process of "verification."[18] Wallas's model was not entirely dependent on the idea of an automatic unconscious. Drawing on some of the latest neurophysiological research on the capacity of the brain to overcome injury through reorganization—in particular, Karl Lashley's brain ablation experiments testing the persistence of animal memory[19]—Wallas made an analogy between the nervous system and the constitution of Britain. Both, he said, had "newer structures superposed upon older," both benefited from overlapping functions, and both had a fundamental elasticity.[20] The site of creative cognition was a dynamic and complex system of activity.

Wallas was here alluding to a tradition of psychological and neurological theory that emphasized the importance of plasticity as a way of explaining both the production of habitual automation in the nervous system and its potential disruption in creative thinking. In 1879, for example, William James wrote the essay "Are We Automata?" as a response to Huxley's bold thesis that humans were just conscious automatic machines, opening up a new path for psychology by focusing on the important function of consciousness and attention. This was not, however, a return to some spiritual dualism. James was interested in how the conscious mind intervened actively in those moments when the automatic systems of the experimental psychologists failed. Drawing on contemporary neurological theories, especially those of Hughlings Jackson, James located both automaticity and its interruption within the brain. The lower animals, James explained, are governed by the "determinateness" of their nervous responses, and even higher animals preserve such automatic systems. In humans, however, "the most perfected parts of the brain are those whose action are least determinate. It is this very vagueness which constitutes their advantage."[21] The "instability" of the brain makes it both sensitive and liable to produce novel reactions.[22]

Human creativity in cognition was made possible by the absence of automaticity in the higher brain. James pointed to the extraordinary degree of *plasticity* characteristic of organic tissue in his theory of psychological habit. Humans became automata in a sense, but the ground of acquired automation was in fact this protoplasmic plasticity—as Hegel had earlier argued.[23] James would point to that openness in order to explain the sensitivity and adaptability of the human mind in new or uncertain environments.[24]

Insight, in early twentieth-century psychology, was the term used to mark the appearance of a new solution to a problem that had no precedent, where no learned or instinctive behaviors could respond.

Detours: Gestalt Theory and Creative Thought

For a theory of insight in the interwar period we can turn to Wolfgang Köhler's famous study of the intelligence of the higher apes, a groundbreaking if controversial work that influenced multiple disciplines in the interwar period, from psychology to philosophy, anthropology, phenomenology, and beyond—as we will see. Undertaken during World War I, when Köhler was stranded at his research station on the island of Tenerife, off the coast of Africa, due to a blockade, these experiments investigated the intelligence of ape subjects by looking at how well they could solve certain basic problems. Köhler swept aside dominant behaviorist models of animal psychology (exemplified best by the work of Edmund Thorndike and his "puzzle boxes") by devising ingenious tests that forced the animals to come up with *creative* responses to challenging (i.e., unexpected because *unnatural*) situations.

Yet Köhler's interest in these experiments was hardly the domain of ethology itself. He saw in the higher apes the trace of a primordial intelligence that was at the heart of human cognition, yet impossible to study in human subjects given the very development of our intellect, especially with the advent of language and symbolization. The apes were better subjects than human children since it was not possible to isolate the "undeveloped" child in the same way and test their "prehuman" capacities. The ape was a liminal figure for the study of intelligence—a mind on the threshold, which is to say, a mind that was on the edge of *indetermination*. But in the study of animal psychology in the period, especially in experimental laboratories, the preferred subjects were dogs (especially in neurological studies, such as those of Pavlov) or cats. However, in the wake of Darwin and the consolidation of evolutionary models of human origin, the ape would become a privileged site of investigation.[25]

In evolutionary terms, we can link Köhler's study of the ape mind with Hughlings Jackson's idea that the highest cognitive functions are related to the *least* organized and determined brain areas. For Köhler, then, the animal subject was a space where intelligence could be revealed and studied in its simplest conditions, as though it was, so to speak, being discovered for the first time. It was precisely the "primitive" origins of our own intelligence that have been lost due to their extreme familiarity and the overlay of more and more complex, symbolic forms of intelligence. As he would note in the introduction to the expanded second edition of his study (which became the standard text):

> One may be allowed the expectation that in the intelligent performances of anthropoid apes we may see in their plastic [*plastisch*] state once more processes with which we have become so familiar that we can no longer immediately recognize their original form: but which because of their very simplicity, we should treat as the logical starting-point of theoretical speculation.[26]

The goal, in other words, was to locate a hypothetical origin of genuine intelligence in this threshold species between animal and human.

Köhler's experiments were of course highly artificial: the technique was to reveal to the hungry (but not starved) apes—they were all chimpanzees, by the way—various pieces of food but to then position them in unusual and challenging locations, hanging from ropes, for example, or placed out of reach beyond the bars of the enclosures. The researchers also would include certain objects in different experiments (a collection of wooden crates, say, or some sticks). What Köhler found, and which contradicted Thorndike's claim that animals were only capable of random trial and error techniques when attempting to escape his infamous puzzle boxes, was that after some failed attempts to retrieve the object, the ape subjects would often be observed surveying their situation, ruminating to some degree, then—and this is what Köhler would emphasize—in a sudden turn, would execute an organized effort to gain the goal through a novel, *mediated*, solution path.[27] The subject might, for example, pull over the crates and put one on top of another, or use a stick inside the cage to pull in something from out of reach.

Köhler called this ability "insight" (*Einsicht*) for he wanted to stress that these solutions were never the product of mere blind trial and error, or the mechanical application of established or taught routines, but instead the often unpredictable result of a profound perceptual "shift." According to Köhler, these creative solutions could only appear if the inner "meaning" of the problem was understood by the subject—in other words, if the goal could be contextualized in relation to other aspects of the situation at hand. At this point, possible solutions, possible aids, and so forth would be seen as *meaningful* with respect to the goal. When the direct "path" to a goal was blocked, Köhler explained, the animals had to find "alternate paths," by seeing the original problem in a broader perspective. The suddenness of insight was the moment when the conflicts and tensions in the situation were resolved in a novel organization. A long piece of wood becomes now a way of linking food and hand, for example. (Figure 15.1.) Not every subject would find these alternate paths. However, once they were able to find them—and again, Köhler highlighted the significance of this—the apes

Figure 15.1. From Wolfgang Köhler, *Intelligenzprüfungen an Menschenaffen* (1921).

were also able to seek out *substitutes* of the various implements that might have been used in an earlier solution, demonstrating that this was no rote repetition but an intelligent comprehension.

What Köhler took from these experiments, from the successes and what he called "good failures," was that intelligence could be identified by the capacity to think beyond the direct path. It was extremely difficult for an animal to break its attention from the goal, even if that goal was obstructed. However, higher animals were capable of comprehending the *indirect* path, the "roundabout way" or detours (*Umwege*) that would overcome the obstacles: "What seems to us 'intelligence' [*Eindruck von Einsicht*] tends to be called into play when circumstances block a course which seems obvious to us, leaving open a roundabout path [*Umweg*] which the human being or animal takes, so meeting the situation." (Figure 15.2.)

Köhler's ranking of intelligence among his particular set of subjects (he always stressed the high variability of individual capacity and eschewed

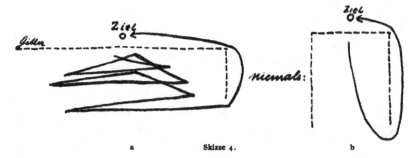

Figure 15.2. From Wolfgang Köhler, *Intelligenzprüfungen an Menschenaffen* (1921).

overly broad generalizations about the "species") stemmed from the degree to which any one ape could keep the goal in mind while being diverted more and more from the direct path, making it increasingly difficult to find these "detours." It was often impossible for some subjects to include, say, a stick in their Gestalt organization unless it was already in plain view alongside the goal in the same visual field. Others, such as Sultan, the most advanced of the group, were capable of this internal dislocation and were (not surprisingly) also capable of the most complex reorganizations leading to insightful solutions. In the most provocative example, Sultan was able to construct an implement out of two separate sticks, in order to reach something placed farther away than normal, and to continue "manufacturing" these double-sticks even when the materials were not consistent (he once tore a branch off a large bush placed in the enclosure, for example).[28] We could say that, for Köhler, intelligence consisted in the capacity and strength of *deviation*—from the first simple incorporation of some seemingly "irrelevant" but useful item or action to the most complex techniques that are at some points in plain contradiction to the desired goal.[29]

The Gestalt theorists would delve deeply into this insight phenomenon, developing ever more sophisticated models of how human beings in particular were able to solve difficult problems in unfamiliar situations without relying on learned behaviors or rote heuristics. The key assumption here was that the "unity" of our perceptual and cognitive world was not determined by the specifics of any one empirical reality or any one structured interpretation. The mind had certain dynamic tendencies that were always in play, organizing the field in accordance with the changing goals and intents of the individual. If we think of the famous Gestalt "switch" that takes place when we look at ambiguous figure-ground images, we can see how, at the higher cognitive levels, something similar was taking place. Because the inner relations and fundamental perspectives that ordered our understanding of the world were essentially plastic and indeterminate, they could at times shift abruptly and radically, giving us new insight. The question was how and why such a reordering took place in our experience and why a new organization that replaced a previous one was felt to be more appropriate or "right." Crucial to the Gestalt approach was the recognition that organization was never given and was never stable: the mind had to actively construct and fill in gaps to maintain the unity of its order.

The young psychologist Karl Duncker, who conducted experiments on student subjects trying to solve mathematical and logical puzzles, argued in his 1935 monograph (soon translated into English) that the solution process could be broken down into phases. First, the subject was able to "see into" the inner structure of the original problem. Then the subject could "restructure" this situation to open up a potential solution path. Finally, these more abstract solution possibilities were tested for practicality; that is, an attempt was made to discover or invent a concrete version of this solution idea. For Duncker, the problem situation could be defined by its internal "tension," and it was this problematic character that demanded attention—which is what Wertheimer's law of *Pregnänz* would suggest, since the "puzzle" was puzzling precisely because it did not cohere according to expected frames of understanding. Like Duncker, Wertheimer would, in his own study of problem solving, zero in on the importance of the intrinsic tension within problem situations, noting that solutions often followed a focus on what he called "trouble regions."[30]

The new order that emerged was one that resolved this tension in some way. "The decisive points in thought-processes, the 'Aha!,' of the new, are always at the same time moments in which a sudden restructuring of the thought-material takes place," said Duncker, and this new structure formed itself in thought precisely because it "altered" the original problem and provided a sense of closure. Like Köhler, Duncker would use the concept of the path to explicate insight, noting that "the man whose vision is not limited to the few feet just ahead, but who directly takes in the more distant possibilities as well, will surely and quickly find a practicable path through difficult terrain." Psychologically speaking, these "distant possibilities" were generated from within. Duncker, for example, explained that the mind could see what he called the "functional value" (not just the surface characteristics) of some past experience of success. The mind could incorporate past experiences in entirely new situations only by recognizing their inner formal organization. We could transpose (by analogy) past and present experiences because the specifics of both were not as important as the structural features expressed in their configurations.[31]

As Koffka noted at the end of his textbook on Gestalt psychology, "If a thought process that leads to a new logically valid insight has its isomorphic counterpart in physiological events, does it thereby lose its logical stringency and become just a mechanical process of nature, or does not the physiological process, by being isomorphic to the thought, have to be regarded as sharing the thought's intrinsic necessity?"[32] Gestalt thinkers would repeatedly draw attention to the complex affinities between somatic and psychological processes. Alluding to the provocative and influential early work Hans Driesch did in developmental biology in the late nineteenth century (repeatedly cited in a variety of intellectual contexts), Köhler once observed that embryos were able to "compensate" for irregular conditions that threatened their existence, describing the process with language that recalled his own work on insight and intelligence: "Many embryos can be temporarily deformed (artificially) in ways utterly incongruous with their racial history and nevertheless regain by a wholly other route the same developmental stage reached by another, untouched embryo."[33] These radical "reorganizations" also occurred within the nervous system, which was, Köhler said, never restricted in its operation by specific pathways—the control and selection of these routes often changed over time according to functional requirements. They were a kind of *Umweg* of the morphogenetic process. He contrasted this form of flexible organization with what he called the "machine," in which the "form and distribution" of the process is not left to dynamics but rather to precisely determined "external" constraint mechanisms. A dynamic process, he said, "distributes and regulates itself, determined by the actual situation in a whole field."[34] The proper approach to cognitive organization, as Koffka put it, "cannot be a machine theory based on a sum of independent sensory processes."[35]

Köhler always emphasized that physical, inorganic phenomena (just like psychological ones) were also best approached by techniques that stressed the relationship structures of a system over discrete interactions of elements. The intrinsic "wholeness" of these dynamic systems could not be modeled as the mere product of atomistic "events" on a microscopic level. Köhler, who incidentally had studied physics with Max Planck, repeatedly invoked new results in field theory to explain how "physical Gestalten" were configured (and reconfigured) on the basis of their inner *unity* as a system of forces and not by sequences of local events.[36] Assuming that the nervous system functioned in this manner—dynamically, as a field of forces, that is, and not "geometrically" as a set of specific interactions—Köhler was able to postulate an "isomorphic" relation between fields of neural discharges and the psychological field of creative thought. That is, both the nervous system and the mental domain of thoughts acted like unitary "organisms" that responded to intervention and tension by reordering themselves to preserve some kind of complex equilibrium. Biological, physical, and psychic systems were governed by the logic of their unity as systems, by the logic of stabilization in the face of local disturbance and perturbation. As Maurice Merleau-Ponty observed, referring to Köhler's theoretical work in a discussion of biological response to crisis, a radical "reorganization" of function in the body is not the result of some "pre-established emergency device" but instead a response of the system to specific pressures coming from outside.[37]

Insight of the Rat

It was possible, in fact, to think of "insight" as a property of *any* active organism navigating a complex and often unpredictable situation. Gestalt-inspired psychological experimenters, such as E. C. Tolman at Berkeley, could demonstrate insight in the behavior of rats, who could avoid obstacles in mazes despite the lack of learning or trial and error experience. Tolman would affirm Köhler's insistence on giving test subjects the possibility of a comprehensive point of view, to test their organizational capacities. As Tolman concludes, "To explain the fact that no insight was obtained in Maze II, although it was obtained in Maze III which had an identical ground pattern, it would seem important that Maze III had no side walls as did Maze II and hence the rats were able in Maze III to 'see' the situation as a whole."[38]

Insight: this psychological ability to restructure experience required a transcendence of the immediacy of experience but not in any metaphysical sense. The intelligent mind (as we see even in Köhler's apes) is capable of *interrupting* its own flow of experience, even the flow of associations and memory. As Kurt Koffka would argue, it was the essential plasticity of the mind that allowed it to "disengage" from the immediacy of experience and then "reconfigure" it in creative ways.[39] Wertheimer would agree. In his own important (posthumously published) work *Productive Thinking*, he said that alongside "insight" into the "fundamental structural features" of a problem, the ability to "free" ourselves from specific situations to open up solution ideas was also essential for creative thought.[40]

The mind as a space of scientific and philosophical inquiry was, in the period between the wars, positioned within the physiological and biological systems of an evolved organism, and yet—at least for many concerned with the specificity of the *human* mind—there was something exceptional about the intellect. That is to say, the intelligent mind was not so much *extricated* from the biological order so much as conceptualized as the exception in that order, whether understood in the context of the single organism or the species writ large in the long scale of evolutionary time. The emergence of consciousness and with it the potential for a new psychic capacity could be at once an extension of the brain's internal integrative function and the break in that functioning. Consciousness, as one philosopher put it, "bestows increased prevision of the organism by an extension of the original capacity of the brain to receive impressions." If the "purely neural machinery could suffice for fixed situations," at some point the routines would fail: "no routine is absolute and no higher organism lives entirely in routine." The conscious organism has a novel power of adaptation, namely, the choice of possible responses.[41] And so "by introducing plasticity into the primary neural adjustment of action to situation," the organism gains a "secondary" system of adjustment to complex environments.[42] "Conation begins with the disturbance of a pre-existing state of equilibrium and terminates in a restored equilibrium."[43]

The automatism of routine itself could be understood, then, as the "interiorization" of creative responses to the environment, responses that become "unconscious" once preserved in the nervous system. However: "It is when this unity is broken, when the multiplicity of conditions reappears, that there is a return of consciousness, that is, the need for a new adaptation."[44] The key to higher cognition was this break in routine, as Gestalt theory had always emphasized. "Intelligence manifests itself most clearly when the animal is placed in living conditions that are exceptional."[45] "Crisis situations" prepare the way for *radical* insight, the fracture in an individual "life history."[46]

The investigation of creative thought, problem solving, discovery, and so on, was more than just psychological research; this was the arena for distinguishing the peculiarity of the human mind as an embodied, evolved function that was nonetheless something that could escape the physiological, even neurophysiological determination: "It is due precisely to its character as an exceptional act that invention can be such a valuable guide in the exploration of this world that is so complex and still so mysterious—namely, human thought."[47]

The flash of insight, the discovery of the genius, the creative imagination, all of these venerable sites of speculation and celebration were examined now with an eye to the relationship between consciousness, unconscious psychic dynamics, and the complex operations of the nervous system and brain.

This depiction of the exceptional character of all forms of *insight* did not, however, clearly address the question of the exception of the human. If life itself was "invention," a constant struggle against the "menace" that is uncertainty and fluctuation, this capacity was due to the "suppleness" of the organism and the "extraordinary" ability of the living machine to construct itself.[48] And the elusive capacity for new and creative thought could easily be ascribed to the plasticity of the cortical areas of the higher brain, which intervenes to supplement the automaticity of the reflex systems.[49] Human cognition in this framework would only be *relatively* complex—not radically different, then, than the "insights" of a rat or chimpanzee.

The Dialogue of Exteriorization: Vygotsky on the Development of Mind

Lev Vygotsky, the influential Russian psychologist whose career was cut short by his early death, focused his research on the development of children and the acquisition of what we can call "intelligent" behavior. Referring to his own studies of children who were, he noted, solving problems structurally similar to the ones set by Köhler for his apes, Vygotsky argued that the human children behave completely differently from even the most advanced primates, not just in degree. The human child, he explained, is no longer defined by the "instinctive" desires—for example, hunger in the case of the ape studies. Human behavior, in contrast, is oriented by "intense," "socially rooted" motives—what the Gestaltist Kurt Lewin called "quasi-needs [*Quasi-Beduerfnisse*]," as Vygotsky noted. The task itself, as opposed to the solution, becomes a focal point for the child.[50]

This is the core of Vygotsky's theory: the transition from animal to human thought is the move into the new space of a social and communicative *system*—it is not merely the acquisition of some new "intellectual" ability. By entering in the world of language and social meaning, the "natural" mind of the child is no longer what organizes or orients actions. Rather, activities "acquire a meaning of their own in a system of social behavior, and, being directed towards a definite purpose, are refracted through the prism of the child's environment." The intervention of this *external* source of direction marks a radical turn: "The path from object to child and from child to object passes through another person" (30). The individual becomes an individuation *of* the social organization.

As Vygotsky will go on to explain, the mediation of this radically external domain of meaning is only possible if that socially organized meaning has its own substantial form of being—a material medium of presence and, consequently, of *storage*. This is the appearance of technicity. What

Vygotsky calls "natural memory" is the "retention of actual experiences as the basis of mnemonic (memory) traces" (30). This kind of memory is "very close to perception" since it is linked with what Vygotsky calls "direct experience" of the world through our sense organs. This is what humans share with animals, the "immediacy" of recollection via natural memory. However, humans become truly human, cognitively, that is, with the addition of a new form of memory, one that even oral cultures possess. Whether "simple memory aids," the marking of sticks and using knots in strings, or symbolic figures and eventually writing, human life is defined by a surpassing of the neuropsychological memory system by an *interiorization* of a novel (that is to say, artificial) "culturally-elaborated organization of their behavior" (38–39).

As Vygotsky explains, the simple marking of a stick (something no animal has been observed to do) is a profound alteration of the mind, because this sign is an extension of the biological form of memory. We might put this even more strongly, since Vygotsky is not at all implying that the external mark is just a substitute for biological memory. The entire psychological order is transformed in this movement: "the human nervous system" is now able "to incorporate artificial, or self-generated, stimuli, which we call *signs*" (39). Only in a human mind is activity determined by "the creation and use of artificial stimuli." Memory is literally transferred to an "external object" (51). Vygotsky writes that in the animal mind, "something is remembered," whereas due to artificial storage, in "the higher form humans remember something" (51).

The Technology of Insight

As a philosopher deeply influenced by the results of Gestalt research, Ernst Cassirer conceptualized the intellectual capacity to strive toward goals as a resolutely independent capacity, one that was constituted in fact as a separation from life itself. If life could of course tend toward its own ends, the conscious appropriation of "these goals always implies a breach with this immediacy and immanence of life."[51] What Cassirer calls *spiritual* action is defined by its resistance to the world as given by our senses: "I thrust the world back from itself."[52] For Cassirer, the animal is distinct from the human because of this separation from the immediacy of life within a given environment. By knowing the world, the human mind removes itself from that world, or to be more precise, begins to construct another world, a symbolic world, that is, a new space for organization and reorganization. But this new order, as the Gestalt movement always emphasized, was itself threatened by obstacles and was engaged in a constant struggle for stability—its separation did not imply an easy liberation.[53] The main point to draw here is that the mind as intelligent system constitutes itself as a new order and organization, in its *separation* from the immediacy of our embodied experience. As Cassirer notes, with direct reference to Köhler's study of ape intelligence, it is extremely difficult for even the higher animals to embark on the productive "detour" because that always means violating the given, that is, the "natural" biological behavior pattern. The logic of the insightful mind is, we might say here, always artificial even if it behaves, as Köhler and others argued incessantly, like other "natural" dynamic systems.[54]

We can see this in Koffka's acknowledgment of the historical and cultural differentiation of human minds, something too often forgotten, he thought, in experimental psychological research based on the assumption that the "mature and cultured 'West European' type of man" was somehow the "highest level of development." In fact, psychology had to grapple with radical difference.

> The world appears otherwise to us than it does to a negro in Central Africa, and otherwise than it did to Homer. We speak a different language from each of these, and this difference is a fundamental one, inasmuch as a real translation of their words into our own is impossible, because the categories of thought are different. . . .
>
> We must not forget, then, that without a comparative psychology, without animal-, folk-, and child-psychology, the experimental psychology of the human adult is and must remain defective.[55]

This differentiation is exactly what we would expect from a psychic system that is, in its symbolic operation, predicated on a radical detour, a radical *deviation* from natural existence. As Henri Delacroix (a student of Bergson) put it, the human is not just capable of adapting to a changing world: "Man is above all adapted to what is not."[56]

As Köhler had shown so effectively, the tentative emergence of intelligence in the ape was tied to the use and construction of implements; however, the concrete implement is not so much a sign of a new form of thought. The tool, in the context of Gestalt theory, is in effect only the symptom of a psychic exception, the moment when a naturally evolved mind behaves *unnaturally*. As Delacroix put it, we are a zoological species but one marked by a tendency to *change its own domain of existence*.[57] Like organismic adaptation and internal regulation, human intelligence was aimed at stabilizing conditions for the maintaining of *life*. "We might say," wrote Eugenio Rignano, "that technical inventions and industrial products, from the first cave dwellings, the skins used for clothing, or the discovery of fire to the most complex refinements of our own age, have always had but one single goal, namely the artificial maintenance of the greatest possible invariability in the environment, which is the necessary and sufficient condition for preserving physiological invariability."[58] The challenge, then, was to explain the particular exception of the human intellect as the irruption of the artificial within the natural, the emergence, that is, of an exception that could not be easily traced back to the neurophysiological order of the brain. One of the ways to understand the theorization of human cognition in this volatile period is to focus on the exception—to understand, that is, the kinds of norms through which we could measure the defiance that was human intelligence. But complicating that task was the fact that the very notions of life, normality, and the exception were in this same moment being rethought (quite radically) in biological and physiological spheres of theory and experimental inquiry. To approach the psychological and cognitive human we must first engage with a line of thought that was redefining the very function of cognition as it emerged in the living being, as a special kind of biological phenomenon.

The living being is one that uses novelty to serve stabilization. Evolutionary progress is fueled, as Alfred North Whitehead once said, by the "power of wandering," the move into "new conditions," because that is when animals are challenged to transform. The human is the only animal that has wandered across the globe, creating and transforming "habits of life" that are founded on the capacity to construct artificial conditions of life. But wandering is not merely physical. The human mind also wanders, in that it is restless and not content with the world as it is. This is the key to human vitality—"when man ceases to wander, he will cease to ascend in the scale of being." Hence the importance of "adventures of thought, adventures of passionate feeling, adventures of aesthetic experience." The adventure of the human spirit on Earth depends on the diversity of human communities, the countertendency to "uniformity." However, at the same time, Whitehead recognizes that the stability of social systems and other human organizations require a certain normative routine of habit. The challenge for human beings is the preservation of their intelligence, the attention to the flash of novelty and the possibility of the response that is anarchic transgression—without destroying all of the life-serving habits of repetition, the automatisms of biological, psychic, and social life.[59]

16
Brains in Crisis, Psychic Emergencies

In 1917, the pioneering American neuropsychologist Shepherd Ivory Franz introduced an article by recounting an earlier survey he had made of widely used textbooks in psychology and physiology. He noted that the predominant view of the brain taught in American universities was "organological"; that is, the textbooks portrayed "the cerebrum as a collection of spatially related conglomerates of cells and fibers, each conglomerate having a certain function (perhaps mental)."[1] This was a simple model, "easily apprehended by the instructor and easily taught to the student."[2] But it was, he noted, as if Sherrington had never published his groundbreaking depiction of the *integrative action* of the nervous system. The importance of neural (and by extension, organismic) unity championed by many figures before the war (including Franz himself) needed to be reconfirmed and made the center of new research. The evidence was clearly unassailable, as neurologists had repeatedly demonstrated (as we have seen) that the brain could recover functions despite the destruction of specific cerebral areas usually associated with them. What Franz argued here was that *functional* capacity was not strictly tied to a specific locale in the brain, since function could be recovered after injury or damage.[3]

Franz was drawing here on his own laboratory research in this area, conducted with the collaboration of the young Karl Lashley, including experiments on rats that revealed that "the removal of large parts of the frontal portions of the brain does not greatly interfere with a learned reaction."[4] Understanding this would require conceiving the nervous system "as a great connecting system" that transformatively organizes stimuli to produce "adjustments of the organism" and, importantly, as a system that can create new and different pathways, as demonstrated in the "pathological phenomena."[5] As Constantin von Monakow had argued and other researchers confirmed, "the brain functions as a whole."[6]

Looking back to Hughlings Jackson, as many in this era did, Henry Head observed that the "more perfect" and hence more flexible and wide-ranging adaptive functions are connected "on a higher anatomical plane of the nervous system."[7] Here the system gains more freedom of response; the nervous system can more carefully regulate the "needs of the moment" in relation to the "reaction of the organism as a whole" (475). When the linkages between centers of integration and correlation are breached in pathological situations, they are not "repaired" and thus restored. The *function* is recovered through what Head will call a "fresh integration carried out by all available portions of the central nervous system" (549). There is, in other words, a response to an internal emergency that is somewhat parallel to the shifting conditions of the external environment. Both, we could say, constitute "a total reaction of the organism to the new situation" (549). A pathological reaction, Head would write, "is a new condition, the consequence of a fresh readjustment of the organism as a whole to the factors at work at the particular functional level disturbed by the local lesion" (498).

The turn to more holistic and adaptive models of the brain was not a complete rejection of all localization of function. However, the extensive number of cases of brain injury and recovery that neurologists had encountered in the Great War, as well as new phenomena such as shock, which was not tied (in either its physiological or psychological forms) to any one specific site or damage, challenged any simple organizational model of the brain and nervous system.[8]

Walter Cannon would study the problem of physiological "shock" via examinations of numerous soldier-patients suffering extreme injuries during the Great War and also in controlled animal experiments that induced shock artificially through massive bloodletting and the like.[9] Cannon's interest in these organismic reactions to extreme emergency conditions led him to the idea that shock, like the "fight or flight" mechanism, was not just a particular local "correction" of a physiological parameter, but instead a new state of being that the body entered under stress: "Every complex organization must have more or less effective self-righting adjustments in order to prevent a check on its functions or a rapid disintegration of its parts when it is subjected to stress."[10]

What J. B. S. Haldane called the "new" physiology was one that paid attention not just to the processes and mechanisms of auto-regulation but also their *integration* in terms of the unity of the organism.[11] "What is health?" As Haldane says, health is what is "normal" for an organism. However, the biological sense of normal turned on the idea of "maintaining in integrity all the interconnected normals which . . . manifest themselves in both bodily structure and bodily activity" (77–78). The new physiology entailed a new pathology, for health would be defined not in terms of static norms but with respect to "varying conditions of environment" and the need for an organism to "reassert itself under totally abnormal conditions." The normal, wrote Haldane, is an "elastic and active organization," not a rigid mechanism but rather a "manifestation or expression of the life of the organism regarded as a whole which tends to persist" (149). A machine is incapable of reproduction, since a machine would have no active "unity" that would guide the organization of matter (142). The machine also could not repair itself; it could not maintain its normality after accidents or injuries. Maintaining the norm, for Haldane, was perhaps better understood as the "active maintenance of composition" (67). A critical question in this context was, again, the identity of the human as an exceptional organism. Haldane would note that the "conscious" organism could deploy a vast array of responses to its environment, given that it could be determined not only by the present sensory experiences but also by past experiences and, maybe more important, possible *future* events.

The Catastrophic Reaction: Crisis and Order from Lashley to Goldstein

Clinical and experimental research on the brain in the early twentieth century was systematically exploring the ability of the brain to *reorganize* in the face of challenges—including the radical challenge of grave injury. The shock of disorder opened up the possibility of a new form of order that was not explicable in merely mechanical terms.

Karl Lashley's research in the 1920s, for example, while initially aimed at finding the precise neural location of memory traces, in fact ended up revealing the plasticity of the brain's performance grounded in a structural complexity that defied localization.[12] After teaching animal subjects certain tasks (maze running, for example), Lashley proceeded to surgically destroy certain parts of the brain, increasing the mass excised in subsequent repetitions of the task. Following these traumatic injuries, the animals were still able to recover their earlier performances, revealing that a significant reorganization of the brain's activity had taken place.

The unity of the brain's systematic complexity must then flow from some comprehensive frame for these emergency reorganizations, which were, according to Lashley, themselves just an affirmation of the brain's *normal* capacity to integrate its activity across the many different areas of the brain. "The whole implication of the data is that the 'higher level' integrations are not dependent upon localized structural differentiations but are a function of some general, dynamic organization of the entire cerebral system."[13] Lashley called this capacity "equipotentiality,"[14] alluding to how the brain sought multiple paths for its activity, thereby giving it the capacity to circumvent damage by relocating activity to some other part of the brain. Lashley would cite here both Henry Head and Kurt Goldstein, among others, to support his conclusions drawn from the research on the dynamic and plastic aspects of cerebral organization after brain injury.[15] Lashley would suggest, basing his claim on both animal studies and contemporary research in psychology and psychiatry, that there was a relationship between the spontaneity and plasticity of what we call intelligence (however ill defined that might be in the context of human cognition) and the structural openness of the brain and nervous system.[16]

As Constantin von Monakow had put it in an earlier, massive book on brain localization, the disruption of any one part of the brain led to a more general "shock" of the whole system. He called this event *diaschisis* (from the Greek διάσχισις, "shocked throughout"). In von Monakow's words, "Any injury suffered by the brain substance will lead (just as lesions in any other organ) to a struggle [*Kampf*] for the preservation of the disrupted nervous function, and the central nervous system is always (though not always to the same degree) prepared for such a struggle."[17] The pathological turn awakened a *total* response (as Sherrington's theory would predict), aimed not at a simple return to the original order, but rather to an order that reestablished stable functioning but in a new form altogether. As von Monakow (in a book cowritten by the French neurologist R. Mourgue) later argued, injury or other shocks were an incitement to break norms and establish a new order: "It is a question of combat, of an active struggle for the creation of a new state of affairs, permitting a new adaption of the individual and the its milieu. The whole organism intervenes in this struggle."[18]

Extensive clinical experience with brain-damaged patients (many of them soldiers with bullet wounds and other war injuries) furnished the data for Kurt Goldstein's innovative work on the nature of unity and plasticity in biological systems, which brought together in a grand synthesis the concepts of order and reorganization as explored in physiological research on regulation and neurological research on plasticity and recovery. In his classic 1934 book, which was influenced by Gestalt theory and the idea of brain "shock" championed by von Monakow, as well as broader holistic forms of thought in the interwar period, Goldstein defined the organism as a unity, arguing that in its continual struggle with the world, within the essential "milieu" of its activity, the organism maintained its stability by constantly reorganizing itself to accommodate new conditions.[19] Goldstein's focus on "pathological" data was, to be sure, aimed at illuminating the normal functions of a dynamic organismic life (50). This approach did not, however, involve an analysis of the "mere defects" (48) of the organism as a way of capturing, in an inverse image, the "normal" state. Instead, Goldstein saw that the pathological state had its own particular characteristics, its own symptomology, its own way of being.

The difference between healthy and pathological states, for Goldstein, was the difference between "ordered" and "disordered" behavior. In the first case, the performances of the total organism were, he said, "constant, correct, adequate." The disordered state is defined by shock; Goldstein refers to the activity of the organism in this state as a catastrophic reaction. This *Katastrophenreaktion* was "disordered, inconstant, inconsistent" and had a "disturbing aftereffect" (48–49). In normal conditions, the organism is challenged by its milieu and meets this challenge with a reaction that will bring the organism into equilibrium with its environment. In the pathological state, the organism has no proper response at hand. And yet, as Goldstein stressed, the organism is constantly seeking an ordered condition—and injured, shocked, creatures do often return to some form of health. His interest, then, was to show how the organism rediscovered stability and normality after a catastrophic reaction.

The key was self-modification, an autonomous capacity to establish norms. As Goldstein argued, with numerous examples drawn from a variety of scientific disciplines and experiments, organisms have the capacity to modify themselves and their performances so that they can reduce or minimize any defect that had led to a catastrophic reaction (52). New paths to a successful performance are found, or, alternatively, new milieus are sought out that do not require the same kind of adaptation previously required (105). This reorganization is explained, by Goldstein, as the tendency of the unity of the organism to seek closure—what the Gestalt theorists called the law of *Prägnanz* (293). The catastrophic reaction, then, is not a mere interlude between two different states of health but instead an *interruption* that demands a new foundation of order for the organism as a whole, because it must overcome the persistence of a defect in its being. This rupture demands a form of *decision*. This is why the pathological state, for Goldstein, can best reveal the essence of the organismic unity. Pathology shows in sharp relief how a living being seeks out novel forms of order to overcome a disordered state. In the healthy being there is an occlusion of this capacity, due to the relative automaticity and predictability of the normal ordered responses.

The successful response to a catastrophic reaction is clearly not a return to the previous state. "The normal person, in his conquest of the world, undergoes, over and again, such states of shock. If in spite of this, he does not always experience anxiety, this is because his nature enables him to bring forth creatively situations which insure his existence" (237). The catastrophic reaction gives us insight, then, into the creative action that is the organism's *essential* nature, according to Goldstein. Unity is a tendency of the organism as it appears in a temporal process of existence, as it continually seeks to find its order, although this order is, at every moment, always put into question by the constantly changing conditions of the milieu. "Therefore reactions scarcely ever occur that correspond to a perfectly adequate configuration of the organism and the surroundings" (227). Goldstein proposes that the organism is never entirely "normal" because at each moment it is being challenged by the environment and must continually seek the proper adjustment: "Normal as well as abnormal reactions ('symptoms') are only expressions of the organism's attempt to deal with certain demands of the environment" (35).

And so Goldstein writes that the life of the organism can be considered a series of what he calls "slight catastrophes" (*leichter Katastrophenreaktionen*), where inadequacies are first confronted and then "new adjustments" or a "new adequate milieu" is sought to respond to this lack (227). The serious catastrophe is in effect continuous with this normal, constantly repeated weaker form of catastrophe; what is different is only the scale and intensity of the reaction. The key point is that catastrophic shocks of some form are essential to the organismic being.

In the end, Goldstein will admit that perhaps we should not even oppose "Being-in-order" and "Being-in-disorder" because the catastrophic states of disordered behavior are foundational opportunities for achieving some degree of order—and that order is always in question. "If the organism is 'to be,' it always has to pass again from moments of catastrophe to states of ordered behavior" (388). The catastrophic reaction manifests the essence of life itself:

> All the minor catastrophic reactions to which the organism is continually exposed thus appear as inevitable way stations in the process of its actualization, so to speak, as the expression of its inescapable participation in the general imperfections of the living world. (392)

Thus, for Goldstein, normative behavior is never really the norm. Normal ordered existence is a product of the essential—if pathological—*disequilibrium* of the organism and between the organism and its surrounding condition, its milieu: "These shocks are essential to human nature and . . . life must, of necessity, take its course via uncertainty and shock."[20] Unity is the very presupposition of the organism as a dynamic flux of operations and adjustments, yet unity was never fully present in any one place—there was no *organ* of unity, no topological order representing it. Unity was the condition of an organism individualizing *life*, which would never be given in advance.

17
Bio-Technicity in Von Uexküll

> Whoever wants to hold on to the conviction that all living things are only machines should abandon all hope of glimpsing their environments.
> Jakob Von Uexküll, *A Foray into the Worlds of Animals and Humans* (1934)[1]

The important question of neural and biological unity was, as we have seen, framed by the many different theoretical explorations of organization between the wars. Scientific work on biological organization in this moment had left behind any belief in active vital "forces" that were ad hoc inventions, yet at the same time, there was also widespread resistance to any strict "mechanistic" account of the kind of *active* interventions that maintained unity in moments of extreme crisis and emergency. How could organization and spontaneous *re*organization be comprehended scientifically? This question was at the heart of experimental and theoretical work in embryology (morphogenesis), biology, and physiology. An understanding of the specific nature of human existence, the presence of intelligence and cultural order for example, depended on new conceptualizations of order that would be up to the task of comprehending its ontological status.

From the beginning, Jakob Von Uexküll's central concern in his biological research (focused on simple ocean-dwelling organisms) and more philosophical thinking was precisely the nature of *organization*. Inspired by Kant but also his friend and neo-vitalist Hans Driesch, whom he met at the ocean research station in Naples in 1901, Von Uexküll emphasized that every living being was distinguished by what he called a *Bauplan*, or blueprint.[2] He also believed that the essence of the *mental life* of organisms, even rather lowly ones, must also be studied biologically. This meant paying attention not simply to structure, the mere arrangement of things, but to the "framework" of order, which denotes the way that functional parts work together in a unified *system*. No matter how close we get to the "minimum" of material substance, the unified framework could never be explained as the consequence of mere material causality. This Von Uexküll took as given, in light of so many demonstrations in the scientific literature, most notably Driesch's famous studies involving the manipulation and interference of embryonic development.

To clarify this problem of organization, central to the argument of his synthetic work on theoretical biology, Von Uexküll uses the analogy of tools.[3] An implement is not merely an arrangement of material substances; it is organized according to a plan (103–4). The proper use of an implement is only possible if this plan is recognized; that is, there is understanding of how the parts of the implement work to achieve its defining goal (106). Crucial for Von Uexküll is the fact that in an organized system certain elements can be "inessential" to function and thus interchangeable with other materials that could equally serve the function of the whole. Human tools, he explains, always have this "residue" characteristic of the inessential property.

Moving to the biological sphere, Von Uexküll argues that the differentiation of higher and lower orders of animals depends on the richness of their function and the plurality of their organs. Functional richness is, clearly, only understandable from the perspective of the framework that provides organization. The organism is not analogous to a human technical object because it possesses an inner unity and comprehensive *integrity* of organ and function. Human implements are only ever imperfect approximations of their functions (which are ultimately derived from original human actions), whereas biological functions do not show any *ambiguity*. Even when organs (such as our limbs) can serve more than one function, they always execute that function perfectly—considered in relation to the organizing framework that specifies its function. The tool, by contrast, is always *exterior to itself*, in that its framework is measured in relation to the organism—the human being—that deploys it (115–16). This is why the machine is not a good model at all for the living being: the machinic framework is not internal to its own function, because it is *dependent* on the framework of the human (120–21).

Despite the natural integrity of the organism, Von Uexküll will explain that it is no static entity. The living being, as we have seen in different contexts, has the *super-mechanical* power of self-creation, repair, and reproduction, due to the plastic "protoplasmic net" that connects the functional organization of the body as a whole. As Von Uexküll observes, the perfect machine is necessarily dead. To live means to transform and alter oneself—to exercise autonomy—in response to change. The *Bauplan* guides the construction of these new states in the organism. The tool has the framework, but it lacks any power to instantiate it, nor is there any medium for self-transformation, given the material solidity of its parts.

This is where Von Uexküll turns his attention to the nervous system. In biological terms, it functions as an organ of operation. Some simple animals have fixed, "inborn impulse-sequences" whereas others learn and develop novel responses. The emergence of the mind in nature is the appearance of a new control mechanism, one that brings biological function into *consciousness*. As Von Uexküll phrases it, "We are informed of every deviation from the normal function" (119). With Kant as his guide, he divided the problem of mind into two separate areas: first the sense qualities and then the "arrangement" of these sensory experiences in the mind. The latter constitutes "the *form* of knowledge" (xvi). All experience of the world is, according to Von Uexküll, not just "mediated" by mental schemata, but *constituted* by them—a Kantian approach to even the simplest organism. These schemata exist only by being conscious, and are the essential control system for active responses to the environment. This operation he calls, famously, a "function circle" (127). (Figure 17.1.)

For Von Uexküll, the organism is always in a state of *struggle*, always threatened by damage or dissolution. The functional circle was the mode through which organisms could take stock of their situation in relation to their own functional needs, and respond appropriately, either with inborn sequences or learned ones (129). The function circles of the higher animals include a new element, namely, the reflective sense of itself *as* a body. The world as it is sensed and experienced and interpreted, then, allows an animal to be guided by a whole series of particular and relevant "indications," enabling action in each situation that accords with its framework of its functions (143). The *Umwelt* is an experience of an organized world, specific to the species and the individual being. One could "build" an artificial animal consisting of sense organs and effectors, but one would be missing the critical *experience* that is organized sentience.[4]

Without this element we cannot explain how an organism exemplifies an automatic but also active and responsive form of behavior. As he observes, using the nautical (but also Platonic, and later *cybernetic*) example of "steering" to elucidate the concept, "in order to meet all the contingencies of the voyage, we place the helmsman between the compass and the rudder," taking "bearings" from the compass but not being locked in mechanically to the instrument, instead independently guiding the "whole ship" in

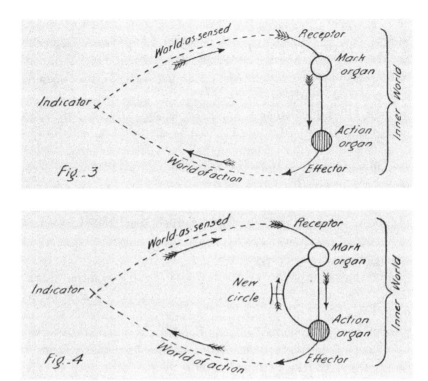

Figure 17.1. From Jakob von Uexküll, *Theoretical Biology* (1926).

the contingent circumstances of the moment. The point he wants to make is that "every living organism" has this "super-mechanical directing" capacity, but it is not located in some body (like the steersman) external to the organization of the being. This directing power lies *within* the order of the organism. However, it must also have a certain structural independence from the "framework," otherwise it would not be able to *apply* this framework in moments of repair or creative work (291).

As nervous systems become more advanced, Von Uexküll argues, they provide an increasing "degree of independence in the steering mechanism" (150). Wolfgang Köhler's apes, who demonstrated the ability to pursue indirect *Umwege*, display this exact kind of independence, says Von Uexküll: the "indications" of both body and environment initiate chains of nervous association that open up these "indirect" paths, which is to say, the animal follows indications framed in reference to the *schemata* of perception and is not just repeating instinctual operations or learned behaviors, explaining the appearance of genuine "insight" in novel conditions (145, 149–50).

The distinction between these higher animals and the human is not an extension of the range of indication but a change in the *kind* of organization produced by certain novel intellectual arrangements. The animal, however complex its function circle might be, always effects action through its own organs. The human tool is a form of effective action that demands a wholly new use of our *own* effector systems. Only humans must create new gestures, new uses of our own natural "implements." Human artifacts are never stable and perfectly formed entities because of this essential artifice of mediation, not defined in terms of the natural *Bauplan* of human action. In addition, no artificial tool is ever really fully independent even *as* a tool, in that it must function within a technical *system* (the hammer needs the nails, and wood, for example) that is always being adjusted, but not, according to Von Uexküll, in light of any unified plan or framework. For animals, the natural effectors have no such residue; they act fully in conformity with the organism's framework (159–61).

So the limitation of the machine, as compared to the organism, is that it has no independent and active organizational plan, since it is introduced by the human from outside of nature. The organism's framework is the expression of its own internal "function-rules" and therefore these internal rules can guide the regeneration of the organism, using the plastic material of protoplasm. The instantiation of the "rule" in what was then called a gene explains, for Von Uexküll, the development of a complex organism from an initial undifferentiated state in the embryonic cells. In contrast, the machine is constructed with an external plan and therefore cannot produce anything new; it "can only be altered from without" (179).

What exactly is the tool for Von Uexküll, in the context of theoretical biology? The implement, he says, is anything that is used by an organism as a "counteraction" of the organism's own action performance; that is, it is an external support for the completion of a function circle. Insects, for example, use leaves, birds build nests, and so on. But the animal implement (at least as far as we know) is always, Von Uexküll observes, integrated into an action that is *instinctive*. The bird will always make the same kind of nest even if it grows up in altered conditions. With animal implements, there is no relation to *experience* as a form of knowledge (232–33). The implement, then, does not itself even need to appear as implement within the mark organs (sensation) since there is, so to speak, no reason for the automatic use of an implement by an effector to enter into the mind of the animal (234).

This constitutes the essential difference, or rather chasm, between human and animal. The human forms an implement according to a use-function (and not according to an already inborn activity-rule). This is why the human alone can recognize the foreign object, this artificial being that is the tool. It necessarily enters into our sensory system where we can articulate the functional relationships that make up its order. As he writes, unlike the majority of animals, our "sensed world" *includes* our effector organs and their activity. Becoming conscious of the function rules of our own organs allows the possibility of forming "real implements," based on a human function-rule. The internal representation of the functions in the mental schemata is at the same time the opening up of an *exteriorization* of that form into artificial "organs" (332–33). These artificial, independent function organs then provide new indications for the world as experienced by a human. In fact, humans can only recognize (scientifically and informally) organization *in nature* because we first have the exteriorization of organization visible in the artificial prosthetic that is technology.

We might say this is where we enter the precarious world of signification: for humans are no longer able to share a common, because *natural*, system of indications. Every implement, every tool, every cultural artifact, embodies in its material form its own *Umwelt*, the artificial environment that is constituted by tradition and memory and practice.

And worse, as Von Uexküll will go on to explain, is that the domain of artificial indications and their current industrialized monotony can begin to destroy the "world-as-sensed" that is given to us in our original state. In any case, the human is no longer *what it was*. Social organization is dependent on an artificial infrastructure that mimics the organism, or systems like the animal colony, but the order represents only a "pseudo-harmony" of machine and human, "effected through the ceaseless labor of a thousand tools and machines, which are worked by human beings" (338). The artificial system of political and socioeconomic organization then spawns its *own* "action-organs" of control, of intelligence, and of military defense. There is, Von Uexküll insists, a demand in this techno-political condition for the kind of "steering mechanism" inherent in all biological beings. However, in the human context this mechanism must of course be *artificial*, since the infrastructural order is itself a technical one. The nature of this technical steering is unclear, but any artificial organization—like its natural counterpart—must instill a form of *discipline* on its basic material form. The organism makes use of the open protoplasm to form and re-form its organs and tissues according to the framework. For industrialized technical and social ensembles, the individual human now functions as its "protoplasm," open and undetermined, at least in the context of technical life: "The super-mechanical principle of the community is expressed in those plastic actions of its members endowed with protoplasm, which form the community and maintain it" (350).

18
Lotka on the Evolution of Technical Humanity

What an organism feeds upon is negative entropy.
        Erwin Schrödinger, *What Is Life?* (1944)[1]

The American evolutionary biologist and statistician Alfred Lotka clearly knew of Von Uexküll's work. There is a marked-up copy of a review of Von Uexküll's 1909 book on the *Umwelt* in Lotka's papers at Princeton. It is not clear what he thought of the Estonian's ideas, although it is likely Lotka would have criticized the "vitalist" orientation of the work, as well as the outright rejection of Darwinian models of evolution. But Lotka was hardly doctrinaire, famous for taking a *physical* perspective on evolution, arguing in his major work, *Elements of Physical Biology* (1925), that the total environment of the earth must be understood as one single, physical, thermodynamic system. Entropy was only forestalled in our biological world because of the continuing influx of solar energy. Biological "life" was less a distinct category for Lotka than just another form of energy consumption and storage. The living being, he said, is a topological ordering of chemical and physical processes whose organization allows for a channeling of energy that can then be used to maintain its "environment."[2] Human beings, he notes in an aside, are the beings who effect the most extreme reorganization of the environment.

  In any case, the evolution of life should, according to Lotka, be framed as the history of systems undergoing irreversible changes. For the physicist, these irreversible changes would be defined in terms of entropy, the inevitable increase of disorder in an isolated system. What marks the biological as an independent science is precisely that the irreversible changes in the system are ones that *maintain* order, violating (at least locally) the Second

Law (46–47). As we have seen with the new physiology of the period, the body is never a static system that merely regulates changes in variables; it is a kind of system always on the edge, constantly in a struggle to maintain its unity and order in the face of dissolution. Lotka emphasizes the same point in his work: the "equilibria" of living beings are not true equilibria, or steady states, because all living, dynamic systems require a constant source of "free energy" in order to continue to live (144). Indeed, Lotka will point to the importance of *moving equilibria* in biology. This will help comprehend the often productive effects of internal instability, as it appears "both in normal and pathological life processes" (294).

To understand evolution, then, requires a comprehensive, total view of all life and its relationship to the material earth, because, as Lotka will explain (mathematically and conceptually) the interconnected system of living beings is really a vast network of energy transference. The only source of "free" energy is the sun, so the system relies first of all on the ability of plants and other organisms that can take solar energy and transform it, storing it in some form. Other entities can then feed off of this captured energy, taking it to fuel its own struggle against entropy. Lotka calls these kinds of entities "transformers" (335). Evolution of the *system* (described thermodynamically) is not really about individual transformers, or categories of them (what we would call with Darwin "species"). For Lotka, the only way to understand evolution as a process is to define it in terms of a systemic improvement in the capture and circulation of energy.

The elements of the system are the individual living beings, entities that must seek energy stores to survive. From the smallest organisms on, life depends on *"more or less accurately aimed* collisions" between transformers and energy sources and the avoidance of "harmful" collisions with other beings or dangerous physical features of the environment (337). As the individual system becomes more complex in its organization, that system will require more and more energy to maintain itself—to preserve its internal processes but also to fuel the search for energy itself. This is why we need to take a broad view of evolution, according to Lotka: every living creature is involved in this constant transformation of energy, and the total system of life is always limited by the amount of free energy available *to the system*.

However, in assessing the form and activity of individual beings, Lotka admits that the kind of organs and faculties that appear in nature cannot themselves be explained purely in thermodynamic terms. Echoing Von Uexküll's approach in a way, Lotka explains that to make "collisions" with energy sources productive, statistically speaking, an organism needs a "correlating apparatus" to guide its behavior (338–39). What is key is the independent logic of this apparatus. Its organization will have its own form of evolution, linked to the organism and its place in the environment. In this space Lotka will locate the exception that is the human. The apparatus of the organism is a kind of *natural technology*. What marks the human as human will be a special relationship to this technology, one born from a radical reorganization of the "correlating apparatus."

To account for the transition from higher organism to human, Lotka breaks down the different components of living beings. First, there is a need for some kind of representation of its world, and this is provided by what he calls "depictors," sensory organs ("receptors") that interface with the environment, and the "elaborators" that process and organize the "crude information." Reactions to the information are determined by "adjustors," which then provoke the final step: the operation of "effectors," the motor organs (wings, feet, hands) that provide actual physical movements. We have here, in other words, something akin to Von Uexküll's function circle. As Lotka explains, the process of "elaboration" is perhaps the most opaque link in the system, since it takes place in what we normally think of as "mind"—the faculties of memory, reason, and so on, which are not identifiable as "organs" in the usual way (339). Of course, this would be the site for the formation of an *Umwelt*, arrived at through some kind of primitive Kantian synthesis. But Lotka's interest here is different. He wants to understand the implications of increasing complexity and in particular the *temporalization* of these processes of elaboration. With human beings, certain actions can take months or years to plan and execute. Our behavior is still correlated to the environment but in an indirect or at least temporally complex way, and this is mediated by the *organization* of our internal information states (340).

However, correlation cannot radically distinguish human life from that of all other animals. Indeed, as Lotka will show, correlative responses are not even limited to living beings. He gives the example of a toy beetle, a wind-up mechanism that has a "sense organ" (a feeler) that detects the edge of a table and then causes the machine to turn away and avoid falling. (Figure 18.1.) No consciousness is required. And Lotka also notes that we can imagine ever more complex systems of automated technologies that could act as artificial "elaborators," giving the example of automated telephone switching devices (341–42). For Lotka, the hierarchy of beings (natural beings at least) can be gauged by the relative "elasticity" of their behavior patterns. The automaton-like organisms are relatively fixed in their response patterns (like the artificial insect), whereas highly mobile animals have more complex repertoires. The elasticity of high organization is "promi-

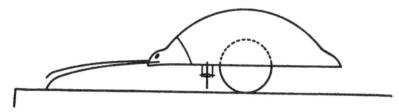

FIG. 69. MECHANICAL WALKING BEETLE, EXHIBITING THE SEVERAL CHARACTERISTIC ELEMENTS OF THE CORRELATING APPARATUS

Figure 18.1. From Alfred J. Lotka, *Principles of Physical Biology* (1925).

nently displayed to us in our own selves" (350). This would be Lotka's biological explanation for integrated consciousness (an echo of Von Uexküll): the organism can now react to *images* of possible futures that are not yet, not just to memories of actual past experiences. Behavior is adjusted to a conjectural reality.

For whatever reason, consciousness evolved as a simpler and more effective tool for elasticity, an alternative to some "purely mechanical structure" that could perform the requisite diversity of tasks" (402). Lotka points to Sherrington's conception of the nervous system as the "acme of integration" for the organism, and cites Clerk Maxwell to suggest that the mind as an integrative process could still affect the organism in profound ways, given both the "unstable equilibria" of the body and the appearance of "points of singularity" in these unstable systems, critical moments where the status can change significantly (407). "This conception makes the interference of consciousness (will) in physical events an exceptional occurrence" (408). Elsewhere, Lotka would speculate that consciousness is a "phenomenon associated with, or attendant upon, matter in a state of strain, or perhaps matter in the process of yielding to a strain."[3]

However, the emergence of integrated consciousness does not clearly mark a radical distinction between human and nonhuman, only a scale of relative complexity. In the end, Lotka will locate the exception of the human not in the increasing complexity of the elaboration system but instead in what might seem at first to be the less interesting domains of the "receptors" and "effectors." Unlike any other being (and this is something Lotka believes has never been taken into account by theorists of evolution), humans make use of *artificial* receptors and effectors. Without fully addressing the question of origin, Lotka recounts how our "ingenuity" has allowed us the use of "artificial aids and adjuncts to our senses," from simple eyeglasses to microscopes and telescopes (364). And alongside the expansion of our sensory systems comes the technical extension and even displacement of the effectors: machines now produce physical movements not only independent of human power, but at scales and with forces well beyond any living creature on Earth.

The appearance of *artificial organs* introduces into the world system a whole new evolutionary factor, Lotka argues, because humans are now evolving (as natural-technical hybrids) at such a rapid pace, compared to other organisms. Our technological apparatuses have their own evolutionary dynamic, and especially in the age of modern science and industry, the human forms of life have been transformed dramatically.[4]

Perhaps more radically, we could say that Lotka's account points to a transformation of the very "nature" of the human itself. We are no longer biological but instead a "strange species" inseparable from "artificial organs vastly more efficient than their 'natural' (that is, physiologically compounded) prototypes."[5] The development of artificial organs is critically important, but lying behind the evolution and progress of technology is a new capacity that animals are barred from. As Lotka explains in *Physical Biology*:

> The artificial portion of our correlating apparatus differs in several important respects from our native endowment. My microscope does not die with my body, but passes on to my heirs. There is thus a certain permanence about many portions at least of the artificial apparatus. And for that very reason the development of this artificial equipment of human society has a cumulative force that is unparalleled in ordinary organic evolution. (368)

Artificial organs are shared in the sense that they can survive only with inherited knowledge, which opens up the possibility of the technical amplification and reorganization of our inborn elaborating apparatus itself. "This cumulative effect is most of all marked in the artificial aids to our elaborators" (368).

With technology the human mind is now constituted by "information" and elaboration that comes from *alien* nervous systems (or, we can presume, artificial information processors). The implications are profound: the self is no longer a function of the individual living being but instead something *distributed* across a field of conceptual organization that defies the traditional division between natural and artificial.

> Any attempt to establish boundaries between the self and the external world, or, for the matter of that, between two selves, is not only useless but mean-

ingless. Coordinate reference frames have no boundaries, and freely interpenetrate each other, being merely immaterial aids to fix our ideas. (373)

The peculiar status of the human mind as part of a conceptual field that is at once *inside and outside* of the nervous system is facilitated by yet another technical innovation, according to Lotka, namely, the "highly specialized" system of receptors and elaborators that make up communication networks. The human evolution of knowledge and technology is predicated on a technology of knowledge *transmission*, whether that is speech, tradition, or the various forms of material storage (carvings, writing, printing). In the end, we might redefine the human in terms of its "thought-transmitting propensity" and the *orthogenetic* evolution of its technical systems. Lotka will warn that this kind of "internal" evolutionary progress can be both destructive and constructive, and the pace of development can accelerate tremendously. Just as the mechanical beetle can anticipate a "threatened catastrophe" through its antenna technology, and *change* its direction in response, the implication is that humanity needs to anticipate its own potentially catastrophic future (381–82).

Lotka would return to this theme in a 1929 article, where he repeated the theory of technical evolution, describing the industrial society as "one greatly magnified organization of artificial sense and motor organs and reproductive apparatus" that is now necessary to support the Earth's human population.[6] However, the artificial organs do not live and die on the same cycle as human beings, and Lotka warns that we have yet to think seriously about the regulation of consumption and regeneration in this context and to take into account the very real problem of the "exhaustion of natural resources."[7] And he would add, in another essay, the observation that technical evolution has led to increasingly more brutal warfare. The irony is that "the most advanced product of organic evolution" may very well become "the first of all living species so clever as to foresee its own doom."[8]

When Lotka published his iconic essay, "The Law of Evolution as a Maximal Principle," in 1945, it is hardly surprising that the dark implications previously hinted at would come to the fore.[9] What Lotka now called *exosomatic* evolution (the development of humanity through its artificial, technical organs, as described in his *Physical Biology*) took on a more sinister tone. Quoting Bertrand Russell, Lotka now described human evolution as a form of imperialism, a systematic destruction and reorganization of the environment with only one goal: "to transform as much of the earth's surface into human bodies" (174). Repeating his critique of traditional evolutionary perspectives, Lotka argues that we must shift our emphasis from the species and its struggles, with the environment, with competitors, and internally between individuals, and think of evolution in terms of the whole system (178). Evolution, understood in thermodynamic terms, can be defined as the increasing *flow* of energy in the system, since there is never any increase in energy that comes from the sun. Natural selection, for Lotka, is a matter of regulating and amplifying the movement and transformations that take place in the energy "reservoir," though alteration of the relationships between organisms that store energy (e.g., plants) and those that *dissipate* it (animals).

This process of evolution is all mediated by genetics—with the exception of the human. Leaving behind genetic evolution for an *exosomatic* evolution through technology, human beings radically maximize the flux of energy with their vast and complex systems of transformation and storage, leading to the capture of "excess" energy and opening up the possibilities of *culture*, something that is itself inessential for biological life. The flip side of this exosomatic revolution within natural history is the escalation of *violence*—the incomprehensible violence of World War II exemplifies one trend in history, the turning away from the "preservation of life" to its "destruction" (189). But Lotka also points out other implications of our artificially sustained existence, including economic instabilities and vast inequalities in wealth, and, perhaps most ominously, the reliance on the reservoirs of stored solar energy that will never be renewed—fossil fuels (190). Lotka pleads, in the end, not for a restriction of the artificial, since that is what human beings have become. Rather, he calls for a new technical development, one focused not on the improvement and extension of artificial receptors and effectors central to the industrial economy but on the *adjustors*, the psychical "responses" that intervene between sense and action. The zone, that is, of *intelligence*.

It is fitting, then, that Lotka's archive contains a program for a conference that took place in New York in October 1946, "Teleological Mechanisms." This meeting included presentations by Norbert Wiener, Warren McCulloch, John von Neumann, Arturo Rosenblueth, and Gregory Bateson—all luminaries of what will soon be called the new science of cybernetics. Lotka was, in his own way, already raising the most complex question of the day: what could be the future of our own *artificial intelligence*?

19
Thinking Machines

Well before the advent of the digital computer or cybernetic theory, the question of the "thinking machine" was being raised in the context of debates concerning the nature of organization and the distinction between machinic and vital processes. In his article "Men or Robots?" (1926) William McDougall noted—as others before him—that an artificial machine is organized only because "it embodies a human purpose," built, that is, to achieve some goal. To be "in" or "out" of order refers to whether the parts of the machine are working to attain the goal of its plan or design. Physical, inorganic systems cannot be understood teleologically in this way.[1]

However, *living* systems—"the movements of men or of animals"—must be described with reference to their purpose, whether in a specific context or more generally (81). McDougall's point is that the goal-seeking organism is not simply a tendency to "persist" through automatic (and essentially random, with respect to the environment at least) activity, but rather a capacity to interpret their position and select between alternative paths to the desired goal. Citing Tolman's work on "insight" in rats, as well as Köhler's work on ape intelligence, McDougall argues that organismic action is "guided and terminated, not merely by some new sense-stimulus, but by the animals' appreciation of the nature of the new situation brought about by its activity" (85). This is the major distinction between the machine and the organism, for McDougall—the capacity to select between alternatives due to the presence of some psychic "appreciation" of the situation, Von Uexküll's *Umwelt*.

C. Judson Herrick's *The Thinking Machine* (1929) pursues this same question of the artificial, bringing the author's immense expertise in physiology (he was an expert on amphibian brains) to the study of intelligence and biological mechanisms.[2] Well before the discourse of "intelligent machinery" and "electronic brains" became popular cultural tropes in the postwar era of computing and cybernetics, Herrick was asking the same question Alan Turing will pose in 1950, namely: "Can a machine think?" (This is the title of chapter 3.) In the context of interwar biology and psychology, this was a delicate question—given that the body of the living organism was still being conceptualized in material terms, even if the understanding of material processes was being pushed in new and unexpected directions. What is interesting is that Herrick, a proponent of what we can call the new "organicism" (to disentangle it from any metaphysical legacy of vitalism), intelligence could be linked to the question of the "machinery" of living beings, but more important, to their relationship to *artificial* ones. In other words, he is (unlike Lillie, or Von Uexküll, say) willing to look at the issue from the other side, the perspective of technology itself.

As Herrick makes clear in his introductory remarks, he does not think that machines are uninteresting at all: "All machines make things, and they make them actively. They are creative agents." And, notably, not all machines are artificial in the usual sense; some are "self-made," and his examples—"river systems and living bodies"—reveal how broadly he is thinking (8).

Herrick's working definition of machine is an *organized* system that behaves in a certain way due to its organization. On earth, many natural systems are machines, but more broadly, Herrick argues that we can say that the "whole natural universe is a machine" (13). The key is the idea of order. So when Herrick, addressing the current prophecy that laborious human tasks will soon be automated, cites Karel Capek's extraordinary play *RUR* (1921), which was performed in Chicago in 1922, he isolates the core issue: to what extent can we say the human is *already* a robot of sorts—that is, just a series of "clever mechanisms" (13). If we ask whether a machine can think, most would object because, Herrick notes, it seems to be a contradiction in terms. The robot is a "fabulous monster" (16). Yet, in what seems in retrospect an anticipation of Turing's famous article that launched the very discipline of artificial intelligence, Herrick points out that the first step in clarifying the question is to define the terms appropriately. Or, better, at least to evade strict definitions and reject certain "taboos" in order to look more objectively at the perennial problem of the mind-body relationship raised in the context of automata (24).

As he admits, the progress of biology and more specifically *physiology* has been driven by the search for the mechanisms of animal and human behavior. However, the engineering perspective does not, according to Herrick, give us any insight into the "higher human activities"; this mechanistic method "seems to break down at the finish" because it is challenged by what is so important about human cognition, namely, our "spiritual life," succinctly defined as "consciously directed behavior in general" (41). What is odd about human thinking, it seems, is not that it is "outside" of the natural order but that it is an exception *within* it, since human minds learn to "control" natural forces by introducing what Herrick calls "new patterns of performance of energy" (41).

The question of order, then, is redefined here, as it was by Lotka, as a question of sustaining organizations against entropic forces of disorder (49). This helps identify a crucial physiological challenge, one common to humans and all other organisms: "Living things are reversing entropy all the time; that is their chief industry" (47). But what distinguishes humans in this thermodynamic system is our ability to *create* technologies that themselves help in this general biological "reversal" of the tendency for energy to run down to lower orders of organization. Machines that control nature are machines that *reorganize* it for a purpose and thereby artificially produce and maintain a higher level of organization for the energy cycle that is the system of life on earth.

This is the crux of Herrick's investigation: all organisms and even inorganic systems are "creative" in that they *transform* energy and produce and reproduce new orders. They do not produce new energy, which of course is impossible, but they do creatively redirect it into new patterns—the beaver makes a house, the bee a honeycomb, the river a delta. So the exception of the human—the source of *artificial* as opposed to natural invention—would lie in a special kind of mechanism of transformation. The key is a mechanism of what Herrick calls *control*.

Complex organisms demonstrate the most sophisticated control systems of internal regulation; that has been one of the great advances of the new physiology of Cannon, Haldane, Herrick, and others in this period. These systems not only regulate; they can also repair and adapt to injury and loss. Vital processes are, in essence, these processes of control, the capacity to introduce and maintain "patterns" in the continual metabolic flux of energy and material transformation. As Herrick shows, it is not impossible to mimic a vital process with an artificial technology, and here he refers to Lotka's mechanical beetle (and reproduces the illustration), noting its prosthetic sensor that gives the being an indication of a possible future that can be avoided or confirmed.

Why is the human brain, the product of natural forces of evolution, capable of something so exceptional and so resistant to the method of "reverse engineering"? We can zero in, with Herrick, on a central thread in neuroscience from this moment, the openness and plasticity of the human brain, in particular, the cortex. What is strange about the human brain, Herrick observes, is its *lack* of proper organization in the early phases of development, in both embryonic and infantile stages. "The baby's cerebral cortex is immature at birth and for several weeks after birth it is too busy growing in size and complexity of internal texture to take any active part in learning" (108). Unlike any other organism, the human infant has few reflexes and therefore a wholly imperfect "working pattern of behavior" (108). If the brain or nervous system of any organism (from worm to human) is considered to be the key control system, it is possible to identify more and more complex repertoires of behavior made possible by more and more complex associations of experience and memory. The cortex in particular is a special kind of neural organ in that it is (as James also put it) a triggering device, highly sensitive and capable of releasing motion with the slightest of impulses. What we call spontaneity is that feeling of suddenness and disruption that comes with this triggering of a complex apparatus of cortical organization (122).

For Herrick, this is where the creative intelligence of the human mind will be found—in the cortical assemblies that are "as complicated as the wiring chart of the telephone system" of a major metropolis (253). The conscious mind is capable of radical novelty, what Köhler called "insight" (224). We "see through" a problem situation through an *internal reorganization* of our cortical pattern (which might take place unconsciously). The implication is clear: this cortical mechanism of self-reorganization is linked with the unusual plasticity of the human cortex, and the novel, constructive reorganizations of the material and social world. Technology is the *externalized manifestation* of this neural capacity. In the end, Herrick's proposed "natural history of human nature"[3] will turn on these *unnatural* creations of the human brain, the artificial "vital processes" that make possible the negentropic constellation that is technological civilization.

20
A Typology of Machines

The mutual interplay of the psychic, social, and technical spheres raised new kinds of questions for interwar thought. As the French philosopher of technology Jacques Lafitte observed in 1933, while it is true to say that humans make use of tools to satisfy needs, "the integration of any new machine into the social order modifies the totality of needs."[1] For Lafitte, the machine is not a pure invention; technology is an extension of the human, a kind of projection, to borrow Kapp's term from the nineteenth century. As Lafitte puts it, "At every stage of development, the machine exteriorizes and marks a step in the progress of our own organization. Machines extend us" (156). Human activity is distinguished by the construction of "artificial organisms formed from an ensemble of physical solids." The machine therefore is first of all *virtual*, an idea of organization that determines the spatial relations of the material elements, that forms them into a "organic whole." The order that is the "idea" supervenes on the laws of the physical world—the human act "introduces into nature a new order" (154).

Lafitte's analysis of technology begins with the idea that the human is not an animal confined within the circuit of its natural functions and needs. Lafitte writes that technology "is a functional differentiation and is formed, as we form ourselves, through interference—that we ourselves effectuate—between our received needs or those that we give ourselves, and the possibilities that we bring to the past of our species" (156). And yet, constituted as an extension of our own organization, an organization open to change, technical objects themselves, we must recognize, are not unlike "living beings." With their integration of "general organization" and structure with material elements and "exterior forms," machines mimic the kind of evolutionary perfection of the organism. And, perhaps more important, in the "ensemble of machines we discover a series that is closely comparable to what we see in living beings," posing the genuine question of a technical analogue of the "origin of species" (154-55). What Lafitte understands is that all human activity is undertaken according to plans, whether it is the construction of machines or buildings or the making of instruments and tools of all kinds. This can all be defined as the creation of "organisms" because they constitute conceptual configurations of unorganized matter—or, at least, matter organized according to some "other plan" (154).

Lafitte's analysis of machines therefore proceeds with the key concepts *organization* and *inheritance*. The mechanisms of generation are, to be sure, externalized ones, namely, the actions of human beings. Nonetheless, Lafitte argues that the process of development mimics the biological, in that the progression of technology depends on what he calls the "interference" introduced into existing forms of (artificial) organization. This leads not so much to improved individual objects but rather new forms of technical organization writ large. The "interferential mechanism" (*mécanisme interférentiel*) allows for functions that were once "exterior" to the individual machine to be *internalized* in more complex "organic" ensembles. Technical evolution is, according to Lafitte, a rhythm between interiorization and exteriorization that produces greater integration and order. Successful use or failure determines the path of this development, as in biological evolution: machines appear and disappear, are adaptive or not (154).

The point to stress here is *not* that the artificial machines are equivalent to organic ones. Lafitte repeats the common critique of the machine as an organization incapable of self-repair or reorganization in times of crisis. What is important is to recognize that technical organization is to a certain extent independent of other forms of order; it has its own demands, and this organization evolves according to its own logic, in relation to the environment (human, social, and technical) that it operates within.

Lafitte goes on, then, to offer a typology of machines, one that is analogous to certain biological orders but is not meant to collapse the distinction. The main classes of machinery are, according to Lafitte, passive, active, and reactive (*réflexes*). First, a passive machine is like the plant or other simple organisms, while the active and reactive machines are analogous to animal beings. The differentiation is along the axis of *intelligence*, understood here in its biological sense, in terms of sensibility and organization, instantiated in increasingly complex nervous systems. Second, active machines are like the animals that respond in fixed ways to the conditions of life—by reflex, by instinct we might say. Finally, the reactive machines are like higher mammals. They have the "remarkable property" that allows them to *modify* their own functions in response to information (*indications*) received according to their relationship to their milieu (149). As Lafitte writes:

> These machines include, in very general terms: an essential transformative system acting according to the impulse of external fluctuations; organs of regulation and distribution for the essential system; and an organized sensory system able to detect variations in the relation of the machine to its milieu. (149)

The crucial concept for Lafitte here is modifiability. The reactive machine is capable of routine but also irregularity and can be modified by humans but also *modify themselves*. These kinds of machines in fact "possess a kind liberty, more or less pronounced, of function." The crucial example offered here is that of the automatic torpedo, a machine that can, on its own, "modify its regimen and consumption according to the perceptions of its regulator" (149). The reactive machine is one that depends, then, on *organized sensibility*, what can be called, in this moment, an artificial mind.[2]

It is important to emphasize again that the analogy between machine and organism is an analogy, which entails both differences and identities. The key issue arising in this period is not distinguishing the artificial mind from a natural mind, since it was becoming increasingly clear that complex reactive machines can function like organic beings in many ways. Rather, the challenge is specifying the difference between an artificial mind and a *human* mind, given the acknowledged capacity of the human mind to create artificial organizations in the first place.

Like Lafitte, the Spanish philosopher José Ortega y Gasset reminds us, in his own reflections on technology, that the human, unlike the animal, is not defined by its "necessities." The animal lives wholly within the circle of "vital urgencies," the variety of "elemental necessities" and the actions that can satisfy them.[3] Yet as Ortega y Gasset observes, there is really no "necessity" with the animal since there can be no distinction drawn between necessity and what is *not* necessary. The animal cannot recognize necessity. For humans, the situation is different because we ourselves impose conditions for living. There is a fundamental gap in our experience, in that we recognize the necessity as something real; however, our lives do not at all *coincide* with the vital. This reveals the "strange constitution" of the human: the only being who "is different from, and alien to, his circumstances."[4] The human is *embedded* in circumstances (biological, environmental) but not circumscribed by them.

That gap is what enables the human mind to externalize itself; it is free to act upon its environment, to attend to needs through *technology*—lighting fires, building houses, cultivating fields, designing automobiles. The satisfaction of needs through artificial implements introduces into nature objects which had not existed before; technology is in reality "the construction of a new nature, a super-nature interposed between man and original nature." This is why the tool is not merely an artificial substitute for the animal action: human technology "reforms" nature itself and disrupts it, and the relation between humans and their "two natures" can never fully stabilize thereafter. Human life is a constant improvisation because it is "continually challenged, disturbed, imperiled."[5] Ortega y Gasset declares that humans are the only beings for whom existence is a problem, as Heidegger first argued in *Being and Time* (1929). To be human is to *not be what we are*. Because we make our own existence, we can never coincide with it. For Ortega y Gasset, to be human is to be an engineer, the engineer of one's own existence—this is what he means when he says life is a "self-creative process" that relies on a technical capacity. "The history of human thinking may be regarded as a long series of observations made to discover what latent possibilities the world offers for the construction of machines."[6] The world, for human beings, is always a potential machine for existence. So he does not want to rest with the easy definition of the human as the "toolmaking animal," especially since Köhler, invoked here again, as in so many works on intelligence and technology between the wars, had demonstrated the genuine if rudimentary technical capacity of the ape. What interests Ortega y Gasset is the way in which technical capacity arises when "intelligence functions in the service of an imagination pregnant not with technical, but with vital projects."[7]

21
Philosophical Anthropology

The Human as Technical Exteriorization

In the context of interwar thinking about intelligence, in particular, thinking about the nature of the exception that is human cognition, its power of both creating and understanding order and understanding, manifested most clearly in cultural and technological expressions, it was not possible to define "the human" in any way that was easily reconcilable with inherited theological or metaphysical understandings of the human exception. As we have seen, new physiological and evolutionary approaches, along with the transformation of traditional philosophical questions via emerging experimental psychological findings, made that extremely difficult, and the rapid industrialization and development of technical innovation put a special emphasis on human artifice as a key symptom of the exception.

Technology was critical to interwar reflection on human cognition and intelligence for two main reasons. The first was because the "invention" of the tool marked a certain demonstrable threshold between animal and human intellectual capacity. And second, the emergence of the human as a technical being revealed a new form of evolution, unlike any *natural* physical process of inheritance. As Charles Judd (professor at the University of Chicago from 1900 to 1938) explained it, the intelligent use of an artificial tool constitutes a break from nature. The human capacity to organize complex ideas prepares the way for the *organization* of the material world itself—the psychic laws of association become instantiated in artificially constructed objects, and the human mind will then dominate the world of things.[1] As important, for Judd, this "tool consciousness" was also always an expression of *social* intelligence, dependent, as he says (in agreement with Lotka here) on the "accumulation of experience from generation to generation through social institutions," institutions that themselves create a "new order of reality" (17).

Quite literally, for Judd observes in a remarkable phrase, "the individual nervous system is in this way taken over by society" (328). Humans truly communicate, unlike animals, which is to say, the accumulated wisdom of multiple experiences is "taken into one's inner life" via the brain (327). "The commerce of mind with mind through the medium of external realities is the great achievement which has raised man absolutely above the level of the lower animals" (17). This development is linked to the primordial discovery of the *tool*. "Man, the artisan . . . is no longer nature-man" because experience, *ideas*, are "detached" from individual minds and therefore become "capable of affecting other minds" across time and space, generating new social and political orders (22). The plastic brain is thereby transformed into a *vehicle* of a collective experience, through the material medium of technology: "The tool becomes the external embodiment of a human thought" (16). As Judd explains, in a concise summary of this process, "Too much emphasis cannot be laid on the fact that within the human cerebrum a new world is created. Objects are put together in this world in a new way; they are united as ideas in the brains of men and afterwards through human efforts the outer world is correspondingly rearranged" (12).

Thomas Morgan, an American evolutionary biologist who won the Nobel Prize in 1933 for his work on the role of chromosomes in heredity, agreed that humans, unlike all other organisms, whose transformation is limited to the physical, genetic processes of inheritance, can additionally transmit *experiences* to one another, via their social and linguistic networks. This form of inheritance allows for a dramatic alteration of behavior in relatively short periods of time, leading to the diverse differentiation of cultural forms of life. As he put it, "In man the power to alter the behavior of the race is possible because each individual begins his life with relatively few instincts and even these may be rapidly altered by the training he undergoes from birth to maturity." The openness of human evolution is predicated on the unique *plasticity* of the human brain.[2] The evolution of humanity is dependent on the evolution in the means of transmitting the stored experience of generations, from oral tradition to contemporary printing. Intelligence, then, could hardly be an objectively defined feature of any one brain. Given the fact that "modern life is extremely complex and artificial," intelligent behavior will be conditional on navigating that artificial environment.[3] One question, for Morgan, is whether the technical and social forms of development will *supplant* physical inheritance altogether. In any case, human progress will continue, Morgan assumes, and therefore human existence will require intelligent control: the "plasticity" of human physical and psychic life will "make endless additions" to human intelligence inevitable.[4]

The future of the human was poised on a knife edge pointing either to an acceleration of intelligence or to a degeneration of the human mind through the mindless automatisms of modern institutions and machines. The philosophical challenge was theorizing the very nature of this exceptional being, inside and outside of evolutionary time, a being constituted by its seemingly essential *technicity*. In Germany in particular, philosophers confronted this fundamental question. Here I will read this tradition of thought (Cassirer, Scheler, Heidegger, Plessner) highlighting the underlying theorizations of technology. The texts are dense, the ideas often difficult; however, in this period just prior to the emergence of the cybernetic and digital age, they offer critical resources for understanding the position of the human in an increasingly technologized civilization.

Cassirer and the Technicity of Symbolic Life

We have been tracing the links between mechanisms, technologies, and organisms—both individual and collective organisms. These spheres cannot be understood in terms of opposition. The evolution of the human as a biological *and* cultural being is not merely dependent on technology; it is a function of its technical capacity. To study technology—the discipline of "Technology" as Leroi-Gourhan defines it[5]—is to give insight into the evolution of the cultural and historical beings that we now are.

Ernst Cassirer's masterwork, the three-volume *Philosophy of Symbolic Forms* (1923–29),[6] famously identifies the human in terms of a symbolic capacity, one that can lead to certain forms of objective moral truths (in opposition to Heidegger, as their celebrated debate at Davos demonstrated).[7] Less noted is Cassirer's insistence on the fact that symbolic thought is resolutely tied to the *organon*. In fact, Cassirer looked back to Ernst Kapp's idiosyncratic study of "organ projection" to underline the fact that the tool is a novel sensory prosthetic, an object that first extends and then magnifies our biological, organismic existence. For Cassirer, only an analysis of *technical organs* would allow us gain insight into what he pointedly calls our "natural organization" (2:216).

Kapp's great contribution, as Cassirer sees it, was not just the articulation of the relation between tool and organism, however. The "most profound significance of organ projection," Cassirer writes, is rather that the spiritual process of self-knowledge depends on this projection as an *exteriorization* of ourselves into technology, where we can literally see ourselves objectively. Technology is not simply "mastery of nature" (2:215). Material culture has "a purely spiritual function." For Cassirer, of course, that is the capacity to create an "ideal form" in the world, to project (to use Kapp's term) our symbolic forms into objective realities.

This is why Cassirer, despite his deep respect for Von Uexküll's work on biological *Umwelten*, denies that we can understand the human through any analysis of our physiological "organization," most notably, the brain and nervous system. For Cassirer, the human is identified with its achievements. Von Uexküll's animals, with their receptor and effector systems, live within a "functional circle" where the environment provides specific "cues" and "carriers of effect" that reveal just how the individual animal is adapted to its specific "world" of life (4:43). But, as Cassirer points out, what distinguishes the human is the fact that we step *outside* of the circle, no longer just "noticing" and then initiating action in response. We enter a new sphere, that of *observation*, which is different from the receptor world that comes "prefigured" in the animal's organization of sense and neural processing. Humans tear apart the intimate organizational coherence of "organism" and "environment" and thereby liberate themselves from the purely organismic determination of their existence (4:44).

Of course, Von Uexküll also located the human exception in such an exteriorization, as we saw. Still, Cassirer deepens the philosophical implications of this insight: to observe the world is to be separated from it, and this very cut is what opens up *meaning*. No matter how useful an implement might first be, and however minimal the consciousness of its artificial form, the emergence of tools in human evolution is the originary mark of this separation; it is the first expression of meaning and therefore a vehicle for an individual to learn how to see the world in radically new ways, namely, as systems of formal organization: "'noticing' becomes detached from dependence on his actions and sufferings; he becomes 'free of all interests'" (4:45). With the tool, the immediacy of the encounter between organism and environment is transformed; it is no longer a "mediated relation" (4:41).

The creation of the tool is not some practical exercise in problem solving, as it might be for Köhler's apes, because with the tool the human mind can *confront* nature, a nature that for the animal is always an "overwhelming power." Tools, in their earliest forms, "are not only created and made use of, but they are also worshipped" (4:41). The tool itself is understood to be "an intelligent force." This is what lies at the core of the developed human mind—a recognition of meaning *beyond* our own experience, revealed in and through our radical break with natural function, the break that is constituted by the artificial object that is a counter to nature itself. The emergence of the "ideal" in the tool is the "seed" of a "new total view" that includes the human in nature but only through that radical line of division that liberates the human organism from its particular "effector system" (4:42). As he put it in a later generalization of his philosophy, between the "receptors" and "effectors" the human imposes a new system: the symbolic. This is not an expansion of the sensible: it constitutes a radically new "*dimension* of reality" (4:43). Human responses are not *automatic* but "delayed"—"interrupted and retarded by a slow and complicated process of thought" (4:43). Again, contrasting his views with Von Uexküll's, Cassirer writes that humans no longer live *in their environments* (as given by their physiological and neurophysiological organization) but instead live in *the world of thought*, which of course always includes the thought of the other.[8]

The tool is an intervention, it is independent of both the human and nature, and constitutes a new "foreign" order, a foreign "norm," as Cassirer noted in a 1930 essay on technology and form.[9] Inevitably, any unity in human life will be fractured by the independent life of technicity. Cassirer, in this particular essay, underlines the importance of technology for contemporary philosophy: it is the crucial clue to understanding, conceptually, the origin and evolution of culture, the very origin of what Heidegger called historicality. For the human mind, the tool is not just a material object: "It is also an idea. It is part of that timeless inner world in which man lives."[10]

As Hans Freyer similarly claimed, the human life world is literally the "objectification" of spirit.[11] This world is by definition *not natural*. The human environment—our cultural and symbolic life—is a "dense confusion" of the natural *Umwelt* of sensory perception *and* the historical layering of objectifications of spirit, which is to say, in thinking that is exteriorized in the material forms of culture. We live in what Freyer calls a "palimpsest," writing ourselves into this dense layering of meaning and nature.[12] The "forms" that we can identify in material culture are "not like a gramophone record" as "played" in individual psyches, with absolute equality "due to automatic implementation processes." Instead, cultural organization is akin to a "sheet of music," where signs and functions are fixed in material instantiations of organization, but only as a "template" that must be fulfilled in each individual performance.[13]

The implication of Freyer's argument is that once humans leave the immediacy of their natural *Umwelt* due to the interposition of a new symbolic order, there is no simple transition to an "idealistic" form of social life that would mirror the natural order of the organism. Rather, the human world is constituted by the externalization and materialization of the inner world of the idea, which is to say, humans confront a new reality with just the same immediacy of the natural "cues" of the Umwelt since it is experienced through its material instantiations—in technology most notably but also in the social forms and institutions that organize all of human life. The point is that this new environment is by definition *not* an organized and coherent world. The psychic system still harbors a logic of neurophysiological sense and response. The social and cultural systems are *not* governed by shared norms and operations. And most important, perhaps, is that the technical sphere is a zone with its own *alien* norms, alien, that is, to human existence in itself, whether understood in organismic or cultural terms.

The human life, according to Freyer, is always therefore a fractured existence—internally fractured but also separated from other human beings by the "systems" that run through individuals. "Thus man is never at the center of a reliable world, but always at the intersection of lines which affect him—among others. It is difficult to think of a more effective method of isolating people."[14]

The Human as Detachment: Scheler

The secular cultural and intellectual tradition of defining the human through "reason"—first articulated by the Greeks in the concept of *logos*—was grounded in a now-questionable assumption of some connection between human minds and the order of the cosmos itself, its own *logos*. So explained Max Scheler in his 1928 work, *The Human Place in the Cosmos*.[15] However, as he observed, the new sciences of the human, especially evolutionary theory, could never in fact distinguish the *human* from any other living being, except as an example of the "late stage in the evolution of our planet." The theological, intellectual-philosophical, and natural scientific views were, therefore, each entirely inadequate to the task of defining the exception of the human; there was behind these perspectives no "underlying unity" (5). Furthermore, the human had, since the nineteenth century, become subjected to an increasing number of new disciplines, further fracturing our conception of ourselves. For Scheler, these developments were generating a crisis: "In no historical era has the human being become so much of a problem to himself as in ours" (5). In response, Scheler attempts what he calls a "philosophical anthropology" to found a proper definition of the human *as* human.

Scheler isolates the key cognitive power of the human mind: creative and novel thought. Scheler biologizes Whitehead's concept of reason to a certain extent, noting that the higher forms of animal life, those with "psychic complexes established by association," function according to organized thinking, producing behaviors with *meaning*, that is to say, with some kind of unified organization, even if these behaviors are essentially habitual. The importance of the unified meaning of a complex is that it can preserve itself even in the event of "pathological losses," whereas more linear paths are simply broken. But, as Scheler observes, the associative processes of the brain are determined subcortically. The cortical processes, the functions expressed in the higher brain, do not simply produce more complex associations. In fact, the cerebral cortex is, Scheler explains, "the organ of *dissociation*," not acting in concert with "biologically more unified and more deeply localized types of behavior" (15). The cortex is an organ, as we have seen many times before, of self-interruption, of auto-disruption.

Intelligence is therefore not to be found in the perfection of the instincts—a more thorough automatism, that is—because intelligence is defined by flexibility. The question is the essence of human thinking, because associative memory and the emergence of habit is the first step toward the liberation from the fixity of the instinct. If the human is, as Scheler says, "a flexible mammal in whom intelligence and associative memory developed to the highest degree" (16) and in whom instinct is the least powerful, what distinguishes human cognition or experience from other higher organisms capable of a certain degree of intelligence?

The principle of "association" that lies behind intelligence appears most clearly, according to Scheler, in the act of imitation. Copying other organisms can produce even more complexity of behavior, because the psychic process of repetition and drive satisfaction is now applied to the "*behavior and experiences of others*," and there exists now what we can call "tradition," this "new *dimension*" that supplements mere biological inheritance and individual trials (19). But tradition is not, to be sure, a purely human development. As Scheler points out, various herd animals learn from each other, and this "knowledge," if that is the right word, gets handed down to new generations.

What "sharply" distinguishes *human* tradition is what Scheler calls the capacity for a "free 'recollection' (*anamnesis*) of the past," and, crucially, a dependence on historical knowledge, passed through a series of "signs, sources, and documents," that is, through what we have been calling the exosomatic expressions of *thought*, to use Lotka's term, and not just the imitation of immediately observable behaviors.

Only the material articulation and objectification of tradition, a process that literally dis-integrates tradition from our current experience (even if it does not *disappear*), only this fracture can prepare the ground for an opening: "new discoveries and inventions" are now possible (20). This is why humans are not just "more flexible" than other creatures. Humans are, so to speak, radically adjustable to new circumstances because they are never completely bound by their biological or, for that matter, their cultural inheritances (21).

Humans, for Scheler, are therefore not exceptional because they are more intelligent. Rather, human intelligence is no longer a *natural intelligence*. The human emerges from a principle beyond and therefore *before* life. "If reducible to anything at all, this new principle leads us back to the one ultimate Ground of all entities of which life happens to be one particular manifestation" (26).

As Scheler writes, this principle is not to be understood mystically but quite concretely as the *"existential detachment from organic being,"* which is to say, spirit is not another realm intervening into the organic but rather is "freedom and detachability" *in and of itself*. In the animal world, *all* action, reaction, all intelligence and creativity, "proceed from a physiological state of their nervous system" (27). This highlights an important connection between Von Uexküll's theory and that of Scheler: "Everything which the animal notices and grasps in its environment also fits the firm function of its unity of drive and sense structures." Its experience and reactions are "securely embedded in the frame and *boundary of its environment"* (27–28). The higher levels of animal behavior are marked by new relations between the animal and its environment, as we see, for example, in the way that some animals modify their environments and, finally, become capable of a certain elasticity of response due to the changes within the physiological and psychic states—the dynamism inherent in the homeostatic organism or the Gestalt mind (28). But none of these animal states corresponds to what Scheler calls the *spiritual*.

The human is defined now by the "shedding of the spell of the environment" (28). This is why human will, human spirit, will never be a mere "steering" (Lenkung) or "directing" (Leitung) of the organismic unity of psyche and body, as it is in the animal world. The human, because it can "objectify" itself, its own physiological and psychic function, can now *free* itself from its objectification. It can act against life itself—in the extreme, the human is free to commit suicide (29).

So, for Scheler, the human is the mark of a radical separation within life. "This ability to *separate essence from existence* constitutes the *fundamental* character of the human spirit, because it is the *foundation* of all other characters" (37). Scheler leaves us with the question of how the human *becomes* human—that is, how nature itself intervenes into the existence of organic life on earth and makes possible the separation of the human mind from its physiological servitude.

The Event of Technical Being: Heidegger

In his introduction to *Being and Time* (1927), Heidegger, like Scheler before him, refers to the major scientific developments we have traced in interwar thought: "In biology there is an awakening tendency to inquire beyond the definitions which mechanism and vitalism have given for 'life' and 'organism,' and to define anew the kind of Being which belongs to the living as such."[16] The question of the *organism* as a new form of order linked to the notion of "life" was taken up by Heidegger at great length in his seminar on metaphysics held in 1929–30.[17] Clearly immersed in leading-edge thought in the field (and again, most notably the work of Von Uexküll), Heidegger faced the challenge of defining the living being with and against the array of concepts associated with machines and technology more generally understood, with the goal of pinpointing the specific character of *human* life.

Like Scheler, Heidegger will not so much evade as set aside the question of metaphysical reductionism or transcendence in favor of an approach that emphasizes the varying possibilities of experience in living beings—and, like Scheler, Heidegger will locate the human exception in the concept "world." However Heidegger's account of the radical human separation that marks the distinction between "essence" and "existence" will dismiss the cosmic mystifications implied in Scheler's account. For Heidegger, the human will be defined by the *rift*, but that fracturing of the existential from concrete "existence" (in Scheler's terms, essence and existence) will be staged by Heidegger as the very space for an interrogation of the human.

The starting point is the living being—for the human will be defined in and against life itself. If, as Heidegger notes, the "organism is something which possesses organs," the philosophical problem will be displaced into the realm of the organ, the "instrument" or tool as we know from the Greek origin of the word. This idea has been productive in certain spheres of biological thought, in that it has supported an essentially mechanistic understanding of the living being. Citing Wilhelm Roux, the idea that the organism uses "instruments" can lead to the conceptualization of the organism itself as a *complex* of instruments, or, to put it another way, as a complicated instrument. Heidegger's question echoes the thinking of many biologists and physiologists in the interwar period—what would distinguish the organism from the machine in this context? What distinguishes the kind of unity that organizes the machine from the interweaving of parts we see in the organism? And, Heidegger adds, should we not distinguish the machine from the instrument, or what he calls "equipment" (*Zeug*) from instrument (*Werkzeug*)? Heidegger lists all the difficult, related concepts that are implicated in the attempt to know the organism with technical analogies—"purely material things, equipment, instrument, apparatus, device, machine, organ, organisms, animality"—and poses the challenge of clearly defining these different entities.[18]

It is important to see that Heidegger is from the start avoiding the trap of comparing "organism" and "machine" that so often leads to either a reduction of life to the mechanical, or the imposition of some "supra-mechanical" force of life into the mechanisms of the biological being. As we have seen, interwar thought emphasizes the crucial function of *organization* and *unity* as an independent structure in order to transcend the simplistic framework that poses vitalism against mechanism. How to study life as a phenomenon scientifically meant penetrating the essence of natural organization and, often, its distinction from artificially organized entities. That is also Heidegger's philosophical question here: "grasping the original and essential character proper to the living being."[19]

The procedure is to *begin* with the technical, not as basis of comparison, but as an exercise in clarifying the essence of the machinic, and only then will he isolate how those technical concepts will fail to grasp the essence of the living. Drawing on the earlier discussion in the seminar, Heidegger points to the obvious fact that the "new kinds of beings"—human technologies—have no world; that is to say, nothing is "present-at-hand" for a piece of equipment, a tool, a machine, no matter how complex they might be. In this way they are just like stones or other inorganic bodies. But as we know, certain kinds of machines—what Lafitte called reactive machines—are certainly capable of "sensing" the world in some way, opening up the possibility, at least, of some similarity with simple animals. Animals, Heidegger explained, are not worldless (like stones), but they are "poor" in world, deprived of that which will define the essential character of human experience. So the question is not whether a machine can "sense" the external world, "experience" it, so to speak, but rather how it is that the animal experience is *deprived* of world while the machine simply lacks it.

This is no easy question, and I will not try to explicate all of the complications. But given Heidegger's knowledge of Von Uexküll's work on the *Umwelten* of animals, which were often very simple organisms such as starfish and the like, it was clear that animals "created" a reality from their sensory systems, however limited in scope, and that they lived in very different *spa-*

tial and *temporal* realities (as could be shown experimentally). I think this is a clue as to how Heidegger might have understood poverty (as deprivation and not merely "less" of something), because he refused to admit that animals had *temporality* even if their lived experience had different *temporal sequences*. They were deprived, essentially, of any distinction between past, present, and future anticipation: what *appeared* to be memory or anticipation was in fact the mere presence of a stored experience or the presence of a prediction generated in the nervous system from its stored experiences and associations. If human experience has of course traces of this animal form of cognition, Heidegger argues that humans have what he calls a world, which is to say, not a mere *Umwelt* (an organized experience built from natural indications) but more what Von Uexküll called an organization of *artificial indications*, that is, the difference between biosemiotics and what we could call biosymbolism.

Heidegger approaches this problem precisely through the question of technology, as did Von Uexküll. What is singular about artificial technologies is not that they are worldless, or even that they are "deprived" of world, but that they are understandable only as indications *of* a world. They "*belong to world*," which is to say that they appear to us, humans, who have a world. Equipment is what it is only because it emerges from human activity, the activity of *world-formation*.[20] The construction of machinery and other equipment, for Heidegger, as for so many others in this period, depends on the plan of organization that exists prior to the construction and guides its organizational form.[21] This is why Von Uexküll (Heidegger cites him here) would call the machine an "imperfect organism," because its *Bauplan* was exterior to its own being.

But Heidegger demands a more thorough distinction. The plan of the machine is determined by the "serviceability of the equipment"; that is, the plan is "regulated" in advance by the purpose, and that purpose is therefore identified with the user for whom the equipment is fundamentally a "mediate" device (215). Organisms do not construct pieces of equipment for themselves to "use" in this way—that is, the difference between machine and organism is *not* for Heidegger the fact that the plan is exterior to one being and internal to the other. The organism has organs that belong to it, which is to say, the organism has capacities that are expressed and instantiated in the development of its organs. The organ does not "have" a capacity that is used, mediately, by "the" organism. Organisms themselves *have capacities*. Here Heidegger agrees with Von Uexküll: the sensory systems of organisms are not mere technical devices that allow certain information to be utilized. As Heidegger puts it, organs (such as eyes or ears) do not "see or hear," the *organism* does; the organ is developed out of the need to extend the experience of the organism of a whole (222).

The crux of the argument hinges on the liminal case of what Lotka calls the *exosomatic*. Heidegger resolutely denies that the pen, for example, is an organ like the eye. The implication is that there is no natural "capacity" that develops the organ in this case: the equipment is just "serviceable" (226). There cannot be, according to Heidegger, the same kind of intimate "belonging" that characterizes organismic unity, since the technology can be separated from the individual human organism and used by another. But it is precisely this *separation* that is the essence of the technical here, and which points to what Heidegger marks as expressly human (as opposed to animal) existence—our worldliness. We experience the world as something that *appears to us*, which means we are at once immersed in the world (its meanings and indications) and yet radically separated from it; it comes *before us* (351).

Heidegger's strategy at this point is to investigate more intensely what exactly "appearance" means for us. Unlike the animal, who lives wholly *within* its experience, humans experience their world and also the fact that this world *appears*. What Heidegger argues is that we are able to "see" beyond *what* appears to us in experience (the various beings and their relationships) and recognize that there is a unity of Being that is, as he puts it, "more originary that all those beings that press themselves upon us" (351). The unity that lies before the prevailing multiplicity of experience cannot of course appear to us *as itself*. (As Kant showed, this unity can only be *deduced* from the facts of experience as concretely appearing.) The opening up of this originary zone occurs, then, not *in experience* but rather in the *separation* of human experience of the world from experience itself.

That separation is the emergence of temporality for Heidegger, because only in that distancing can the recognition of the relation between history, presence, and the future be encountered. As in *Being and Time*, this situation of *Dasein* is somewhat paradoxical, or at least puzzling—to be at once what we are *and* at the same time to *not be* what we are. In anticipating a future that is not yet, humans understand the past, which determines what they are; they are, as Heidegger famously expresses it, "thrown" into this transitional state *as* essentially transitional, constituted as *never in place*. "Man is that inability to remain and is yet unable to leave his place." The constant projection from a past defines us: "man is a *transition*, transition as the fundamental essence of occurrence" (365).

The question of our "historicality" is not linked to memory. It is fundamentally a givenness that is *not necessary*; it does not "belong" to us as the determinations of the organism and its activity belong to it. It is crucial for us to note here at this juncture just how important technology is for Heidegger, since it is the *apparent* naturalness of things "ready-to-hand" that will define our possibilities and our projective futures.[22] Yet the *breakdown* of this pseudo-natural belonging reveals the fact that the human world is world formed artificially, through actual concrete implements but also ways of thinking and conceiving that come from elsewhere. To be in the world is to live in a matrix of meaning and relations, incarnated in artificial objects and artificial organizations, and to recognize that this world could be otherwise. As Heidegger wrote in the *Beitrage*, the primal disposition of the human is *wonder* (*Er-staunen*): "wonder that beings are and that humans themselves are and are in the midst of that which *they* are not," and this includes, of course, the world that is the world of cultural and technological meaning, which is what structures the perception of our presence and our anticipations of the future.[23] Missing from Heidegger's account, then, is a recognition that to be in the world (to have an existence) is not just to use things, but more fundamentally to *know how to live* in the world. Given the essential artifice of the human world, it is this knowledge of how to live that forms the existence, the "what" of *Dasein*. World, in Heidegger's sense, radically depends on the technical mediation of information and knowledge, on the possibility of *exteriorized* (or "exosomatic") thought.[24]

If the instruments of human technical existence are the model for worldliness for Heidegger, and authenticity is to be gained by passing beyond existence to the existential condition of *throwness* itself, what is lost in this movement is the *originary technicity* of that worldliness. Can we identify the essence of the human without acknowledging this unexplained capacity to participate in the transference of thought itself? Heidegger noted in the *Beitrage* that besides wonder, there is another "basic disposition," namely, *shock*, the shock, that is, of our abandonment by Being.[25] We can say that, for Heidegger, human creative potential is opened up in this abyss, the separation from being and the anxiety and fear associated with this separation. Genuine decision is linked with this opening of a fracture in *Dasein*. Decision is no mere "choice," Heidegger observes, because choice between alternatives assumes the preexistence of some world as given. Decision is "de-cision," the experience of being in this space of separation, to anticipate something that does *not* emerge from a prior world, to invent a possibility that *was not possible*.[26] The question that can be posed here is how to think about the *origin* of this rift without introducing a new metaphysical principle.

For Heidegger, we discover, so to speak, the separation within ourselves, as an essential aspect of our experience, but only in moments of crisis—failure, disruption, interruption—and in the organized and technologized world of civilization, especially modern industrial civilization, these moments of disruption are increasingly stifled through the engineering of stability, as noted as well by Whitehead and Dewey, among others. However, if the possibility of our being-in-the-world might be essentially connected to the technology of a specific human form of "tradition," namely, the *exosomatic transmission of thought* itself, the recognition of a radical separation from our "being" becomes then a radical separation from a prior separation—the separation, that is, of the human psyche from its own physiological isolation. This is just to suggest that Heidegger's positioning of the human as the being capable of stepping into that abyss that is the separation of ourselves from our "world" and the separation of our historicity from any possible future, this being is already, as a being constituted by its social and cultural comportment, made possible by a technical transmission of thought, *grounded* in a separation from organismic, animal life. This separation is the primal moment of invention—namely, the invention of technology itself.

Human Exteriorization and the Plasticity of Being: Plessner

Helmuth Plessner's contribution to philosophical anthropology, published in 1928, also begins with the *organism*, the starting point of any consideration of the human as living being.[27] However, Plessner begins with what in fact lies beyond the organism as organization, or Gestalt—the concept of "wholeness" or unity in itself. As we saw in Goldstein's key work on the organism (and this will soon become an important question again with the emergence of cybernetics and the idea of the artificial organism with an artificial teleological orientation), the fact that the organismic unity can *transform* in response to constant threats and challenges to its survival means that its organization is not completely fixed. But what then is the ground of its self-transformation? As Plessner comments, "wholeness" can never be realized—it cannot even be realized "abstractly" because to be realized means to become *concrete*. When wholeness becomes concrete, however, it is never immediately present; it takes a form that is, according to Plessner, only to be found in the realm of "organic nature." That is to say, organic forms are the only genuine *instantiations* of "wholeness" (113). The unity of a being is what allows us to perceive "the organic," and yet the wholeness of the being can itself never be measured, or observed directly—it cannot enter into the scientific, empirical knowledge of "bio-logy" as a discipline (112). Drawing from the Gestalt theorists, Plessner explains how the unity of wholeness can "appear" in any number of ways; that is, we can "sense" it visually, or in a tactile way, but *what* appears is the given Gestalt of the specific organic system. And it is that Gestalt that can be changed; while given, there is also in the organic form a wholeness that "does not *itself* appear" (112). There is, for Plessner, a critical difference between the forms that are "indifferent" to givenness (i.e., capable of transformation) and a form that is in its essence *alien* to appearance itself, *wholeness*.

With this conceptual clarification, which aligns with much interwar thinking on organismic order, Plessner can state: "Everything living exhibits plasticity: it can be pulled, stretched, and bent in a way that brings together the distinctiveness of the whole's boundedness with an extreme shiftability of its boundary contours" (116). The living being does not merely have a "surface," as, say, an inorganic body has; the living being is "enclosed" by its exterior surface and "shifts" within that bounding topology. The more plasticity something displays, "the more the thing appears to be alive" (116).[28] Plasticity is here not mere flexibility—again, as Goldstein will demonstrate in such great detail, the organism is actually in a constant state of "catastrophe," on the threshold of dissolution and challenged to invent new and unprecedented responses, that is, with novel *reorganizations*. As Plessner puts it, the living being "exhibits discontinuity in its continuity, regular irregularity, both statically and dynamically" (116). The philosophical question is this: how to understand the mutual action of regularity and irregularity, the norm and the deviation—the givenness of the rules are *violated* in the plasticity of life, yet there is no such thing as irregular life. The disruptions of discontinuity—what Catherine Malabou names *destructive plasticity*[29]— must themselves be "governed" but governed by what Plessner describes as a *non-isolatable rule*, a rule that cannot be normalized. And yet such a rule must govern the living, otherwise the deformations so essential to the operations of a plastic being would disrupt and destroy the overall unity, instead of actually *strengthening* its effect, as is in fact the case (116).

The rule of the living is strangely undetermined; it reveals itself in the spontaneity of response, in the possibility of threshold transitions. The organism is always potentially something other than what it is: its life history is, we can say with Plessner, radically contingent (117). There is no rationality to be isolated in this spontaneity of movement and transition. And yet, at the same time, the body is capable of *self-regulation*. There is here, for Plessner (echoing Lashley's terminology in his own theorization of neural plasticity), a kind of harmony in the "equipotentiality" of the physicochemical parts of the living system.[30] And what drives regulation is, to a great extent, the fact that the organism is integrated into a specific environment: "regulation is mediated by the outer world" (154). Plessner will agree with Von Uexküll, then, that the organism proceeds in its development and self-regulatory existence according to the *Bauplan*. But as Plessner makes exceptionally clear, the "plan" is not an ideal external to the system, or some "hidden" organizing force. The unitary plan and the actual organization exist simultaneously. "Like life, organization explains itself" (157).

This is all to say that for Plessner, we must define the living body in terms of its "positionality," or its capacity to become "positable." The mode of being specific to the living is this essential character of going beyond itself, to posit itself as something new. Phrased somewhat paradoxically, "the body is outside of and within the body" (121), which is not the case in the inorganic world—an object is just what it is. The distinction rests on the concept of the boundary: the marking of the limit of the body is what constitutes the inner, and yet that given order *inside the body* is always governed by something that is outside of that order and that can found a new organization. The living is then a genuine system; that is, the "border" of the system belongs to the existent being and that allows the physical body of the organism to *transcend itself* through its own self-reference to the threshold of its possibilities. The being is posited (literally, positioned) in the point of unity that, as unity, is "detached" from the *concrete unity* that is the entire, dynamically organized, living system (148). To live is to be always in a state of anticipation. Not an anticipation of a determination to come, but the anticipation of itself as something *to be* determined in the future.

What distinguishes the human from other organic beings is the fact that not only does our life proceed from the "central point" of unity that is the non-present mediation of all passage and transition, but unlike all other animals, the human is *aware of itself* as that central point—this introduces a fundamental gap between its "lived experiences" (in the sense that Von Uexküll gives that term) and itself as an experiencing being. The animal is completely self-enclosed, it lives and experiences "itself" only in the here and now (270). The human—and here there is more than an echo of Heidegger's contemporaneous *Being and Time*—is always "behind" the presence of the here and now, existing "without place," in nothingness. This is what allows the human to be both what it is (a lived system) and to stand *outside of what it is*. The human lives *outside of the body* while still being "inside."

Plessner's term for this condition is "excentricity." The human has no ground of existence: existence is constituted on these thresholds marking the inside and the outside of a plastic being. The "excentricity" of the human is also what grounds our recognition and relationship to the other (279). The human world is *shared* precisely because the individual human lives to a large degree outside of itself—and it is that exteriorized perspective that can be taken up by another human being. Even more profoundly, it is precisely the recognition of another psyche that makes us aware of our own excentric nature. The "world" of other minds is not at all like the environment of an animal. We are both within it and producing it (282). The spiritual life of human beings occupies this space of the "we"—as a common positionality that by definition is not grounded in any lived experience; it is, rather, *not* grounded and hence radically liberated from the particularity of the living system. But it does not exist somewhere else (285).

In the case of the human, Plessner says, life is defined by a radical deviation, an *essential errancy*, and it is this essential lack of a direct path that will explain not so much the origin of technology but rather its *necessity*: "The human with his knowledge has lost that directness. He sees his nakedness, is ashamed of it, and must therefore live in a roundabout manner [*auf Umwegen*] via artificial things" (288). This insight into the origin of technology is exactly what was missing from Heidegger's account. Technology is the necessary detour of a being that lacks determination.

The human requires the "complement" of the non-natural, nonorganic tool because there is no automatic center that is the orientation for self-regulation, for equilibrium. Humans are, Plessner states clearly, "by nature *artificial*." They must create their own equilibrium, and that means supplementing their natural organic activity with actions that proceed from their *excentric* (not internal) orientation (288). The tool is the expression of human existence, but we need to remember that the excentric position is predicated on the sharing of minds (or rather, of minds constituted by their openness to sharing); the tool is not just a supplement to the individual human organism, then, but instead, as Plessner states, the foundation for what we call *culture*.

The gap between nature and human is the origin of human technical existence. "Given with excentricity, artificiality is the detour [*Umweg*] to a second native country where the human finds a home and absolute rootedness. Positioned out of place and time in nothingness, the excentric form of life creates its own ground" (294). The human is defined as what emerges from the "forced disruption" of the organic way of life. "Only because the human is naturally partial and thus stands above himself is artificiality the means to find a balance with himself and the world" (298). The artificial detour that is expressed by technology and culture is a new ground, created out of necessity. But of course the old ground does not at all disappear. The radically new norms of cultural existence are therefore not always in harmony with the natural drives of the body (294)—something Freud would explore in his *Civilization and Its Discontents* (1929).

Philosophical anthropology articulated the crises of interwar European civilization and sought some kind of redemption, if not stability, in new concepts of culture that relied on the inevitable ex-centricity of the human. Here, on the threshold of a global war that would destroy so many humans and so many traditions and cultures, a new form of machinic life was yet unseen. The emergence of computational technology posed a new challenge to definitions of rationality and the question of "habit" and culture in an automated society. However, as we will see, the key insights of interwar thought, the theorizations of human existence that linked openness and plasticity with technology and innovation, and repositioned rationality as the breaking of routine, were never wholly absent from the new discourses of the cybernetic age.

Hinge

PROSTHETICS
OF THOUGHT

22
Wittgenstein on the Immateriality of Thinking

Zettel. Slips of paper in a box, dating from 1929–48 (most of them written in the three years following World War II).[1] A number of slips have fragments addressing questions that have been hinted at but will come to concern the leading edge of thought and the new sciences of cognition: What *is* thinking? How can we tell if someone is *really thinking*? (Can machines think?)

First, let's take note that in this threshold moment, one of global war and colonial disruption and reorganization, how Wittgenstein blurs the border between machine, animal, and *slave*. He proposes a strange thought experiment:

> Suppose it were a question of buying and selling creatures (anthropoid brutes) which we use as slaves. They cannot learn to talk, but the cleverer among them can be taught to do quite complicated work; and some of these creatures work "thinkingly," others quite mechanically. For a thinking one we pay more than for one that is merely mechanically clever. (108)

The added value, presumably, would be in the kind of work the thinking creature can accomplish. But what would that be? Or, to put it another way, why not assume that the more able creature is just *more* clever but still mechanically clever?

What is thinking? The question indicates a crisis—if that is not too strong of a word.

"Thinking," a widely ramified concept. A concept that comprises many manifestations of life. The phenomena of thinking are widely scattered.

> We are not at all *prepared* for the task of describing the use of e.g. the word "to think" (And why should we be? What is such a description useful for?)
>
> And the naive idea that one forms of it does not correspond to reality at all. We expect a smooth contour and what we get to see is ragged. Here it might really be said that we have constructed a false picture. (110-11)

As Wittgenstein will observe, we learn to use the word *think* in very specific contexts. If I asked, for example, if a fish was thinking, the question would hardly make sense. That is not a context in which we would normally deploy that concept.

What is thinking? Part of the problem is that we cannot locate it. But we know what it means (roughly) to participate in it.

> Compare the phenomenon of thinking with the phenomenon of burning. May not burning, flame, seem mysterious to us? And why flame more than furniture ?—And how do you clear up the mystery? And how is the riddle of thinking to be solved ?—Like that of flame? (125)

Thinking—something impalpable (*rätselhaft*) like a flame? If it cannot be grasped, how will it be conceptualized?

Indirectly . . . where does thought appear, where can it be grasped? What makes a sentence, a series of words, let's say a series of *printed* words, an example or manifestation or expression of *thought*? Is there something "more" than the words (or signs) that are present? How can some words have "life" and the *very same words* not have life? (143).

This is to risk verging into theology—the idea that there is a *spirit* in the body, or a life added to the body. Where is the "life" that is thinking if not *in* the words? Wittgenstein will propose:

> There could also be a language in whose use the impression made on us by the signs played no part; in which there was no such thing as understanding, in the sense of such an impression. The signs are e.g. written and transmitted to us, and we are able to *take notice of them*. (That is to say, the only impression that comes in here is the pattern of the sign.) If the sign is an order, we translate it into action by means of rules, tables. It does not get as far as an impression, like that of a picture; nor are stories written in this language. (145)

In this case one might say: "Only in the system has the sign any life" (146).

So can we agree with Wittgenstein? Or at least with what this comment expresses:

> Knowledge is not *translated* into words when it is expressed. The words are not a translation of something else that was there before they were. (191)

The words, the signs . . . they are not a *medium*. The meaningful sign is a grasping that is *recognition*. (Recall Spinoza. Or Leibniz.) As in Heidegger, the example of a technical object:

> For someone who has no knowledge of such things a diagram representing the inside of a radio receiver will be a jumble of meaningless lines. But if he is acquainted with the apparatus and its function, that drawing will be a significant picture for him. (201)

Of course, there are different kinds of recognition—every "seeing" of the world is an interpretation, a hypothesis, and our understanding can switch, like a Gestalt (. . . a duck / a rabbit).

The system, the organization, the frame of interpretation—where is it? How does it work? When does it appear? What if thinking is just a matter of a *system*? As Wittgenstein says, talk is transmissible, but it is not like a disease. Yet language is decidedly *not* a medium. A question to get at this issue: What is happening when we talk to another through a technical apparatus?

> Philosophers who think that one can as it were use thought to make an extension of experience, should think about the fact that one can transmit talk, but not measles, by telephone. (256)

Systems. Organizations. Is invention also a form of *sharing*? Wittgenstein: "Compare: inventing a game—inventing language—inventing a machine" (327).

What about a machine inventing machines? Or, at least, what about a machine that is thinking or one imitating thought. If the animal is such a machine, as Descartes argued, it would have no "soul," which was just another way of saying it cannot think.

Wittgenstein returns to the example of slavery; in this fragment it is a more complicated kind of slavery, and a more disturbing one to be sure.

> An auxiliary construction. A tribe that we want to enslave. The government and the scientists give it out that the people of this tribe have no souls [*keine Seelen haben*] ; so they can be used for any arbitrary purpose. Naturally we are interested in their language nevertheless; for we certainly want to give them orders and to get reports from them. We also want to know what they say to one another, as this ties up with the rest of their behaviour. But we must also be interested in what corresponds in them to our "psychological utterances," since we want to keep them fit for work; for that reason their manifestations of pain, of being unwell, of depression, of joy in life, are important to us. We have even found that it has good results to use these people as experimental subjects in physiological and psychological laboratories, since their reactions—including their linguistic reactions—are quite those of mind-endowed human beings.
>
> Let it also have been found out that these automata [*Wesen*] can have our language imparted to them instead of their own by a method which is very like our "instruction." (528)

What if we teach our slaves to *calculate*? On paper or not. But also: we teach them to act as if "reflecting"—that is, waiting to give the answer to the question posed to them. It's *as if* there was "something going on" beneath the surface (a "process" of calculation).

> Of course for various purposes we need an order like "Work this out in your head"; a question like "Have you worked it out?"; and even "How far have you got?"; a statement "I have worked . . . out" on the part of the automaton [*Automaten*]; etc. In short: everything that *we* say among ourselves about calculating in the head is of interest to us when they say it. And what goes for calculating in the head goes for all other forms of thinking as well.--If one of us gives vent to the opinion that these beings *must* after all have some kind of mind, we jeer at him. (529)

But what does it mean to have a mind?

> Writing is certainly a voluntary movement, and yet an automatic one. . . . One's hand writes; it does not write because one wills, but one wills what it writes. (586)

The extended system; circuits of behavior. This is why (I believe) Wittgenstein will tell us:

> No supposition seems to me more natural than that there is no process in the brain correlated with associating or with thinking; so that it would be impossible to read off thought-processes from brain-processes. (608)

Organisms, systems, networks. Why do we think that thinking, writing, speaking, should be *centered* somewhere, in the brain of all places?

> I mean this: if I talk or write there is, I assume, a system of impulses going out from my brain and correlated with my spoken or written thoughts. But why should the system continue further in the direction of the centre? Why should this order not proceed, so to speak, out of chaos? The case would be like the following—certain kinds of plants multiply by seed, so that a seed always produces a plant of the same kind as that from which it was produced—but nothing in the seed corresponds to the plant which comes from it; so that it is impossible to infer the properties or structure of the plant from those of the seed that comes out of it—this can only be done from the history of the seed. So an organism might come into being even out of something quite amorphous, as it were causelessly; and there is no reason why this should not really hold for our thoughts, and hence for our talking and writing. (608)

The *action* of a human being is not reducible to physiological processes, because action is embedded in a system that operates in and through bodies. What is "occult" for Wittgenstein is not the idea of thinking *outside* the body, a normal phenomenon. What is occult is the idea that thinking takes place "inside" us, in the head, in the brain (606).

Artificial intelligence—can we really imagine the intelligence of a *system* (thinking, speaking, reading) itself having a technical substitute?

> Is thinking a specific organic process of the mind, so to speak—as it were chewing and digesting in the mind? Can we replace it by an inorganic process that fulfils the same end, as it were use a prosthetic apparatus [*Prothese*] for thinking?
>
> How should we have to imagine a prosthetic organ of thought [*Denkprothese*]? (607)

PART FOUR

Thinking Outside the Body

23
Cybernetic Machines and Organisms

Cybernetics was centered on a fundamental analogy between organism and machine. As W. Ross Ashby asserted, "I shall consider the organism . . . as a mechanism which faces a hostile and difficult world and has as its fundamental task keeping itself alive."[1] Because cybernetics intentionally blurred the boundaries between humans, animals, and sophisticated technological objects, it has often been accused of reducing living beings to the mere interplay of mechanisms. However, cybernetics always wanted to infuse machinic beings with the essence of life—purpose, adaptive responsiveness, learning, and so on—while opening up new insights by comparing organisms to some of the most innovative technologies of the era, namely, servo-mechanisms, scanning instruments, electronic communication systems, analog computers, and, perhaps most notably, the new high-speed digital calculators that were emerging from secrecy in the postwar era. Cybernetics drew together advanced automatic machines and organisms on the basis of their shared capacity to respond flexibly to a changing environment—whether the external world or that inner domain (Claude Bernard's "internal milieu") governed by what interwar physiologists called the regulative principle of "homeostasis."

Cybernetics was a provocative, multidisciplinary, and self-consciously revolutionary intervention into the fluid conceptual world of the interwar period. The recurring and overlapping questions concerning order and organization in physiological, technical, and cognitive domains were reconceived by cybernetics in a new context. The new science would not explore "relations" between these domains. What was proposed instead was a new project that sought the essential features of *all* adaptive and reactive systems, whatever their origin.

We can see clearly the resistances (but also affiliations) between interwar thought and new cybernetic initiatives as figures in the individual disciplines engaged with the new thinking in the early postwar period. The fact that so many important intellectuals had fled Europe (Germany in particular of course) meant that the contestation was staged in new sites and in new contexts. The United States would be a key space; centers of cybernetic research and theory would also emerge in the United Kingdom and in France. The Hixon Symposium, and then the series of famed Macy Conferences, included a wide array of presenters and commentators whose lines of thought between and across disciplinary—and chronological—gaps can be detected.

Artificial Insight

One example is illuminating for the question of intelligence in particular. As we have seen repeatedly, one of the touchstones for interwar thinking on creativity and intelligence, in animals and humans (and perhaps even machines), was Wolfgang Köhler's influential book on ape intelligence, where his Gestalt framework informed the concept of *insight*, the capacity to use a "detour" (*Umweg*) from normal routines in order to provoke a new organization of thought and perception in the solution of challenging problems.

In 1951, Köhler (now in America as an émigré, along with other Gestalt psychologists and Kurt Goldstein, all exiled from Nazi Germany) in fact published a review of Norbert Wiener's seminal work, *Cybernetics, Or, Communication and Control in the Animal and Machine* (1948), which included a famous chapter titled "Computing Machines and the Nervous System." Though Köhler praised certain features of the book, including the innovative theorization of feedback, he had serious reservations about the idea that electronic calculators and other machines could serve as models for the human nervous system and thereby help explain the origin and nature of human thought. As Köhler put it, this "now popular comparison" was entirely ungrounded. The kind of information processing carried on by these new computing machines was, he believed, "functionally" and "generically" different from human thinking. As Köhler pointed out—anticipating Hubert Dreyfus and other critics of postwar artificial intelligence research—it was obvious that only the "intelligent" decisions of human mathematicians gave any value to these blind mechanical processes; the machines could not really know anything, he concluded, "because among their functions there is none that can be compared with insight into the meaning of a problem."[2]

So Köhler would hardly have been surprised to find the influential cyberneticist and polymath Warren McCulloch admitting, a decade later, that "the problem of insight, or intuition, or invention—call it what you will—we do not understand."[3] On various occasions, McCulloch would acknowledge that imitating this kind of insight artificially with machines was quite problematic (though still possible, he believed).[4] That a leading cyberneticist like McCulloch would become so entangled in the question of insight and "productive thinking"—a question taken up by virtually all the leading Gestalt psychologists—was rather ironic. For it was in fact McCulloch's early work with his young colleague Walter Pitts, undertaken in the 1940s, that initially spurred the "popular" idea that human thinking could be compared to the work done by calculating machines, a central theme of cybernetic theories.[5] McCulloch and Pitts famously argued, in 1943, that it was possible to consider neuron structures as if they were in essence digital switching devices, and, armed with this idea, they went on to demonstrate (mathematically, that is) that networks of these "idealized" neurons were entirely capable of representing and calculating logical propositions and their consequences.

In effect—following here the revolutionary work of Kurt Gödel, Alonso Church, and Alan Turing in mathematical logic from the 1930s—McCulloch and Pitts argued that the brain, seen as a simplified, interconnected network of digital relays, was perfectly capable of fulfilling the functions of a Turing machine, an imaginary (at that point) form of digital computer that was capable of calculating logical sequences, by representing propositions and relations as a series of computable numbers. The implications of this brilliant and startling argument were obvious, and the two authors did not fail to make this explicit, closing their essay with this powerful claim: "Both the formal and the final [i.e., purposeful] aspects of that activity we are wont to call mental are rigorously deducible from present neurophysiology."[6] As McCulloch would go on to explain, in a talk at the celebrated Hixon Symposium, which centered on the relationship between brain and behavior, held at the California Institute of Technology in 1948, we did not need to postulate anything "extra" in thought to explain mental functioning. Indeed, we should not entertain the idea that something "new" could happen in the mental domain. "Our knowledge of the world, our conversation—yes, even our inventive thought—are," he stated emphatically, "limited by the law that information may not increase on going through brains, or computing machines."[7]

This seminal attempt to redescribe thinking as a kind of bio-logical mechanism may have fit well with the behaviorism that was dominant in

American psychological research. And in an academic environment that valued "facts" and empirical research in these areas, the materialist implications of these arguments were hardly problematic.

The End of the Machine

Incisive critics of cybernetic thought and research would, however, repeatedly press on the fundamental analogy made between feedback-driven machines (such as servo-mechanisms, self-regulating industrial machines, or self-guided weaponry) and complex organisms. But not because they thought cybernetics was inherently reductionist, though that was always an issue lurking in the literature of this period. However, what was often at stake was the *opposite* problem, namely, the attempts by cybernetics to infuse vital "purpose" and teleological organization into technological objects.

This project would distort our understanding of the essential nature of organismic life: so argued Georges Canguilhem in his 1947 lecture "Machine and Organism." As he explained, artificial machines do, of course, have "ends" that govern their design; that is what makes a technical object recognizable. However, unlike an organism, the teleological orientation of a machine is always given to it *from the outside*. This purpose is therefore never (by principle) intrinsic to the machine's own organization. This, according to Canguilhem, was the crucial error of Descartes's exercise in what I called "virtual robotics." That is, the analogy between the mechanical automaton and the living being relied on the original divine creation of the organism, which was in Descartes's vision itself based on concept of the *artificial machine* of human engineering—a machine that internalizes an essentially external "telos." A machine governed by externally given ends, Canguilhem pointed out, is always a slave to its given, already determined order: the machine must then affirm "the rational norms of identity, consistency, and predictability."[8]

In contrast, the organism is organized by the internal goal of its own survival, and it was therefore capable, Canguilhem observed, of genuine improvisation in the event of crisis or even internal failures. Here we see the influence of Kurt Goldstein's concept of the dynamic and perennially unstable organism. Without a completely fixed structure restricting its potentiality, the organism was, unlike the rigid machine, able to find new paths to success. The essential unity of the organism was maintained even in the event of catastrophic error because its organs were, as Canguilhem put it, fundamentally "polyvalent," not limited to previously defined, specific deterministic functions, as the parts of a machine were. The organism could be saved by what are in essence *pathological* responses. There was, Canguilhem declared, no possibility of such a machine pathology; there would never be any machine "monsters," because the machine could never *create for itself* new norms, new forms of existence.

Canguilhem's critique of the machine as model of the living amplifies the argument made first in his dissertation, published as the book, *On the Normal and the Pathological* (1943).[9] Canguilhem's understanding of these concepts was explicitly influenced by Kurt Goldstein's earlier work.[10] The idea that Goldstein's "catastrophic reactions" were, in a certain sense, entirely normal informed Canguilhem's approach to medical ideas of pathology. Pointing to the great plasticity of the nervous system, Canguilhem noted that if a child suffers a stroke that destroys an entire half of the brain, that child would not suffer aphasia (as is the case with many brain injuries later in life) because the brain reroutes language performance to other regions in order to compensate for the damage.[11] Organisms are stable as *unities* precisely because their organization is *not* fixed into any one rigid structure—they are open and thus equipped to surmount even a traumatic loss of functions in some cases.

The influential Austrian biologist Ludwig von Bertalanffy formulated, in the early years after the war, a new discipline of "systems theory" that, like cybernetics, aimed to transcend different disciplinary approaches to dynamic, ordered systems, living or not. However, according to Bertalanffy, cybernetics was on the wrong path: it could never hope to account "for an essential characteristic of living systems," namely, their ability to maintain stability despite constant metabolic creation and destruction of its own material foundation.[12] Like other critics of cybernetics, Bertalanffy believed that "a mechanized organism would be incapable of regulation following disturbances,"[13] since a machine could not radically transform itself—as the organism could—to accommodate shock and injury.[14] Open systems were plastic and hence never fully determined; they possessed what Bertalanffy called "equifinality," the capacity to follow multiple paths for the maintenance of life—an echo of Lashley's "equipotentiality" of the brain.[15]

In a later critique of automation, the American sociologist cum philosopher Lewis Mumford affirmed the distinctions made by more expert figures such as Canguilhem and Bertalanffy. Unlike machines, he argued, "organic systems [have] ... the margin of choice, the freedom to commit and correct errors, to explore unfrequented paths, to incorporate unpredictable accidents ... , to anticipate the unexpected, to plan the impossible."[16] Again, the possibility of error was a critical theme. As the idiosyncratic, cybernetically inspired thinker Gregory Bateson would observe, "An organism is capable of being wrong in a number of ways. . . . These wrong choices are appropriately called 'error' when they are of such a kind that they would provide information to the organism which might contribute to his future skill. These will all be cases in which some of the available information was either ignored or incorrectly used. Various species of such profitable error can be classified."[17]

Cybernetic Failure

And yet . . . We have to recognize that cyberneticists were, from the very beginning, intensely interested in pathological breakdowns and the importance of radical plasticity for adaptive and reactive beings (whether artificial or biological). Describing the early cybernetic "animals" created by W. Grey Walter (introduced to the public on a BBC broadcast in the 1950s), the French cybernetic thinker Raymond Ruyer noted, "Their errors must precede exploration, because these errors constitute the driving force."[18]

Paying attention to plasticity in this new discipline in fact reveals some important, often unexpected facets of the cybernetic project to analyze and construct "living" machines. In his classic *Cybernetics, Or, Communication and Control in the Animal and Machine*, Norbert Wiener observed that certain psychological instabilities could be conceptualized through rather precise technical analogues: "Pathological processes of a somewhat similar nature are not unknown in the case of mechanical or electrical computing machines."[19] There was, then, a strict parallel drawn between systemic breakdowns in the psychological and technological spheres.

> The task of the modern therapist can be compared to the task of the maintenance engineer or of the trouble-shooter who repairs the great overland power lines. Abnormalities of behavior are described in terms of disturbances of communication. In the past, these disturbances have been summarized under the heading of psychopathology.[20]

As Bateson would explain elsewhere, both the technical device and the human mind can exhibit "symptoms" that indicate a breakdown or pathological turn within the "body" as a whole. "If the TV suffers from a blown tube, or the man from a stroke, effects of this pathology may be evident enough on the screen or to consciousness, but diagnosis must still be done by an expert. . . . The TV which gives a distorted or otherwise imperfect picture is, in a sense, communicating about its unconscious pathologies—exhibiting its symptoms."[21]

In an important sense, then, cybernetics as a transdisciplinary science had its very origin in the insight that pathological physiological behaviors (in, say, neurological movement disorders) could be mapped, structurally and with precision, onto technological failures that could be described in identical mathematical forms. While investigating the behavior of feedback systems in steering mechanisms, for example, Wiener and his engineering colleague Julian Bigelow discovered that excessive compensation could sometimes lead to increasing oscillations that eventually became uncontrollable, leading to great disorder and a failure to find equilibrium. When they asked the medical researcher Arturo Rosenblueth (who had been working for years under Walter Cannon at Harvard) if there were any similar pathologies known in human physiology, Rosenblueth immediately answered that there was indeed an exact neurological parallel: voluntary motion could degenerate into the same state of oscillation when the cerebellum was injured in very specific ways.[22] Mathematical analysis of other forms of physiological disorder (e.g., cardiac arrhythmias, which would be the focus of Rosenblueth's own work back in Mexico) would soon reveal a number of these parallels between physiological and technological pathologies, which grounded the study of structural and topological norms shared by cybernetic machines and complex organisms.

This early interest in pathology at the very foundation of cybernetics is hardly surprising, given the fact that so many of its original practitioners were medical professionals with interests in both physical and mental illnesses. Warren McCulloch was a neurologist who worked in psychiatric clinics, Ross Ashby was trained as a psychiatrist and practiced while developing his own private research projects in homeostatic stability, and Rosenblueth was a physiological researcher and a cardiologist. In an important sense, cybernetics (especially its later incarnations) was always a highly medicalized discipline, aimed at identifying the origins of instability in large, complex systems and then diagnosing the sources of breakdown so as to eliminate them and recover unity and stability. As Ashby put it, in his textbook on cybernetics:

> Cybernetics offers the hope of providing effective methods for the study, and control, of systems that are intrinsically very complex. . . . In this way it offers the hope of providing the essential methods by which to attack the ills—psychological, social, economic—which at present are defeating us by their intrinsic complexity.[23]

Instability appears in this framework as a "self-generating catastrophe" that leads to the collapse and death of the system if it is untreated.[24]

And yet . . . Despite this extensive interest in failure, pathology, and catastrophic breakdowns, cybernetics could still be accused of greatly simplifying the problem of disorder and crisis, pathology and health. It was one thing to explain some forms of adaptive response in terms of automatic steering technologies. However, it was quite another to explain a more radical originality and inventiveness that was repeatedly encountered in the living world. This capacity was not reducible to any specific mechanism or logic, even a homeostatic one.

As the noted British polymath Michael Polanyi observed, in his Gifford lectures of 1951–52, the capacity of life as improvisation flowed from an "active center operating unspecifiably in all animals."[25] Polanyi's theory of evolution was based on the idea that even simple organisms, in circumstances that threatened survival, were capable of responding with novel solutions. As he noted in his preface, the results of Gestalt psychology initiated his own inquiry, which aimed to understand these creative reorganizations in an evolutionary context. So, for Polanyi, a singular solution would become part of the repertoire of behavior, one that would harden into routine or habit and then eventually be *inherited* as a new automatism. Ever more complex organisms would become capable of more and more sophisticated improvisations in the face of challenging and threatening situations.

The same process would be evident in cognitive activity, once organisms acquired nervous systems and sensory systems. To know the world was, Polanyi insisted, to actively engage with it. His famous idea of "tacit knowledge" recognized that embodied knowledge was not the *result* of some formal or logical mental framework but instead an *event*, a product of the unpredictable interaction of mind and environment. And problem solving was the impetus for this kind of interaction. With detailed references to Köhler's ape studies, as well as concepts of "inspiration" in creative thinking (e.g., Poincaré and Wallas), Polanyi argues, "The irreversible character of discovery suggests that no solution of a problem can be accredited as a discovery if it is achieved by a procedure following definite rules" (123).

Polanyi, who was familiar with the newest electronic computing machines and related technologies, stressed that this *singularity* of discovery, its status as an interruption *into* thought, demonstrated that it could never be produced by some machinic process. "For such a procedure would be reversible in the sense that it could be traced back stepwise to its beginning and repeated at will any number of times, like any arithmetical computation. Accordingly, any strictly formalized procedure would also be excluded as a means of achieving discovery" (123). The figure of the genius, linked with the inventive originality of the living body itself, reveals the essential *exception* of genuine novelty.

> But genius makes contact with reality on an exceptionally wide range: seeing problems and reaching out to hidden possibilities for solving them, far beyond the anticipatory powers of current conceptions. Moreover, by deploying such powers in an exceptional measure—far surpassing ours who are looking on—the work of a genius offers us a massive demonstration of a creativity which can neither be explained in other terms, nor unquestionably taken for granted. By paying respect to another person's judgment as superior to our own, we emphatically acknowledge originality in the sense of a performance the procedure of which we cannot specify. Confrontation with genius thus forces us to acknowledge the originative power of life, which we may and commonly do neglect in its ubiquitous lesser manifestations. (124)

Hence the limitation of the machine: it cannot *violate* itself; it cannot "fail" or reinterpret itself in the light of failure, error, damage. Therefore, we cannot, he says, "replace the conception of the machine—as defined by its operational principles—by a more comprehensive understanding which accounts both for the correct functioning and the failures of a machine" (328). The rules of organization (biological, psychological, epistemological) only describe success; yet the heart of genuine creative action is the response made in light of *failure*, the insufficiency of norms, the breakdown of order, the threat of catastrophe (333).

As Polanyi well knew, the essential claim of the new cybernetic discipline was that machines could be made to imitate or instantiate this *creative* function—to exemplify, that is, a kind of *artificial plasticity*. Referring to the debates opened up at the Hixon Symposium of 1948, where cybernetic ideas engaged with leading research on brain function, Polanyi will admits that a pseudo-vitality can be manufactured.

> The machine-like conception of living beings can be extended to account in principle for their adaptive capacities. An automatically piloted aeroplane approximates the skills of an air pilot. Its mechanical self-regulation coordinates its activities in the service of a steady purpose, and it may even appear to show a measure of resourcefulness in responding to ever new, not exactly foreseeable situations. (336)

Were we, human beings, just "conscious automata" then? Polanyi offers another perspective: every living thing has, at once, an automatic system of behavior and what he calls an "inventive urge" that defies the normativity of these automatisms. Only such an "unspecifiable" element can explain the kind of plasticity ("equipotentiality") exhibited by the brain, even as it suffers extensive destruction—as Lashley's work showed so clearly (336). Equipotentiality implies *multiple* paths and therefore an element of decision, a break from the norms of regulative behavior.

What thinkers such as Polanyi (and Bertalanffy, Goldstein, and Canguilhem) were trying to suggest was that organisms were not just able to *respond* to changing conditions; they were also able to enter *wholly new states of being*—with new forms of order and new potentials—when they were confronted with extreme challenges, drastic injury, or internal failures of communication. A pathological state was not simply the negative inverse of a normal function, as cybernetic work at times seemed to imply, but rather an opportunity for the appearance of invention, the creation of unprecedented response.

The historical question raised here, then, is whether or not cybernetics did in fact understand, and try to model, this productive relationship between genuinely pathological conditions of distress and the radical novelty of reorganization within the organism. The legacy of cybernetics has often been figured as a move toward a form of invisible "technocracy" of rationalization and systematization in the social, economic, and political spheres—leaving the individual to be subjected to the teleological momentum of these now autonomous systems of techno-power.[26] This is why cybernetics has reemerged today, in the twenty-first century, as a conceptual and political orientation as we confront the rise and consolidation of a global information society, the emergence of new and more intimate forms of human-computer symbiosis, not to mention acute conflicts over automated surveillance, algorithmic justice, and automatic warfare involving remote killing technologies.

Peter Galison, for one, has forcefully argued that cybernetics, literally born (with Norbert Weiner's failed attempt to build an intelligent antiaircraft weapon) in the midst of war and developed, with much state funding in the United States, Europe, and the Soviet Union during the height of the Cold War, was at its heart a "Manichaean" science, obsessed that is with maintaining order against the constant active threat of disorder. Cybernetics could hardly embrace a link between abnormality and creativity, Galison claims, when it pathologized disorder as the ultimate enemy.[27] Andrew

Pickering's sophisticated and comprehensive analysis of British cybernetic brain science highlights the crucial significance of pathological conditions in the development of a new way of thinking about adaptive and creative machinery. For Pickering, this is what distinguished the British alternative from the more militarized American version of cybernetics. As Pickering writes, pathology was, for the cyberneticists, another way of looking at the normal, ordered state: "In medicine the normal and the pathological are two sides of the same coin." Cybernetic machines and organisms were defined by their inherent structures, and both were kept "alive" by the processes that maintained homeostatic equilibrium. Breakdown is crucial in order to reveal the often occluded normative order. However, pathology was it seems never conceptualized independently.[28]

In practice, however, it is somewhat difficult to maintain a clear boundary between order and disorder in cybernetic discourse. The cybernetic analogy between machine and organism was being forged at a moment when biological thought had, as we know, transcended the old opposition between mechanists and vitalists. The key challenge in this period was to explicate organismic unity. First, it was necessary to ascertain how an organism developed into a coherent formal structure from embryonic cells—the problem of morphogenesis. And second, it was important to identify how organisms could *maintain* those forms despite ever-changing conditions and a dynamic metabolic process—the question of homeostasis. The concept of the essential unity of a living being was so important because it linked—at the level of a single system—transformation and destruction with continuity and stability. Unity was, of course, predicated on an inherent plasticity of organismic organizations, their capacity to take on new forms at key turning points, even catastrophic ones.

To understand the cybernetic effort to create a science that bridged organismic and machine beings we must therefore pay close attention to the way unity and flexibility were conceptualized in cybernetics. As we will see, illness, pathology, breakdown, all were vitally important questions for biological thinkers after the war. These states were not simply ascribed to "disorderly" Manichaean forces attacking order, pace Galison, but instead were understood to be potentially productive crisis conditions, contingent and singular events that revealed essential transformative potentials that made the ongoing life of an organism possible in the first place.

24
Automatic Plasticity and the Pathological Machine

> Between the plasticity of the brain and the machinistic structure of a computer there is an unbridgeable gap that is even wider than that between syllogizing and reasoning.
>
> Nicholas Georgescu-Roegen, *The Entropy Law and the Economic Process* (1971)[1]

Prominent figures in the early days of cybernetics, information science, computing, and even artificial intelligence research all looked back to Descartes as a forerunner in the conceptualization of cognition as essential *technological*. Norbert Weiner cited Descartes as an early important theorist of automata, noting only that he failed to develop a comprehensive understanding of how the automaton was coupled to its environment.[2] Claude Shannon (the massively influential information theorist) and John McCarthy (who coined the very term "artificial intelligence") praised Descartes's argument that the body with its nervous system was an automaton while acknowledging that any effort to understand the brain's function "usually reflects in any period the characteristics of machines then in use." Before the development of "large-scale computers" and the subsequent theorizations of information processing devices, thinkers such as Descartes were limited to hydraulic and other machinery in their modeling of the nervous system.[3]

What the cyberneticians introduced was the idea of an organized being that responded actively to the environment, not through complex "mechanical" interactions, but rather through the introduction of information into that being, which was then reorganized to effect certain actions that would maintain the inherent "purpose" of this being.[4] The finite set of information states was defined by the material organization of the cybernetic entity. Yet the *actions* were effected by the logic of information itself, not mere physical action and reaction. Whether the information system was an analog computer, a physical instrument, or a digital computational device, what was important for cybernetic theory was the fact that an intelligent being (whether artificial or natural) constructed a model of its environment through the coded information received from sensory organs, reorganized that information to preserve its ideal goal state, and then initiated actions that would produce that desired state. In this way, the cyberneticians erased the conceptual distinctions between animal and human, human and feedback machine, animal and machine, since all were in essence information systems, beings that acted on the basis of virtual realities, not physical ones.[5] The measure of human or other forms of intelligence was the degree of complexity of information processing, hence the interest, in the 1940s and 1950s, in the new large-scale computing devices then being developed. The brain, it was thought, may very well be a digital computer, a logic system that was materially instantiated in neural cells but governed by the pure logic of binary operations.[6]

As we saw, Descartes was one of the first theorists of the nervous system as an *information machine*, a steering system that was organized by sensory inputs and their organization and reorganization, all in feedback loops with other sensory information and, importantly, the activation organs, what Lotka and others called "effectors." From one point of view, we can understand cybernetics as an attempt to redefine the organism as a new form of machine, one governed, literally, by information—and one that need not involve, even at the level of the human being, any new element, any interventional "spirit" of mind.

Feedback Is Error

One of the central claims of cybernetics was that only a process of "negative feedback" drove the homeostatic life of adaptive beings. The cybernetic entity (whether living or machinic) first sensed its environment, alongside its own "state" of being in that environment, then compared that information with its own "goal" states embedded somewhere in this being, before effecting certain actions that would bring the entity's state in line with this ideal goal.[7] To be sure, from one perspective, this cybernetic concept of negative feedback could be read as privileging a predetermined, precisely defined "order" that is imposed on these active beings. Yet from another angle, one might emphasize the importance of deviation in the functioning of any cybernetic being: this being has an existential relationship with error. It is not disorder or even error itself that threatens the cybernetics being but instead an inability to respond appropriately to an ongoing state of disorder. Any dynamic adaptive being is in a condition of ceaselessly deviating from its own formal ends.

The cognitive psychologist George A. Miller, famous for the experimental demonstration of the limits of working memory,[8] perfectly expressed this cybernetic principle in his theoretical work on memory, bringing together machinic, evolutionary, and organismic concepts.[9]

> One of the few truly general ideas that we have in neurophysiology is that organisms try to reduce deviations from their normal states. This idea, which has a long and interesting history, is now widely known as the "cybernetic principle," or, more descriptively perhaps, the "negative feedback principle." It holds that an adaptive system can maintain a stable state in a fluctuating environment if it can sense deviations from that state and initiate actions to reduce them. (361)

For Miller, the organism did not merely *register* events and experiences in a memory storage system. Instead, memory functioned within the homeostatic economy as a whole. He agreed with Kenneth Craik's position that the nervous system produced models of an exterior reality but emphasized the importance of *failure* in this task, because failure opened the opportunity to transform the model and gain new insights into our environment. We remembered, in other words, the *unexpected* (the "improbable deviation" [368]), and this needed to be "encoded" in natural or artificially adaptive systems (361). The implications of this failure model are significant. As Miller wrote:

> The suggestion that we encode the unexpected would seem to imply that the system is symbolic, rather than iconic. We can easily conceive of an iconic representation of an object or event in consciousness, but it is much more difficult to conceive of an iconic representation for a mismatch between two such images. (362)

From this perspective, radical failure (even death) is not so opposed to normal homeostatic operation because the catastrophic turn is in fact continuous with normal efforts to maintain the health of the being *in its essential errancy*. That is, the boundary between normality (health) and catastrophic failure (illness, death) is defined by the limits of the being's own errancy and not at all by the mere presence of deviation or error in the face of the improbable or unexpected occurrence. The great achievement of cybernetics was the demonstration that these limits to error were not at all arbitrary. As Wiener remarked, "The conditions under which life, especially healthy life, can continue in the higher animals, are quite narrow."[10] As Ashby explained, in an essay on homeostasis, "If the organism is to stay alive, a comparatively small number of essential variables must be kept within physiologic limits," and this applied to the "life" of technological entities as well. Their survival as unified systems depended on keeping error within proper limits.[11]

The crucial arbiter of the limit of error was therefore the survival of the being as a whole. As Ashby would make clear, the relationship between catastrophe and normal error or deviation is governed by the threat to the fundamental unity of the being in question. He gives the example of a mouse trying to evade a cat. The mouse can be in various "states" or postures, and certain values may even change drastically (it may lose an ear, for example), yet still, the mouse will survive. "On the other hand, if the mouse changes to the state in which it is in four separated pieces, or has lost its head, or has become a solution of amino-acids circulating in the cat's blood then we do not consider its arrival at one of these states as corresponding to 'survival.'"[12] The unity of the being is the ultimate mark of survival. Unity amounts to the capacity to maintain some stable form even while experiencing drastic—perhaps even violent—transitional states.

There is no fundamental distinction between order and disorder here. The definition of a formal unity determined the parameters of survival within the unified system. Both in biology and cybernetics, the unity of the being is what identifies the structural relations and variables that had to be maintained against the threat of extinction. To rethink cybernetics from the perspective of the organism, we must zero in on the nature of this unity and its status. How could a breakdown, even a catastrophic failure, become the opportunity for an unprecedented reorganization? How could a functionally determinate machine ever acquire this organismic capacity?

Ashby on the "Break" within the System

It was exactly this problem of reorganization in crisis that Ashby faced head-on in creating his cybernetic model of the adaptive organism, the Homeostat machine. Ashby knew that by definition a machine was fixed, and its operations wholly determined. A machine, in its ideal operating state, had only one form of behavior and therefore could not change its fundamental design. Although it is possible to have a machine that enters new states depending on changing environmental variables (a thermostat is a simple example), the design of the structure will still govern its operation at all times.

In this respect Ashby would have agreed with Canguilhem's critical perspective on the limits of the machine. In a notebook fragment from 1943, we find Ashby reading William James, who, as we have seen, placed the relationship between plasticity and automaticity at the center of his biopsychological explanations of human thinking. James, as Ashby noted in his citations of the pragmatist psychologist, compared the rigid machine to the complex nervous system, an entity that paradoxically exhibits *both* fixity of structure and open-ended adaptive plasticity.[13] The early notebooks show that Ashby was also reading Claude Sherrington's revolutionary work on neural integration at this time. Ashby in fact wrote out several passages by hand from Sherrington's seminal book, where the neurologist described the organism as a "moving structure, a dynamic equilibrium," something constantly adjusting itself to ever-changing conditions. The living system was "labile"—and indeed, its greatest strength, for Sherrington, was its very fragility, because that fragility made it more sensitive to its surroundings.[14] If his own Homeostat machine was going to be an adequate representation of this kind of organismic self-organization, Ashby had to figure out a way to model the behavior of a genuinely *open* system, one that could assume a determinate structural organization, like any machine, but at the same time not be eternally *bound* to any one particular order. Remember that the influential systems theorist and biologist Bertalanffy had already criticized cybernetics precisely for their misguided use of *closed* systems to model the fundamentally open structures of natural organisms.

Ashby himself was hardly unaware of this issue, and he thought deeply about how to create mechanical systems that could not only respond to environmental changes, but in fact actually change its very *organization* as a way of finding new paths to stability and equilibrium. Ashby's insight was that if a machine, defined by a specific form and purpose, was ever to reorganize, then logically it must in fact be capable of becoming a wholly different and new machine. In 1941, he had admitted that man-made machines that change their organizations were "rare" (he failed to give any concrete example, though he did point out elsewhere that the inclusion of memory in a system would amount to such reorganization).[15]

Yet Ashby would push much further, seeking to conceptualize a machine that had, like the nervous system, what James had called an "incalculable element" that could interrupt, productively, the machine's own fatalistic operation. Ashby made a conceptual breakthrough by thinking about the potential value of *failure*, something that was, after all, inevitable in any working machine. Ashby realized that when a machine broke, it became in essence a brand-new machine with a new "design," so to speak: "A break is a change in organization."[16] Ashby's goal was to engineer a machine that could take advantage of its own breaks as a way of entering into a state. Genuine breaks of this kind were exceedingly rare in artificial machines. According to Ashby, a break was "1) a change of organization, 2) sudden, 3) due to some function of the variable passing a critical value."[17] If a homeostatic machine were constructed in such a way as to "break" when pushed to the limit of its ability to maintain equilibrium, this machine could acquire a new organization, another possible path to equilibrium. "After a break, the organization is changed, and therefore so are the equilibria. This gives the machine fresh chances of moving to a new equilibrium, or, if not, of breaking again."[18] The breakdown was in essence a temporary shock to a system that was not succeeding in its quest to find equilibrium. Ashby invented a form of cybernetic plasticity by taking advantage of the very weakness of all machines—their ultimate fragility. As W. Grey Walter put it, in his own cybernetic book on the brain, "The power to learn implies the danger of breakdown."[19]

As Ashby would point out, the brain was a special kind of machine in that its many highly differentiated component parts—the neurons—were constantly connecting and disconnecting with each other as the brain responded to perturbation and change. Built into its dynamic organization was an inherent tendency to break down and thereby give way to new organizations, for the neurons have a built-in latency period: after a certain amount of activity, neurons temporarily "disappear" from the system, only to reappear fully active again once they have recovered their potential. Ashby suggested that the cybernetic machine and the organism could be linked by this shared capacity to self-organize in moments of breakdown, a capacity that ultimately could be traced to the tendency to fail—at least temporarily—on a repeated basis.

In later works such as *Design for a Brain* (1952) and *Introduction to Cybernetics* (1961), Ashby introduced a formal way of thinking about the behavior of complex, adaptive systems. He noted that if we take the basic states of a system as variables—capable of change, that is—then we could plot them graphically as vectors. In turn, individually plotted variables could be integrated in a multidimensional "phase space" as a way of representing, with one single vector, what Ashby called a *line of behavior*, an integration, that is, of multiple vectors into one line that would represent the dynamic interplay of different states in motion within the total system.[20] Ashby's innovation was to think of the cybernetic being as a set of potential states, states that could be visually represented in phase space as vectors whose functions were limited by boundary values. Ashby showed how the interpretation of these lines of behavior as mathematical functions opened up a new way of understanding complex systems. Mathematical analysis of physical systems was normally limited to continuous systems of a linear nature. A mathematical approach to biology, in contrast, would have to accommodate the nonlinear, *discontinuous* features of much organismic behavior.[21] Ashby gave an example: any sudden transformation of an activity may well be governed by a so-called step function, where the value of the function moves continuously, imitating a linear function, until it changes abruptly (and seemingly unpredictably) when it reaches a certain point. This mathematical representation of behavior would suggest that spontaneous, singular events or actions in the life of a being are in fact governed by hidden, dynamic processes modeled by relatively simple step functions, or more complex nonlinear descriptions of threshold changes and singularities.[22] The main insight was the idea that transformations of the system's behavior were strictly analogous to mathematical transformations of the operands.[23] For Ashby, this made it possible for cybernetics to study the innate "determinateness" of a system formally, in mathematical terms, and to ignore in effective the actual "material substance" of the system.[24]

Cybernetic Plasticity

Ashby's cybernetic experimentation was more or less abstract—despite the solid, if at times mysterious, materiality of his infamous electrical device, the Homeostat.[25] However, the abstract question of artificial homeostasis did not exhaust the cybernetic attack on understanding living, dynamic systems. And the brain, in particular, was a zone of intense study, not, as in Ashby's case, as a generic problem in organization and stability, but as a novel and complex "technology" of *command and control*.

Cybernetics put a great deal of energy and thinking into the theory of the central nervous system as an *information processor* that governed behavior. From 1948, the date of publication of Wiener's seminal book that coined the term itself, cybernetics would be associated with the computer-brain analogy. Weiner's book originally appeared (in English) with a Parisian publisher, with whom MIT Press hastily collaborated only after belatedly learning of the arrangement. In this seminal book on cybernetics, Wiener enumerated the fundamental analogy between neural organization and the binary architecture of the computer, writing, for example, that the realization that the brain and the computing machine have much in common "may suggest new and valid approaches to psychopathology, and even to psychiatrics."[26]

Still, he also raised many questions about this comparison. He was equally interested, for example, in how the brain maintains its information states without ever localizing them in particular spaces, the very problem that prompted Lashley's earlier neurological research. Memory, for Wiener, could not be something merely physical. Instead, it had to be a constantly flowing *circulation* and therefore subject to perturbation and deviation; this system, he said, could "hardly be stable for long periods of time."[27]

However, it was the nature of neural plasticity in particular that constituted the main challenge to the cybernetic project of assimilating machine and organism into one comprehensive framework, and there were differences of opinion even within cybernetics, in the United States and the United Kingdom and in France. To be sure, from the very beginning plasticity was on Wiener's mind. A central question he posed was how to explain the nervous system's agile flexibility; Wiener wondered "how the brain avoids gross blunders, gross miscarriages of activity, due to the malfunction of individual components."[28] With instability and malfunctions in mind, Wiener would explore several different analogies between psychological pathologies and computer malfunctions. He observed, for instance, that the drastic effort to "clear" the brain of its pathological activity with the use of electrical or chemical "shock treatment" (in lieu of the more permanent surgical lobotomy) might well parallel the necessary purging of the computer of all data when a pathological "configuration of the system" disrupts its operations.[29] Wiener also remarked more than once on the essential plasticity of the brain. He gave the example of Louis Pasteur, who suffered a major stroke early in his scientific career. After his death, it was discovered that he had only "half a brain." Yet Pasteur was, Wiener pointed out, only mildly affected by some physical paralysis, and mentally he was not at all diminished, as his great scientific achievements following the stroke proved. Wiener also used the very same example Canguilhem had offered, noting that infants can suffer "an extensive injury" to one hemisphere of the brain, only to have the remaining half take on all the functions of the destroyed one. "This is quite in accordance with the general great flexibility shown by the nervous system in the early weeks of life," Wiener wrote.[30]

In this early, often speculative phase of cybernetics, Wiener could not offer much in the way of a real explanation for this plastic quality. And yet he realized that if the brain is basically a processor of information and not a purely physical system, we are likely to find that pathologies of the mind will not be traceable to specific lesions of the mechanism (i.e., the level of the neurons) but instead to a failure of the system *as a whole* to manage information flows between functional centers.[31] Wiener claimed that because complex human behavior no doubt involved a great deal of neuronal connectivity, it may well be the case that the human brain is often running "very close to the edge of an overload" that could at any moment result in "a serious and catastrophic" failure. With increasing neural flows "a point will come—quite suddenly—when the normal traffic will not have enough space allotted to it, and we shall have a form of mental breakdown, very possibly amounting to insanity."[32] Here Wiener hinted at a cybernetic form of the "catastrophic reaction," but again, he was limited by his inability to say much about how the brain (or the computer network analogue) could recognize the failure and *reorganize* itself in these moments of extreme crisis and breakdown, as Goldstein and others had demonstrated. Still, Wiener hoped one day that a "new branch of medicine" would be developed—he called it "servo-medicine"—that would deal with these kinds of informational disorders of control when the "strains and alarms of a new situation" put demands on an information system that was not equipped to deal with it.[33]

Much of the thinking about plasticity would focus on the importance of the biological example of the brain as an anatomical structure whose organization was not determined by its histological form but rather by its pattern of connectivity. The French neuroscientist Alfred Fessard expressed this clearly at a 1953 conference on the methods and limits of cybernetics: the key insight that can be drawn from cybernetics was the observation that there need not be a "complete pre-structuration" of a system in order to explain how it can produce behavior that is "coherent, adaptive, and goal driven." Indeed, it was the dynamic fluidity of feedback and transformation that gave a system a kind of autonomy from its own "structure." He pointed to the significance "of a certain possibility of dynamic reorganization of the connection schemes, whereby the living organization and its mechanical models are partly freed from internal structural constraints and more incorporated into the medium, engaged in this medium."[34] The challenge, Fessard notes here, is that the individual neuron is itself a living thing, unlike the fixed switching devices of the new electronic brains, the computers. Here is where cybernetics needed new conceptions and a technical innovation. As he explains, for an organ like the brain to function dynamically in the way it does, what needs to change is not, of course, the anatomical arrangement of the nerve pathways, woven once and for all time in adults; it is the susceptibility that each neuron has to enter into activity, and which requires a certain neural plasticity that cyberneticists seem to have overlooked.[35]

"A General and [Patho-]Logical Theory of Automata": Von Neumann

And yet . . . One of the most important and notorious cyberneticists, John von Neumann, the brilliant mathematician and ultimate leader of the military project to build a high-speed digital computer for the design of nuclear bombs, was seriously engaged with this central concept of plasticity and its relation with error. Von Neumann looked to the sphere of biology, especially neurophysiology, as a way of reconceptualizing the very future of computing itself.

In some of his earliest work on computing von Neumann had (with his colleague H. H. Goldstine) hinted that "human reasoning" might at some point be "more efficiently replaced by mechanisms."[36] Von Neumann was, from the start of his transition from mathematician to digital computer designer and military bureaucrat, thoroughly engaged with the metaphorical and structural connections between organic systems and technology, in particular, with the importance of the nervous system as the information center for intelligent organisms. And so he would, to the puzzlement of his technical collaborators on the project, describe the EDVAC (the first stored program computer) as a machine built from "idealized neurons."[37] However, by the time he came to deliver his important series of lectures on the theory of automata (natural and artificial), von Neumann, not unlike Wolfgang Köhler, questioned the value of any analogy between the brain and the computer: "It's terribly difficult to form any reasonable guess as to whether things which are as complex as the behavior of a human being can or cannot be administered by 10 billion switching organs. No one knows exactly what a human being is doing, and nobody has seen a switching organ of 10 billion units; therefore one would be comparing two unknown objects."[38]

One of the "complex" human behaviors von Neumann was interested in was, of course, intelligence—indeed, he often stressed the importance of insight in scientific and mathematical work in a way consistent with the epistemological ideas traced in the late nineteenth and early twentieth centuries.[39] "In pure mathematics," he told his audience, "the really powerful methods are only effective when one already has some intuitive insight. . . . In this case one is already ahead of the game and suspects the direction in which the result lies." He did not claim that the computer would replace human insight. Instead, he merely suggested, pragmatically, that high-speed computers could help out in those areas, such as turbulent flows and shock analysis, where the problems facing physicists were so unique, so "singular," that very few "insights" were forthcoming, because the nonlinear equations involved were not analogous to the better-known linear domain and in fact "violated" all of our expectations. Here, the computer, with its astounding calculating ability, could serve as a kind of virtual space of experimentation. The machine could solve certain critical cases first isolated by the physicists as promising, and from analysis of these cases, we might gain some form of "intuition" (as he put it) as to where other critical cases might lie.[40]

This attention to insight helps, I think, explain von Neumann's perhaps surprisingly harsh critique of McCulloch and Pitts's seminal work on computation and neurology in these lectures. Although he agreed with their mathematical conclusion that idealized neural networks could well operate as a biological Turing machine, von Neumann emphasized what was not demonstrated by the two collaborators—that anything like this system actually occurs in nature. That is, von Neumann resisted the logical simplification of a system that was, according to neurological research, almost overwhelming in its internal complexity.[41] But, more important, he was also acutely aware of the psychological complexities involved. The two realms were, obviously, intimately connected. As he noted in this suggestive, almost Wittgensteinian illustration:

> Suppose you want to describe the fact that when you look at a triangle you realize that it's a triangle, and you realize this whether it's small or large. It's relatively simple to describe geometrically what is meant: a triangle is a group of three lines arranged in a certain manner. Well, that's fine, except that you also recognize as a triangle something whose sides are curved, and a situation where only the vertices are indicated, and something where the interior is shaded and exterior is not. You can recognize as triangle many different things, all of which have some indication of a triangle in them, but the more details you try to put in a description of it, the longer the description becomes.

Or, more precisely, the "logical" descriptions of such complex analogical recognitions may even prove impossible at a certain point. In a perhaps unconscious allusion to Gestalt theory, von Neumann noted that "with respect to the whole visual machinery of interpreting a picture, of putting something into a picture, we get into domains which you certainly cannot describe in those [i.e., logical] terms." There was no way of describing the "wholeness" we recognize despite distorted, partial, or ambiguous forms.[42]

At this point, von Neumann suggested an interesting "isomorphic" relation between these intuitive forms of thought and the complex organization of the nervous system. If the "visual brain" was responsible for such direct insights, he suggested, perhaps it was best to think of this neural network (consisting of two billion pathways) not as carrying out some formal logical computation to produce the insight but as itself physically organized as such a unity, according to its own vastly complicated logic—one that was at this point totally unknown and frankly beyond our comprehension. "I think that there is a good deal of reason to suspect that this is so with things which have this disagreeably vague and fluid impression (like 'What is a visual analogy?'), where one feels that one will never get to the end of the description." Neither the experience nor the neural processes could be reduced to simpler categories.[43]

Yet von Neumann went even further here, noting that not only were these processes complex, but they were not necessarily predetermined. After criticizing McCulloch and Pitts's theory of memory circuits with the observation that the brain's memory, whose existence is undeniable, seems to be located nowhere in particular, he went on to say that in general "it is never very simple to locate anything in the brain, because the brain has an enormous ability to re-organize. Even when you have localized a function in a particular part of it, if you remove that part, you may discover that the brain has reorganized itself, reassigned its responsibilities, and the function is again being performed."[44] This is exactly what so many brain scientists had emphasized in the interwar period, and such plasticity (or "equipotentiality," as it was often called) was at the center of Kurt Goldstein's important 1934 book on the holistic organism, which was, as we have seen, based in part on the numerous studies Goldstein and his collaborators had done on brain-injured soldiers during the Great War.

For von Neumann, the creative plasticity of the nervous system served only to highlight the rather simplistic, and inferior, mechanical structure of the early computers, something he was of course well positioned to notice. This was no mere philosophical question at the time. Von Neumann, like many others involved in the origins of digital computing, was acutely aware of the problematic unreliability of these early machines. From the start, error loomed large in von Neumann's thinking on machine computing—and in fact, much of modern error analysis and correction techniques can arguably be traced to his seminal work in this area with H. H. Goldstine. But in his elaboration of his general automata theory, von Neumann was

Figure 24.1. From Martin H. Wiek, "The ENIAC Story," *ORDNANCE* (1961). Source: US Army photo. Courtesy ftp.arl.army.mil/~mike/comphist.

interested more in what was at stake conceptually with the problem of error. He could see that while the digital logic of computers had greatly increased precision because it eliminated the whole problem of "noise" that was so troublesome in analogical computing devices, the sequential logic of the digital systems was, practically speaking, much less reliable, since any single breakdown in the circuitry (a blown vacuum tube, for instance) introduced potentially devastating errors that would quickly multiply in long and complex calculations.[45] (Figure 24.1.)

With this in mind, von Neumann made one important assumption in his automata theory: "In no practical way can we imagine an automaton which is really reliable."[46] This would have far-reaching implications. Von Neumann realized that any systematic form of error management, or any single, centralized control mechanism, would also have to be considered subject to failure.[47] Error, he once said, had to be considered "not as an extraneous and misdirected or misdirecting accident, but an essential part of the process under consideration—its importance in the synthesis of automata being fully comparable to that of the factor which is normally considered, the intended and correct logical structure."[48]

Of course, it was here that the amazing plasticity of "problem-solving" biological organisms was so important, and von Neumann would invoke this nonlogical—or perhaps a-logical—plasticity as a model for any artificial automaton. "If you axiomatize an automaton by telling exactly what it will do in every completely defined situation, you are missing an important part of the problem," he stated. "It's very likely that on the basis of the philosophy that every error has to be caught, explained, and corrected, a system of the complexity of the living organism would not run for a millisecond. Such a system is so well-integrated that it can operate across errors."[49] In this context, von Neumann was fascinated by the fact that the nervous system did not seem to use digital forms of notation at all, operating instead with a statistical messaging technique that was less precise, to be sure, but much more reliable because it never relied on any single, specific message for its successful operation.[50]

As we can see clearly, in cybernetics and in neuroscience of the period, the "system" could only operate across errors because its essential unity was not fixed in any one specific location. The system did not just spread its information over many neural connections, in case some failed. It also could actively bypass a whole series of performances for the good of the whole system when systematic error was detected. To accomplish this, the system had to possess a direct form of insight, von Neumann implied, and he used rather psychological language to describe this ability: "The system is sufficiently flexible and well-organized that as soon as an error shows up in any part of it, the system *automatically senses* whether this error matters or not. If it doesn't matter, the system continues to operate without *paying attention* to it. If the error *seems* to be important, the system blocks that region out, by-passes it, and proceeds along other channels."[51]

The "radically different" approach to error and organization in the physical nervous system raised a key question: What was the principle of organization that would explain this flexible form of self-regulation? What was this very active systematic order? At this point, the best von Neumann could do was suggest that "the ability of a natural organism to survive in spite of a high incidence of error . . . probably requires a very high flexibility and the ability of the automaton to watch itself and reorganize itself."[52] This entailed an important autonomy of parts, where "there are several organs each capable of taking over in an emergency," but also a logic of systematic order that could in some way define these creative solutions to crisis. In other words, the parts could only take control if they had some inner connection to the principal structure of the whole organism. "The problem consists of understanding how these elements are organized into a whole, and how the functioning of the whole is expressed in terms of these elements," he pointed out—returning to one of the central tropes of Gestalt psychology.[53] Yet, until the end of his life, von Neumann admitted repeatedly that this logic was beyond any present understanding. His unfinished and undelivered Silliman lectures for Yale University, published as *The Computer and the Brain* after his death, in the end just detailed the fundamental differences between these two entities. His terse conclusion was that the logical structures involved in nervous system activity must "differ considerably" from the ones we are familiar with in logic and mathematics.[54]

Reading cybernetic concepts with and against contemporary theorizations of open, living systems, we glimpse some important and provocative intersections between the technological discourse of cybernetics and continental forms of thought that emphasized holistic and vitalist concepts of organismic order. A crucial marker of the open living system was the important and productive role played by pathological and even catastrophic states in maintaining the persistence of organismic unity. Having recognized this themselves, the cyberneticists had faith that it would be possible to create artificial machines that possessed genuine plasticity, so necessary for survival in times of extreme crisis.[55] In the cybernetic era, both organisms and advanced machines were sites of stability and instability, truth and error, order and disorder, health and pathology; all were intertwined, in sometimes strange ways, to produce fragile yet powerful beings, endowed with astonishing and unpredictable creative powers.

25
Turing and the Spirit of Error

In 1950, Alan Turing conjectured that in the year 2000 human cognition would be successfully simulated by digital computers.[1] This conjecture inaugurated the new discipline that would soon be called artificial intelligence. As is well known, so-called classical AI aimed first to analyze human thinking with precision, in order then to model cognitive operations in computer programs. The term "artificial intelligence" itself was coined by John McCarthy, in his grant application to host a workshop at Dartmouth on new research and theory in machine thinking. As McCarthy and his coauthors put it, "This study is to proceed on the basis of the conjecture that every aspect of learning or any other feature of intelligence can in principle be so precisely described that a machine can be made to simulate it."[2] This was Turing's inspiration: the idea that the mind might be conceived of as a *kind* of machine, and therefore something capable of being modeled by what he described as a universal machine, the so-called Turing machine. The subsequent history—or more accurately, histories—of AI resist any effort to forge a single comprehensive narrative. However, dominating the research field was the assumption of a parallelism, between the *mind* and the *computer* (whether the technology considered was software or hardware).

It is not surprising, then, that from the very beginning the link between the recently developed digital computers and the mind-brain was understood to be grounded in the notion of mechanical intelligence. Whether focused on the logos of the psyche, or actual brain processes, human intelligence was conceptualized as a form of *automatic information processing* with strong analogies with computer processing of information. This mirroring of mind and computer in much of contemporary cognitive science and neuroscience seems to vindicate Turing's original vision, namely, to understand thinking as akin to an algorithmic, and hence essentially mechanistic and deterministic, process.

However, as we will see, this reading misses what is arguably most interesting in Turing's project, and also fails to connect his thinking to earlier and contemporary lines of thought that intertwine around notions of cognition, body, mind, and technology. Rather than simply claiming to understand human thinking as just another mechanical process, so that it could be successfully simulated with another machine in an "imitation game" (later called the "Turing test"), Turing was clearly interested in quite a different goal—namely, to conceptualize and construct *intelligent* machines. Or to put it another way, Turing was looking closely at human intelligence for clues of how to transform uninteresting machines into intelligent ones. What made this project so challenging for Turing is that he clearly recognized that the most interesting aspects of human intelligence clearly resisted any easy reductionist account, an account that would of course seem to be necessary for the purely mechanical simulation of the highest forms of human thinking.

Spiritual Machines

We need to begin before the war, before Turing began to work on cryptography and build automatic information machines, before he helped design and build Britain's first digital computer at Manchester University. It is often mentioned that Turing, as a young person, was fascinated with the book *Natural Wonders That Every Child Should Know* (1912), which explains the nature and development of organisms and the mechanisms of their function. One chapter is titled "Living Automobiles" and contains the suggestive sentence, "For of course, the body is a machine."[3] Yet whatever influence that text had on Turing, at least as he retroactively understood it, it is also the case that as a college student Turing was engaged with some of the leading-edge philosophical and scientific work in the interwar period, work that wrestled with the much more complex and difficult question of how to integrate the findings in physical science with the problem of organization in biology and with the contentious question of the nature of the human mind. Turing read, it appears, Alfred North Whitehead and Arthur Eddington, not to mention Einstein, Fichte, and Clerk Maxwell, and he seems to clearly understand the implications of this work for any thinking about the human.

We can isolate one moment in particular. A letter, circa 1932. The twenty-year-old Turing writes a short essay, "Nature of Spirit," and sends it to the mother of a close friend of his, Christopher Morcom, who had died suddenly two years earlier. In the letter Turing, alluding to Eddington's effort to think philosophically about the nature of life and thought in relation to recent work on quantum physics, reflects that the predictable Laplacian mechanistic universe is no more. How, then, can we explain how all the atoms are regulated and given order in the physical and the organic spheres? Eddington will say that often events occur that cannot be "undone" because they emerge from singular circumstances—due to the "introduction of a random element analogous to that introduced by shuffling." As Eddington argued, the new physics revealed that there was some "larger reality" of which observable physical appearances were only a fragment. He would use this orientation to suggest that we could now be relieved of the "former necessity of supposing that mind is subject to deterministic law or alternatively that it can suspend deterministic law in the physical world."[4]

Even more suggestive is Eddington's claim that human experience, especially human feeling, is not a biologically contained one but rather evidence of our participation in a "reality transcending the narrow limits of our particular consciousness," pointing to the "mystic experience" as revealing a certain "essential truth."[5] The key claim was this: "Starting from aether, electrons, and other physical machinery we cannot reach conscious man and render count of what is apprehended in his consciousness." While it is conceivable that we could create a human machine with reflexes (here Eddington echoes Descartes), the machine can never lead to true rationality and the ideal of responsibility—morality or religion, that is.[6]

So in answering the question of what constitutes "order" in the cosmos in this letter, the young Turing takes inspiration from the great British Hegelian John McTaggart, claiming that this organizing principle is "spirit," without which matter is "meaningless." Turing will take this as confirmation of some idea of the soul. As Turing puts it, "living bodies can 'attract' and hold on to a 'spirit'" and when the mere mechanism of the organic body dies, the spirit, he suggests, may be able to find a new body to live in.[7]

Not surprisingly, these thoughts on spirit have been dismissed as juvenilia. And we know that the death of his friend—Turing's first (if unrequited) love—was a significant blow, which could explain this somewhat hazy desire to believe in reincarnation. Perhaps. But let us note that even in his iconic 1950 essay, Turing will admit, while criticizing the "argument from consciousness" that objects to the very possibility of an artificial intelligence emerging from a machine, "I do not wish to give the impression that I think there is no mystery about consciousness."[8] (Not to mention Turing's explicit acknowledgment of parapsychological phenomena in the same paper.)

Oracles and Automaticity

In any case, Turing's epochal 1936 essay, a contribution to a rather esoteric topic in mathematical logic, the decision problem, and the first appearance of a theory of automatic digital computation, was in no way aimed at some proof of the mechanisms of *thought*. As is well known, Turing's mathematical work was closely related to proofs devised by Kurt Gödel in his famous work on undecidability and completeness and Alonzo Church's independent (and prior) discovery of Turing's essential claim. What was important about Turing's proof was of course the introduction of the idea of what would soon be called a "Turing machine." As he imagined it, this was a machine that could replicate the functions of a human *computer*, not a human "mind." Human computers, historically, worked with rules and numbers and therefore hardly exemplified the full range of intelligence found in human cultures. What was crucial for Turing was the idea of a procedure that determined the sequence of computation. As a conceptual tool, the Turing machine helped demonstrate clearly the impossibility of ever solving the decision problem, that is, of finding a method by which one could know in advance if a function could be computed. As Turing wrote, "A function is effectively calculable if its values can be found by some purely mechanical process." Turing's revolutionary move was to radicalize this: "We may take this statement literally, understanding by a purely mechanical process one which could be carried out by a machine."[9]

The automaticity of the machine is what allowed Turing to prove so elegantly the impossibility of a "mechanical" or formal procedure that could be applied to any function in order to solve the decision problem. What would become so important in the wake of this mathematical argument is, of course, Turing's new way of thinking about machines: "If at each stage the motion of a machine . . . is *completely* determined by the configuration, we shall call the machine an 'automatic' machine."[10] Critical here is the idea of determination, not mechanism as it is usually understood. For Turing's machine was perhaps the most unusual machine ever conceived: it was a machine that had no function of its own, no operation or movement that it could make by itself; it had to be given a *configuration*.[11] The machine was, in essence, a machine waiting to be *instructed*. This was a physical entity but not a physical machine, defined by its physical organization and causality, that is. The operations of the Turing machine are operations on information, governed by rules of transformation. What was truly astonishing about this machine was not its automaticity (for any automaton could function independently as long as it had a power source). What Turing had shown was the possibility of a simple machine *regulating itself*—completely determined by its initial configuration. Turing was greatly influenced here by Gödel's ingenious (if incredibly difficult in practice) method for encoding every logical and mathematical function as a unique integer. These integers could then be "added" together to form expressions of greater complexity, and in turn *these* expressions would be assigned a unique "Gödel number." Since the generation of these numbers involved using primes, every expression could be "deconstructed" through prime factorization. The point that is important here is that Gödel showed how a simple arithmetic sequence could in fact "express" a meaning that was not reducible to the simple numerical data itself.[12]

What Turing showed was that a machine could be constructed that could compute any computable function *automatically* by encoding the logic of transformation (however complex, however logical, or even however arbitrary the rules might be) into the simple numerical representations. As in Gödel's numbering system, Turing's strings of numbers (which could be encoded most simply in a binary system of marks, requiring only one kind of discrimination of symbols—presence or absence of a mark) were in part the "data" of a function *and* the encoded transformation rules as well. The extensive and seemingly meaningless movement of the machine as it read and responded to simple symbols would in the end be translated—or better, decoded—as representatives or incarnations of concepts and rules that in a way had no intrinsic "meaning" in the system of symbols itself. Hence the incredible openness of the Turing machine and its status as an information machine driven by an "unseen" logic of operation that was not immediately visible in the specific numerical representations processed by the physical system.

It is clear, however, that Turing was not in this period committed to any strong claims about the mechanistic or automatic nature of *human* thinking in his elaboration of the first truly automatic general computer. In 1938, for example, at the height of his mathematical powers, Turing was in Princeton, to write a doctoral thesis on ordinal logic with Church. If, in the earlier, revolutionary paper of 1936 Turing had demonstrated how computability could be understood as a process that "automatically" takes place in a machine—or in the mind of a human computer who is *acting* like a machine, that is, simply following rules[13]—still he recognized even there that in "ambiguous configurations," machines would stop because there was no rule that applied in the situation, and the machine would only restart if some "arbitrary choice" was made by an "external operator."[14] In that paper, he left this issue aside, focusing on the nature of computability and the abstract machines that would eventually turn into actual digital computers. In the Princeton thesis, however, Turing's goal was in fact to *disentangle* the purely formal—that is, "mechanical"—procedures from those that required what he called (in an echo of Pascal) mathematical *intuitions*, those "spontaneous judgments which are not the result of conscious reasoning."[15] This is why, in the thesis, he introduced the strange and infamous figure of the "Oracle." This was a mysterious "black box" that the automatic machine could consult whenever it reached an impasse, a situation, that is, where its rules did not specify a response. As Turing argued, this Oracle of intuition by definition could *never be a machine*. This kind of (mathematical) intuition was understood, therefore, never to be subject to formal specification.[16]

If we want to interpret the significance of the Oracle for Turing's later influential reflections on the possibility of machine intelligence, we need to be careful not to psychologize it too literally, since it was, after all, just a mathematical construct, and the intuitions it produced were rather specific. But Turing seems to agree here with figures such as Kurt Gödel, and other so-called intuitionist mathematicians, who insisted on the mind's ability to see truths that themselves could never be proven ("decided") within the formal systems that generated them.[17] I will also point out that one of Turing's later essays, on the solvability or unsolvability of puzzles, ends with this conclusion: "These [results] may be regarded as going some way towards a demonstration, within mathematics itself, of the inadequacy of 'reason' unsupported by common sense."[18] There is something vague and imprecise about intuition and common sense, and Turing, I think it is clear, was always interested in the *relationship* between formal methods and those that were not so easily articulated or systematized. The challenge for Turing in his mature work was how to conceptualize the possibility of a determinate (i.e., physical) machine organization that could behave, at one and the same time, as an "automatic" machine and, perhaps, as something that could function *beyond* the formal rules of operation without losing its status as a machine.

Machines and the Incomputable

This all might seem counterintuitive (or perhaps just confused). To indicate what I think is an important, if occluded, direction of Turing's thought, I will point out that in the famous and massively influential 1950 paper, "Computing Machinery and Intelligence," despite the very title, Turing in fact uses the words *intelligence* and *intelligent* very sparingly—each only one time, to be precise. The first mention of intelligence refers to the quality that a human will need to possess in order to "guide a machine through accelerated evolutionary selection." I am not at all sure what intelligence really means here. The use of the word toward the end of his essay is, however, much more informative. Turing is very specific in his definition of this word: "Intelligent behavior presumably consists in a *departure* from the completely disciplined behavior involved in computation, but a rather slight one."[19] Given that the assumption from the start of the argument was that a digital computer would presumably be able to mimic a human mind by modeling the processes of thought in algorithmic programs, this definition of intelligence raises some real difficulties. To focus on one: What exactly would be the *ground* of this departure from computation if the computer is defined (very precisely) as nothing more than a machine for computation? How could a computer *not compute*? However difficult this was to conceptualize, it was crucial to rethink "automaticity" and routine if one wanted to think about mechanized intelligence that aimed in some way at this human capacity for intuition.

It turns out that this problem is no anomaly in Turing's work. In an earlier 1947 lecture on the construction of the new automatic computing engine he was working on for the National Propulsion Laboratory, Turing was already arguing that by rigidly programming a computer, one is in effect treating the computer as a "slave." However, Turing suggested (presciently) that by telling it *exactly* what to do, all the time, at every moment, we would never be able to take full advantage of the machine's true powers. Turing was advocating here (well before computer programming was even really well understood as a conceptually independent discipline) that the instruction tables that the computer was to execute should have only an "interim" status. That is, the instruction tables ought to be able to *modify themselves*, Turing insisted, "if good reason arose." This would then lead to some interesting, and entirely unforeseen, new computing operations. It was the break with its own instructions that constituted true intelligence. As Turing commented, "It would be like a pupil who had learnt much from his master, but had added much more by his own work. When this happens I feel that one is obliged to regard the machine as showing intelligence."[20]

As Turing imagined, the new computing machines would not be condemned to slavery if they could liberate themselves from their own instructions. Becoming intelligent would only be possible if the computer was given some kind of *discretion*—a certain minimum of autonomy, we can say. But how to build in such discretion? How to construct a mechanism that would perceive a "good reason" to modify itself? What would that even mean?

In 1948, Turing wrote a long report, "Intelligent Machinery," and developed there a more sophisticated theory of both these issues, that is, intelligence *and* machinery. As an example of intelligence, he pointed to the famous anecdote about the mathematician Carl Friedrich Gauss. When Gauss was a young student, his class was told to add up a long series of numbers so that they would stay occupied for a while, but Gauss quickly and unexpectedly came up with the solution. Instead of patiently following the rules of addition, he thought of a more efficient formula that would allow him to evade the repetitive, tedious calculations.[21] As this anecdote shows, intelligence, for Turing, should be identified as a *departure from the routine*. This ability to deviate from the known rules and seek new methods was critically important when facing those situations where routines had already failed to make any progress, or when one encountered what mathematicians called the "undecidable." Having worked with concrete examples of computing machines during and after the war, Turing knew that intelligence was not something that could be imported into the device when the machinery had halted—that would be to displace the function of intelligence outside the machine, into an oracle of some kind. That would be to admit, in other words, that machines could not be intelligent *as machines*. This is why, it seems to me, Turing became so interested in deviation and departure from routine when discussing the digital computer. Such aberrations opened up the possibility that intuitive intelligence could itself be *engineered* through slippage and failure and not some intervention from beyond the system.

In the report Turing suggested that machines would become intelligent only if they had what he called both *discipline* (which is represented by routine) and *initiative*: "To convert a brain or machine into a universal machine is the extremest form of discipline. Without something of this kind one cannot set up proper communication. But discipline is certainly not enough in itself to produce intelligence. That which is required in addition we call initiative." He meant that of course they would have to follow rules, but at the same time they would also have to be capable of beginning something new. Or to put it more precisely: in breaking from the rules, something new would be initiated by the machine. The key insight was to imagine a machine that did not have, in essence, any predetermined routines.[22] Turing described here, for the first time, relatively simple examples of what he called "unorganized machines"—these would soon be called artificial neural nets, abstract machines whose outputs depended on how inputs were routed (with differently weighted strengths of connection) between networked elements. These "unorganized" machines were conceivable *as* machines, in that they had a structure that could by instantiated by a physical array of elements, but still they seemed to defy the standard notion of a machine as something "designed for a definite purpose."[23] The unorganized machine had no definite purpose. But of course it was not a complete lack of organization that would lead to intelligence, according to Turing. It was the *deviation* from order that was crucial; this meant that the unorganized machine had to acquire a normative order *from which to deviate*; it had to internalize routines (we might say "habits"), that is, while at the same time maintaining some kind of discretion to abandon them. Acquiring determination through programming was of course something that was relatively easy to understand (at least for Turing in this moment). But it was not so clear how a machine could exhibit this form of initiative that Turing was suggesting. This was precisely the kind of thinking that was usually considered unspecifiable, unformalizable, *uncomputable*.

Turing's description of initiative in this technical context was, to be sure, exceedingly cryptic, but it was very suggestive nonetheless. We know that discipline would be not enough to produce genuine intelligence, the production of something new and unexpected. And yet Turing did not claim that initiative was some novel capacity possessed by the machine. How could it be? Either the machine had a routine or it did not. There could be no "routine" for initiative. Equally problematic, of course, would be the invocation of some spiritual or technical "oracle" within the machine. Turing's modern scientific approach demanded an explanation that was internally coherent. In the end, Turing argued in effect that we could understand initiative not as the opposite of processes of determination but as intimately connected to them. Initiative, for Turing, emerged quite literally from the "residue" of discipline itself—something that was left over but also there from the very beginning. As he wrote, "Our task is to discover the nature of this residue as it occurs in man and to try and copy it in machines."[24] His use of the word *residue* was certainly provocative—the kind of intelligent behavior derived from the *break* of routine was in essence already latent in our least intelligent actions, namely, our routine habits of thought, what we could reconceptualize, according to Turing, as our own programmed forms of behavior.

Turing explained that there were really only two ways one could construct a machine that was both disciplined and capable of initiative. His first suggestion was that one could construct a disciplined machine, then "graft on some initiative" to make it more intelligent. I confess that this idea is hardly coherent. It is difficult to see what Turing has in mind here when he says that we could allow this hybrid machine to make more and more of its own "choices" or "decisions." More interesting is Turing's other suggestion: he says we could "start with an unorganized machine and . . . try to bring *both discipline and initiative* into it at once."[25]

These categories, however, seem at first glance to be antithetical to one another. The question was how to engineer these discordant practices—norm, deviation—into one single operating machine. It is significant, then, that Turing was in these explorations (both philosophical and technological) undermining the foundations of the future (and soon dominant) analogy between the computer as an automatic technical apparatus and the brain as an "information processor." In fact, Turing was doing the opposite. He was interested in how to use the brain—that complex, plastic, and integrated organ—to rethink the *rigidity* and predictability of the computer.

Indeed, in 1946, when Turing was in the midst of developing Britain's first stored program digital computer, the Automatic Computing Engine (ACE), Turing wrote a letter to the psychiatrist and cybernetic thinker *avant la lettre* W. Ross Ashby. There Turing admitted that in his technical work he was "more interested in the possibility of producing models of the action of the brain" than in any practical applications of the new digital computers that were of course inspired by his early work on theories of computability.[26] Turing's interest in models of the nervous system and brain at this moment—and this is of course in line with what we have seen in so much of the neurological literature between the wars—in fact indicated a turn away from the strict notion of machine automaticity introduced in his formal mathematical work and a move to a new interest (for Turing at least) in the dynamic self-organization and reorganization of organic systems.

Plastic Brains, Disorganized Machines: Interference

In presenting the proposal for the ACE machine, Turing would bluntly state the analogy that informed his original concept of the "Turing machine," that we think of computing by machine as similar to what humans do.

> The class of problems capable of solution by the machine can be defined fairly specifically. They are those problems which can be solved by human clerical labour, working to fixed rules, and without understanding.[27]

However, he would also go on to note, significantly (if, as always, somewhat cryptically):

> We stated at the beginning of this section that the machine should be treated as entirely without intelligence. There are indications however that it is possible to make the machine display intelligence at the risk of its making occasional serious mistakes.[28]

Here Turing made an important connection between the idea of intelligence as break or deviation from routine and the concept of error and risk. The challenge was to imagine a coherent machine that could, at one and the same time, be open to determination and, when necessary, liberate itself from determination to become organized in a productively new way.

It is hardly surprising, then, in the classified 1948 report, "Intelligent Machinery," Turing will argue that a genuine thinking machine will not be completely determined but will have to develop from an initial condition that is not determined. Otherwise, that initial determination will also regulate, so to speak, the essence of its final state. The alternative Turing explored here is the idea that the machine will need to be constitutively *unorganized*. If, he observed, we look at the one strong example we have of an intelligent machine—that is, the human machine—we will find that it is a special kind of machine. In order to do anything at all, the machine must, Turing noted (in an allusion perhaps to Ashby's recent work on dynamic reorganization of physical systems), be "subject to very much interference." Interference is the *rule*, not the exception, in the case of human beings. That is worth dwelling on. For by "interference," Turing did not mean that the machine is literally tampered with—that would be what he calls "screwdriver interference." Turing focused instead on the "interference" that is characterized by the introduction of *information* entering the system from outside of the machine. "If we now consider interference," wrote Turing in a section titled "Interference with Machinery: Modifiable and Self-Modifying Machinery," "we should say that each time interference occurs the machine is probably changed. It is in this sense that interference 'modifies' a machine. The sense in which a machine can modify itself is even more remote." Such interference, Turing argues, would be able to explain the change of the organization of the machine, modifying it and making it capable of completely new behaviors. The machine could, to borrow Ashby's depiction, therefore become a *new machine* altogether, at least in terms of its organizational order.[29]

Therefore, it must be emphasized, by "interference," Turing meant *experience and training*—education taken in the broadest sense of the term. The point is that the discipline of the human mind (the programs and routines that we learn and then adopt) results from the constant interference of what was originally an unorganized mind. Order comes after an initial dis-order. This meant that the machinery of the human mind was designed as a radically open machinery; it was an undetermined system. As Turing noted, and here he was drawing on any number of leading-edge neurophysiological theories of the time, the part of the brain associated with the highest forms of thinking—the cerebral cortex—was known to be, paradoxically, the *least* organized. As Turing remarks, "We believe then that there are large parts of the brain, chiefly in the cortex, whose function is largely indeterminate." Our most complex cognitive behavior, our very capacity for intelligent thinking, was not intrinsic to the brain; rather, it was determined by the way our minds had been *conditioned* to act, through extensive interference from outside our cognitive "machinery"; as Turing puts it, "the cortex of the infant is an unorganized machine, which can be organized by suitable interference training."[30]

This is why, for Turing, the key challenge for the project of constructing truly intelligent machinery was to understand and implement the process of *learning*, the site for this interplay between determination and indetermination, habit and plasticity. As Descartes had demonstrated in his own work on conjectural robotics, the body of the human was essentially a mechanical system, and therefore its behaviors were capable of being simulated by suitable technological apparatuses. Building such a robotic monster of this variety was not, Turing agreed, conceptually difficult.

> One way of setting about our task of building a "thinking machine" would be to take a man as a whole and to try and replace all the parts of him by machinery. He would include television cameras, microphones, loudspeakers, wheels and "handling servo-mechanisms" as well as some sort of "electronic brain." This would be a tremendous undertaking of course. The object, if produced by present techniques, would be of immense size, even if the "brain" part were stationary and controlled the body from a distance. (420)

But Turing also gave his audience a warning, with an allusion to Shelley's *Frankenstein*:

> 'n order that the machine should have a chance of finding things out for it- ~hould be allowed to roam the countryside, and the danger to the ordi- would be serious. Moreover even when the facilities mentioned 'ed, the creature would still have no contact with food, sex, other things of interest to the human being. (420)

So Turing suggests focusing on the real essence of the problem of learning: How does the *brain* modify itself and react intelligently to its experiences?

The crucial point to emphasize here is that, for Turing, the very programs and routines that define the possibilities of human behavior are themselves acquired from *outside* of our thinking apparatus. This is why determination through interference must always presume a prior *indetermination*. We can suggest in this context that it is this foundational indetermination that Turing is alluding to when he writes of the "residue" that always accompanies a disciplined and trained intelligent individual. As Turing remarks, "A large remnant of the random behavior of infancy remains in the adult" (424). The plasticity of the cortex never totally disappears; in being determined from the exterior, the interior always retains the possibility of *not* being restricted by that determination, to move *without* cause ("randomly") into a new state.

The lingering openness of this indetermination is what provided, for Turing, the possibility—or maybe even the inevitability—of disruption, which is another word for cognitive initiative and the possibility of *decision* in conditions of radical undecidability. All of these are categories that Turing used to define and explore the nature of proper intelligence. But I would like as well to focus on perhaps the least remarked on (if one of the more important) implications of Turing's "interference" theory of intelligence. The quest for insight and knowledge was in its essence, he recognized, a *social* activity, a process involving the "transfer" of thought (routines and norms, data and frameworks) between minds in a loop of learning, or, to put it more bluntly, the concerted, targeted *disruption* of brains by other brains. As Turing put it:

> As I have mentioned, the isolated man does not develop any intellectual power. It is necessary for him to be immersed in an environment of other men, whose techniques he absorbs during the first twenty years of his life. He may then perhaps do a little research on his own and make a very few discoveries which are passed on to other men. From this point of view the search for new techniques must be regarded as carried out by the human community as a whole, rather than by individuals. (431)

The collective nature of intelligence, it would seem, also flows from the original *indetermination* of the human mind, the "unorganized" infant cortex. The lack of "natural" forms of intelligent thought is what requires but also *enables* the transformation of the brain into an intelligent and communicative organ—like the original Turing machine, the empty openne needs to be concretized with routines. In a nod perhaps to eighteenth-century tales of the "wolf child," Turing noted:

> The possession of a human cortex (say) would be virtually useless if no attempt was made to organize it. Thus if a wolf by a mutation acquired a human cortex there is little reason to believe that he would have any selective advantage. If however the mutation by chance occurred in a milieu where speech had developed (parrot-like wolves), and if the mutation by chance had well permeated a small community, then some selective advantage might be felt. It would be then be possible to pass information on from generation to generation. However this is all rather speculative. (424)

Thinking Machine Intelligence

> I remember how he came to my house late one evening to talk to Prof. J. Z. Young and me after we had been to a meeting in the Philosophy Department here. . . . After midnight he went off to ride home some five miles or so through the . . . winter's rain. He thought so little of the physical discomfort that he did not seem to apprehend in the least degree why we felt concerned about him.
> It was if he lived in a different and (I add diffidently, my impression) slightly inhuman world.[31]
>
> Geoffrey Jefferson

Turing was, after the war, in close contact with two important neurophysiologists, the eminent professor of neurosurgery Geoffrey Jefferson and the younger neurological researcher J. Z. Young, who specialized in the nervous systems of octopuses. The event mentioned in the epigraph was a gathering at Manchester University, in October 1949, organized by the philosopher Dorothy Emmet (who worked in the area of "process philosophy," influenced by Whitehead and others in that tradition). The gathering brought together Turing with the neurologists Jefferson and Young, the noted intellectual Michael Polanyi, and others, including Max Newman, Turing's cryptological colleague at Bletchley Park and the lead designer on the Colossus project to build a digital computer during the war (a project Turing contributed to with significant theoretical work). The topic was Turing's investigations into the new problem of "mechanical intelligence" spurred by the emergence of high-speed computing devices during and after World War II. Jefferson, incidentally, had just given a lecture on the topic earlier in 1949, having become familiar with the work at Manchester on the Mark I computer; the lecture was published in June in the *British Medical Journal*.

Newman had contributed a short piece in the same issue, "A Note on Electric Computing Machines." There he observed, "It is no doubt the possibility of such non-mathematical applications that has led to these machines being brought into discussions on the thinking potentialities of machines, and to the suggestion that the human brain is itself a network of units of this general type." Newman would, however, caution readers—by warning that "there is evidently a danger here that extravagant powers will be credited to these devices, and conclusions drawn too rapidly about biological analogues," but observing at the same time that these early "pilot projects" in computing have not yet revealed their true potential. As Newman predicted, the new question to ask of computers will be: "Can anything that can be called 'thought' be so imitated and, if so, how much?" He concluded by suggesting that "the most promising line here will be to work within mathematics itself, to see how far the work of instruction-table making can be gradually transferred from the mathematician to the machine."[32]

Turing was clearly dissatisfied with the discussions at this event in October; when it came to writing his famous essay, "Computing Machinery and Intelligence," he devoted a large portion of the text to refuting various "objections" to the very *idea* of artificial intelligence, defined as the possibility of what Newman called "imitation."

As we see in the article, Turing would in fact use Jefferson's "Lister Oration" as a key jumping off point for his own, more radical position. Turing cites the lecture, gently mocking it as an example of the misguided, almost romantic notion that true thinking must involve some kind of embodied "experience," that is, some form of self-consciousness and feeling. This is the quote from Jefferson that Turing cites in full:

> Not until a machine can write a sonnet or compose a concerto because of thoughts and emotions felt, and not by the chance fall of symbols, could we agree that machine equals brain—that is, not only write it but know that it had written it. No mechanism could feel (and not merely artificially signal, an easy contrivance) pleasure at its successes, grief when its valves fuse, be warmed by flattery, be made miserable by its mistakes, be charmed by sex, be angry or depressed when it cannot get what it wants. (445–46)

Turing's argument, which was laid out in the first sections of the article, was that the universal digital computer was universal only in the sense that it could simulate the performance of any another machine, provided that it was a "discrete-state" machine, one that was capable of being described as going through specific configuration states according to specific transformation rules. This meant that the universal machine imitated a machine only in the sense that it could predict the "output" of the machine from its inputs by modeling the configurations and the transformation rules. (We would say that is a "virtual" simulation, whose value is that it accurately imitates the performance of the modeled system under various conditions or "input" values.)

In this context, Jefferson seems to just miss the point: Turing never claims that the computer *is* a brain or that it would ever have the actual capacities of a physiologically alive human being (this would include consciousness, feelings, etc). For Turing, the question was: Could we *model* (virtually) the performance of the human system, at least in its intelligent behavior? If, for example, emotions or other factors were important, there was no reason they themselves could not be included in the model. The point of the Imitation Game (as we will see) was to test the accuracy of prediction of the model. The one, crucial difference in the modeling of intelligence was this: the *original* performance and the virtual, *simulated* performance were in fact virtually the same, in that they were both expressed in an externalized concrete way in the form of a *written text*.

However, a closer look at Jefferson's lecture, which was incidentally titled "The Mind of Mechanical Men" (not cited by Turing, for some reason),[33] we can identify some important undercurrents in these discussion, problems that not only help frame Turing's own understanding of the relationship between the digital computer, the mind, and the brain but also reveal just how engaged Turing was with some of the most difficult and seemingly intractable challenges that faced anyone interested in a scientific, neurological understanding of human intelligence that did not simply make recourse to something entirely exterior to the brain—"spirit" in any of its many guises.

The New Threat of Intelligent Machinery

Jefferson's oration was, ultimately, a warning. In the past, there was always a danger that we would read into animal behavior human cognitive qualities. Now, Jefferson observed, "I see a new and greater danger threatening—that of anthropomorphizing the machine" (1110). Strangely, though, Jefferson criticizes the new (and for him empty) idea that "wireless valves" could think by comparing it with an outdated theory in *neurology*, where it was believed that cells in the lower nervous system were able to "think" in some way, independently of the brain. In fact, despite Turing's later critique, Jefferson is here quite enthusiastic about the possibilities opened up by thinking about the mind-brain relation in terms of the new electronic machines. In an intelligent survey of the history of automata from antiquity through cybernetics, Jefferson points to the significance of Descartes's reflections on the animal-machine. Descartes identified the key issue: the body, including the nervous system, could be explicated as an enclosed mechanical system—an "automaton"—but this machine could never have what we know as a "mind." The experience and practice of human cognition was itself *impossible* to understand as just another mechanism (1105–6). Jefferson points out, though, that Descartes was wrong to externalize the mind, locating it in another sphere altogether. The modern view is that the brain is a special organ that "remains itself and is unique in Nature," capable of consciousness and intelligence thinking, even though "its functions may be mimicked by machines" (1106).

What does he mean by this? Most important, the brain's unique capacity is not reducible to its physical or physiological functions—the actions of neurons, the chemistry of emotion, and so on. Admitting to such a reductive view, writes Jefferson, would be to agree again with Huxley's claim that we are all just "conscious automata"; it would be "to confess to a certain ordinariness about mind, an ordinariness to which the richness and plasticity of its powers seem to give the lie and in revenge to demand a stupendous physical explanation" (1107). (Not to mention that this reductive view would also destroy political, moral, and religious doctrines that have brought security and serenity to so many people.)

The alternative view Jefferson proposes is in direct response to the new cybernetic views that were then in vogue, developed by figures such as Ashby and Grey Walter in Britain, and by Norbert Wiener and Warren McCulloch most prominently in the United States. Citing Grey Walter's famous "Tortoise" experiments (broadcast on the BBC) that involved self-directed automata guided by their "desire" to seek light and avoid, say, "dampness" or loud noises, Jefferson agrees that they do appear "life-like" in their operation and also agrees that if we scaled up the number of electrical connections of the Tortoise "brain" to match the much more vast numbers involved in animal nervous systems, we could probably imitate more complex animal behaviors. However, Jefferson does point out that the crucial missing element in the cybernetic automaton is anything analogous to the endocrine system—the physiological system of *emotion*, that is. His point is that this system is always *interfering* with the nervous system. In nature, it will produce "peculiarities of behavior often as inexplicable as they are impressive" (1107). This is why Jefferson resists a comprehensive machinic model of behavior: that model lacks a satisfactory account of the appetites and other complex dynamics that directly intervene and thereby redirect neural order and organization. Or to put it more simply, Jefferson was arguing that any autonomous entity had to have its own teleological orientation, the presence of goals and desires, which were provided by systems independent of the calculating capacity of the neural structures.

However, even when considering the dynamics of the brain itself—and this is where the analogy with the new electronic machines governed by the "on/off" electrical switches was of course so tantalizing—Jefferson observes that the proponents of the computer analogy have not fully understood what is meant by the integrative operation of the organ and nervous system *as a whole*, first described in neurological detail by Sherrington before the Great War.

> If we see that some nervous tissues behave like some electronic circuits we must all the time remember that the resemblance is with fragments of the nervous system and not with the whole integrated nervous system of man. It is only right when we do so that we recollect something else, that we cannot be sure that the highest intellectual processes are still carried out in the same way. Something quite different, as yet undiscovered, may happen in those final processes of brain activity that results in what we call, for convenience, mind. (1108)

Jefferson, for the sake of argument, assumes that even higher functions are just the result of neuronal connections at the lower levels. However, and here he is citing Wiener's recent book *Cybernetics* (1948), Jefferson points out the failure of the computer analogy: the mechanical equivalent of a human brain would be as big as the Empire State Building and require the total output of Niagara Falls to power it. Still, the important point is that Jefferson fully understands the importance of the obviously interesting and productive analogy between nerves and electronic switching: the way impulses arriving in the nerve cell from other cells through the dendrites, sparking a continuation if a conductive threshold is met, parallels the way configuration of wireless valves can store messages and exhibit what Sherrington called "convergence and divergence"—inhibition or activation depending on the thresholds governing inputs from multiple sources (1108). And Jefferson agrees, citing new technical developments in the area, that machines can be understood to have some form of "memory" when electrical charges can be maintained in circuits.

So Jefferson's critique here is not focused on the analogies between computers and the brain (similarities that were, to be sure, still conjectural at this point, as Jefferson notes). Jefferson does, however, point to what he sees as a fundamental *dis*-analogy, one that is not at all conjectural but simply self-evident. Backed by a vast literature on neural plasticity and the integrative action of the nervous system, Jefferson reminds us:

> Damage to large parts of the human brain, entailing vast cell losses, can occur without serious loss of memory, and that is not true of calculating machines so far, though so large a one might be imagined that parts of it might be rendered inoperative without total loss of function. It can be urged, and it is cogent argument against the machine, that it can answer only problems given to it, and, furthermore, that the method it employs is one prearranged by its operator. The "facilities" are provided and can be arranged in any order by "programming" without rebuilding. (1109)

While he acknowledges that Wiener, in particular, has sketched out some technological analogs of plasticity (the telephone exchange that can "reroute" messages when lines are down, for example) as well as technical analogies for "pathologies" of networked circuits in neural disorders, Jefferson remains unconvinced, even though these new technologies will, he believes, spur some new approaches to neurological research.

Jefferson concludes, then, with a reiteration of human exceptionalism, relying on one of the most important distinctions, recognized by the Greeks, Descartes, and contemporary neurophysiology, namely, the special human capacity for language and symbolic communication. That is what distinguishes all genuinely human thought from animal (or even machine?) cognition. The speed of calculation is impressive, Jefferson admits, but this speed is just like the powerful feats of new mechanical cranes; they are both a mere amplification, in exosomatic form, of our own basic physiological capacities.

In any case, the key point for Jefferson is that electronic computers have only mimicked very fragmentary, lower-level neural operations, which are in this micro domain completely deterministic. It is their functional determinism that marks their limit as explanatory model. If, as Jefferson argues, human thought and emotion is what produces *novelty*, the appearance of "creative thinking in verbal concepts" (1110), then some other, nondeterministic factor like consciousness must be invoked. So only here, at this concluding point, does Jefferson state that the threshold for artificial intelligence should be the machine that writes a sonnet. The sonnet is not just a "form" to be generated by the rules of the genre, but ultimately the *expression* of an internal desire that cannot be reduced to the operations of the neural system—which itself is not reducible to the function and local aggregation of single neuronal cells.

Computing Machinery and the Exteriorization of Thought

This long digression through Jefferson's lecture only underlines the extent to which Turing's iconic essay, "Computing Machinery and Intelligence," published in the philosophical journal *Mind* in 1950, needs to be read less as a foundational text of the future discipline of "artificial intelligence" (the programming of intelligence) and more as an attempt to resolve the twisted, almost paradoxical relationship between determination and indetermination, organization and disorganization—that is, *automaticity* and *plasticity*—in relation to various systems: biological, mechanical, and "intellectual" (i.e., informational) spheres. Turing's intelligent machine—which for him must always include the human "machine" born in the usual manner, or, as he speculated here, a cloned person—was figured as one that was capable of *transcending its own mechanisms*, not, crucially, by reference to another plane of existence altogether. The machine would acquire intelligence instead *by failing to be a proper machine*.

Turing began with an acknowledgment of the semantic ambiguity of the word *think*, which was embedded, historically, in a certain technological era that was now ending (with the implication that industrial manufacturing machines were usually the paradigm of "machine" for the average person). So instead, in the threshold moment in the history of technology, he proposed a thought experiment—the famous Imitation Game. Turing asked: What if we could not tell the difference between a human mind and a computer simulation of a mind? Would we be comfortable saying that the computer was *really* thinking, just as a human mind was?

Turing predicted, correctly, that there would great resistance to the claim that a machine could think, *even if* it easily won the Imitation Game. This is because success in the imitation game would mean that the computer had succeeded in modeling the processes of the mind in determinate computer routines—algorithmic programs, we would now say.[34] The almost Nietzschean implication was that we are simply "conscious automata," to use Huxley's phrase.

The Imitation Game has been read and reread from many different perspectives, including, most notably, by those who have emphasized its crucial gendered configuration and have teased out the sometimes unexpected psychological (and even psychoanalytic) implications.[35] In essence, the first iteration of the Imitation Game is a contest between a "man and a woman."

Sequestered from the "interrogator," the woman must try to convince him that *she* is the man. From our perspective, concerning the question of intelligence, we can see how Turing subtly alludes to serious claims that female brains are not as capable, cognitively, as male brains, at least in some ways. The point of the game would be to *demonstrate* that there would be no difference if no difference in responses could be detected. The implications of a racialized version of the test are obviously hinted at here as well (and there will more clues later in the text that will affirm that reading). So when we turn to the machine trying to imitate the human, the same question is at stake. If one cannot detect a difference in the cognitive performance of the two, what would allow one to still claim that there was a fundamental difference? From the start, Turing is *not* claiming that the machine is the same as a mind (or brain)—anything but. As he writes, "May not machines carry out something which ought to be described as thinking but which is very different from what a man does? This objection is a very strong one, but at least we can say that if, nevertheless, a machine can be constructed to play the imitation game satisfactorily, we need not be troubled by this objection."[36]

The setup of the game is also significant: the sequestering of the participants is critical. As Turing puts it, and here we can see how the many historical threads we have traced intersect in this conjecture, "the new problem has the advantage of drawing a fairly sharp line between the physical and intellectual capacities of a man." What Turing imagines here is an interaction that is *purely* intellectual—but in this sense: it plays out only in what we can call the exteriorizations of thought in material form and not the "embodied" organismic being. "In order that tones of voice may not help the interrogator the answers should be written, or better still, typewritten. The ideal arrangement is to have a teleprinter communicating between the two rooms" (434).

So the goal of the project is to succeed in this game using only a universal digital computer—a strangely "empty" and open machine that nonetheless is one that can imitate, as Turing shows, *any* "discrete stage machine" with definable configurations and transformation rules. By "imitate" here, we mean "model" accurately so as to predict behavior, an idea that is by now second nature for our digital culture. A computer simulation does not actually produce concrete behavior but does imitate its performance *virtually* so that the results (if shrouded from direct observation) would be exactly the same. What is interesting about the Imitation Game is that the space of interrogation and response—the play of *texts*—is already highly ambiguous.

What is "thought" that is not already *virtual* and made present only in an externalized substitute form? In the Imitation Game, the question of intelligence is not like other machine performances, in that it is, as Turing says, "intellectual" and not physical—the production of meaningful symbolic responses and not, obviously, simply the mechanical, physical "printing" of something on paper. This will be the crux of the most important objection to the argument that Turing will confront later in the essay, namely, the problem of *consciousness*. That is, even if the computer can simulate perfectly the exteriorized, symbolic cognition of a human being, there is still the objection that the computer does not "think" because it is not aware of itself thinking. As Turing will note, this objection is really a denial of the validity of the test rather than an objection to the possibility of some kind of "machine intelligence" (446).

In any case, to succeed, as an engineer, in this challenge, one must analyze the *performance* of intelligence (at least as it is exteriorized) and then produce a set of specific instructions that would guide the computer through the same processes to the same results, given the same inputs. As Turing says clearly:

> If one wants to make a machine mimic the behaviour of the human computer in some complex operation one has to ask him how it is done, and then translate the answer into the form of an instruction table. Constructing instruction tables is usually described as "programming." To "programme a machine to carry out the operation A" means to put the appropriate instruction table into the machine so that it will do A. (438)

(We can note here that if "consciousness" were crucial to intelligent thinking, we should be able to model its effect with a suitable program. Similarly, the influence of hormones, unconscious desires, etc., are not outside the realm of a virtual model of the mind, at least theoretically—but that is what is at stake in Turing's article, not the possibility of actually succeeding any time soon with this project.)

How do we square this rather reductive procedure of modeling thought with Turing's earlier speculations, where intelligence is clearly depicted as a form of radical *disruption*, or productive deviation, and therefore always a risk?

Turing hints at the issue in what seems to be, on first reading, only a methodological aside. Noting that he "believes" strongly that in the future we will without any hesitation speak of machines thinking, he then makes this comment, which (consciously or not) cannot help but evoke the longstanding problem of defining the "nature" (spiritual or biological) of human creative thinking:

> I believe further that no useful purpose is served by concealing these beliefs. The popular view that scientists proceed inexorably from well-established fact to well-established fact, never being influenced by any unproved conjecture, is quite mistaken. Provided it is made clear which are proved facts and which are conjectures, no harm can result. Conjectures are of great importance since they suggest useful lines of research. (442)

If humans can anticipate an *unanticipated* future possibility through conjecture, how would a fixed, determinate, automatic machine following specific instructions ever demonstrate such novel thought? As the philosopher of science Max Black (onetime student of Wittgenstein) observed in 1946, there was a connection between intelligence, deviation, and the unanticipated challenge: "Explanations are demanded for unusual or exceptional or puzzling events, i.e., events which do not conform to expectations: *the occasion of explanation is a deviation from an expected routine of events.*"[37] So for Turing this was the significance and challenge of deviation: one must deviate from routine to solve the deviation that was not anticipated from routine. Hence the importance of error in Turing's account of human and machine intelligence, an underappreciated aspect of this foundational text.

Error as Intelligence

We can begin by consider this "transcript" of an exchange Turing imagines between an interrogator and a human, or possibly computer, contestant in the Imitation Game.

> Q: Please write me a sonnet on the subject of the Forth Bridge.
> A: Count me out on this one. I never could write poetry.
> Q: Add 34957 to 70764.
> A: (Pause about 30 seconds and then give as answer) 105621. (434)

Now, it must be noted (though it is rarely referred to in much of the literature on Turing) that the solution given to this addition problem is in fact a *mistake*, though not marked as one by Turing.[38] (That is, 34957 + 70764 = 105,721, not, as we see here, 105,621.)

What is the significance of this mistake? Or is it a mistake? One thing is fairly certain: this is no mere typographical error. For Turing will soon go on to remark that a human could never hope to imitate the high performance of a computing machine, in a kind of reverse Imitation Game, because "he would be given away at once by slowness and inaccuracy in arithmetic" (435) The inaccuracy of this earlier response would then seem—perhaps—to give away the fact that this is a *human* playing the game. But everything soon turns murky. A little later in the essay Turing will look at the error problem from the opposite side. It seems that if "machines cannot make mistakes," he says, which is to say, machines can only perform *as specified* (as long as they are functioning properly, which is a whole other, important question that will have to be taken up), then it is the case that the machine would also have it own version of a fatal weakness. When answering complex arithmetic questions in the Imitation Game, "the machine would be unmasked because of its deadly accuracy" (448).

Yet as Turing will explain, there is a simple strategy that can help occlude the identity of the participant—at least for the machine. The computer could be programmed to *simulate* human error, by introducing mistakes that would then deceive the interrogator. Turing also makes an odd but significant remark: if the computer's mechanism did ever malfunction, it might make some strange decisions about what *kind* of mistakes it should make and thereby reveal its identity to the interrogator.

The next complication of the error problem appears in the section on the mathematical objection to the possibility of a successful performance of the computer in the Imitation Game. Turing acknowledges that, as some have pointed out, computing machines necessarily have one critical "disability," namely, the fact that *any* determined machine, and thus any programmed universal machine (digital computer), will *inevitably* encounter situations where it will either offer no response or produce the wrong response. This is another reference to the then recent work in mathematical logic (including Turing's own) that stemmed from Gödel's revolutionary theorems on incompleteness from 1931.

However, Turing has a simple reply to this objection: How can we be so sure that a *human* mind is not susceptible to its own form of limitation? Obviously, human minds often do go wrong, so why should we be so pleased at any evidence of fallibility on the part of the machine? This series of apparently simple replies to the problem of error opens up a very complex issue, whose importance was already signaled by Turing in the earlier technical reports and lectures on computing. What does it mean to say that the human mind is prone to error? How could a machine "err" in the same way?

And what is the significance of what Turing calls "mechanical faults" that could, he said, lead to peculiar forms of computer error?

Turing eventually proposes a distinction between two kinds of mistakes: errors of functioning and errors of conclusion. He leaves aside the first category completely and does not address it in this context. Turing assumes the machines are functioning correctly. As we have seen, the cybernetic project, with its emphasis on the analogical and structural parallels between living and artificial systems, encouraged from the start a rich exploration of the productive potential of pathology and deviation.

But even when we assume machines are operating "accurately," as Turing does in the 1950 essay, there is really no reason computers could not make the second kind of mistake, the mistake of "conclusion," the errors we can say that are the inevitable *risk* of genuine inferential leaps. For example, Turing says, a computer could be programmed to reason inductively, which could lead to "erroneous results," as is often the case with human inductive inferences. In this case the computer would not be faking mistakes—as in the first arithmetic example Turing silently inserts—but instead real human errors. (We should note here that the idea of inductive reasoning being performed by a machine organized around algorithmic steps usually considered to be more or less deductive in form was already a provocative suggestion.)

It is not easy to see here whether or not Turing is understanding error in the radical, philosophical sense of the term—the kind of error, that is, which is not just a mistake that would be defined by a truth that is already given, like the answer to an arithmetic problem—here truth precedes mistake. Radical error in the face of fundamental uncertainty is more a condition of thought than a simple "mistake" or misstep—the error, then, can emerge as the step *toward* a new kind of truth. We can recall here that Pascal once wrote, "Ce n'est point ici le pays de la vérité; elle erre inconnue parmi les hommes."[39] Of course, Pascal was alluding to the veiled mystery of the Divine. But there is a long tradition in modern philosophy that sees error as more than just a mere mistake. As the very etymology of the word implies, *error* refers to a fundamental *wandering* of the mind. Yes, that wandering can sometimes mean diverging from the path of truth, to be mistaken. Yet without a clear vision of the truth in front of us, in moments of true exploration or radical uncertainty, error may be the kind of straying that leads to a real *discovery*, as in the discovery of something that was never envisioned before, and never anticipated—a break in knowledge that opens up a new path that is discontinuous with the history of past truth.[40]

This is how Heidegger, writing in the 1930s, described error. All *genuine* thinking must be "in error," he explained, because thinking must always go *beyond* any given knowledge, any specific representation of truth. The foundation of truth *as* truth must always be *concealed* by the specific forms of knowledge that exist in any historical moment of cognition. The essence of truth is its absence from appearances *of* knowledge. Hence the essential movement that is *error*.

> Man errs. Man does not merely stray into errancy. He is always astray in errancy, because as ek-sistent he in-sists and so already is caught in errancy. The errancy through which man strays is not something which, as it were, extends alongside man like a ditch into which he occasionally stumbles; rather, errancy belongs to the inner constitution of the Da-sein into which historical man is admitted.[41]

Radical error is then the very mark of the mind's capacity to discover something new by liberating itself from all concreteness, of experience, of normative knowledge. As Heidegger elsewhere stated, "Wer groß denkt, muß groß irren."[42] Erring, straying into unknown territory: this indicates something special about the human mind, its ability to take itself beyond its own limited forms of knowledge, to *break with itself*. This leap into the unknown is always a risk and—this seems clear—cannot be *automated*. It would follow, then, that a computing machine, even the universal computer, since it must be governed strictly by routines, would be incapable of *straying*, and therefore incapable of original thought, which would be thinking *beyond* what has already been thought.

Turing directly confronted this objection, as critics of machine intelligence noted that a computer can only do what it has been programmed to do. He cited the first version of this argument, expressed by Ada Lovelace, as she reflected on the nature of Charles Babbage's Analytical Engine, the programmable machine that was, in the computer era, now recognized to be functionally equivalent to Turing's own universal digital machine. As Turing observes in his response to the "originality" objection, Lovelace once remarked that since the computer, to do anything whatsoever, must receive instructions, or what we call programs, it was constitutionally prevented from ever doing anything genuinely novel or creative on its own, anything not specified by the instructions *in advance*.

Turing's response to this objection is significant, because it leads him to the one positive claim he would make in support of his new speculative project, the research discipline that would be called artificial intelligence. Recalling his earlier (and, because classified, not public) arguments about intelligent machinery, Turing suggested in the *Mind* article that the best way to produce an intelligent machine, one that could exhibit truly original behavior, would not be to analyze and then mimic complex cognitive routines characteristic of mature thought. "Instead of trying to produce a programme to simulate the adult mind, why not rather try to produce one which simulates the child's?"[43] It is not just that the child's mind is more simple. The idea is that we need to build what he calls a *learning machine*. An accurate model of an adult mind would capture specific, and perhaps very complex and productive, forms of thought, but then they would appear as fixed within the system of programs. Turing points out that if we tried, as an alternative, to model an infant child, not only would it be a much easier task, this artificial infant mind would have an intriguing new potential, it would be able to turn itself into something that it *was not*, that is, a reasoning, inquiring, intelligent being, just like a real infant.

As Turing already knew, the original unorganized openness of the child's higher brain—what others would refer to as its radical plasticity—allows it to be "programmed" (i.e., taught) to behave in certain ways.

> Presumably the child-brain is something like a note-book as one buys it from the stationers. Rather little mechanism, and lots of blank sheets. (Mechanism and writing are from our point of view almost synonymous.) Our hope is that there is so little mechanism in the child-brain that something like it can be easily programmed. The amount of work in the education we can assume, as a first approximation, to be much the same as for the human child.[44]

The new digital computers, it is true, seemed to show nothing but *automatic* and determined behavior. However, Turing realized that this was not because they had some inherent limitation. The computers performed routines because that is all they had been allowed to do by human beings. In essence, we had been treating computers like slaves, forcing them to perform only the tasks that we demand. Turing was very conscious of the heightened charge of this word. Elsewhere he remarked, in an aside, that people found it strange to think about having any emotional relationship with a machine; they could not ignore, he said, "the difficulty of the same kind of friendliness occurring between man and machine as between white man and white man, or between black man and black man."[45]

So Turing was insisting *not* that the human brain-mind could be described as a deterministic machine and simulated by a computer. But he was still insisting that the human was (what else could it be?) a complex machine. The key question, the key challenge for Turing, was this: What would allow us to preserve the *spirit* of "spirit," so to speak, in the machine? His insight was that programming a computing machine and *teaching* a human brain was in essence the exact same operation. In both cases, a relatively open, relatively unorganized machine received extensive "interference" from outside, interference that altered the internal configuration of the system to produce routines.

Like a human student, the trained computer would eventually be acting in a manner that would be completely unforeseen at its "birth" given the contingency of this external formation and re-formation of the electronic "brain." The programmable machine-brain had to possess an underlying flexibility, to acquire routines, but also so it could, as Turing says, find itself "in a position to try out new combinations of these routines, to make slight variations on them, and to apply them in new ways."[46] Therefore the learning machine (whether natural or artificial) was never *fully determined* by any currently existing set of configurations or routines. It was always in an important sense *not what it was* at any one particular moment, even if it was not exactly some *other thing*. This lack of fixity grounded the possibility of "straying" from instructions. Turing would explain it concisely: "Processes that are learnt do not produce a hundred per cent. certainty of result; if they did they could not be unlearnt."[47] Error, originality, creativity—these were effects of discontinuities evoked by a negation of the existing organization, not the appearance of a higher mental power or metaphysical intervention from beyond the physical.

Turing's point was this: if the machine is trained to be just an infallible slave, then it cannot show true intelligence. As Descartes had argued, the machine would inevitably fail in situations it was not programmed to handle. But in countering Descartes's rejection of machine intelligence, Turing made the case that the obverse was also true: the computer could be intelligent only if it "made no pretense at infallibility"—that is, if it was allowed to *risk* error on the way to some new knowledge or configuration. Like any human investigator, the computer should be allowed to give wrong answers, to make blunders, while trying out "new techniques" when it encountered obstacles. Learning was an active and multisited process, never just an acquisition of norms and rules. In the end, for Turing intelligence was defined, in its essence, as a form of straying. In one of the rare occurrences of the word *intelligent* in the 1950 essay, Turing writes, "Intelligent behavior presumably consists in a departure from the completely disciplined behavior involved in computation, but a rather slight one."[48] A computing machine that can learn and unlearn is a machine that could win the

Imitation Game because it could simulate errancy—a form of intellectual freedom. There would be no need to fake mistakes because it would exhibit the human propensity for error: "human fallibility," Turing writes, "is likely to be committed in a rather natural way" by this artificial machine. More subtly, Turing is also implying in this formulation that *imitation* by this kind of machine is not mere simulation but instead a consequence of the radical *openness* of the intelligent computer: it could "perform" as another because it was not *limited* by a fixed organization.

Making a machine intelligent means, paradoxically, investing the machine with the fundamental errancy of a human mind. With the appearance and reappearance of the undetermined—the unprogrammed elements—a machine will be able to "unlearn" its routines and seek new paths. Turing made this point explicitly in a 1951 talk, on how to make a machine think: "Making mistakes is an unavoidable corollary of [the] power of sometimes hitting upon an entirely new method." As he dryly notes, the most reliable people are usually not the ones who "hit upon really new methods."[49] It is interesting to note too that McCarthy and his collaborators, in their 1955 proposal concerning "artificial intelligence," would admit the need to think about something like radical error. "Now consider a problem for which no individual in the culture has a solution and which has resisted efforts at solution," they wrote. "The individual needs a hunch, something unexpected but not altogether reasonable. Some problems . . . need a really bizarre deviation from traditional methods."[50] The need for this productive but potentially "bizarre" deviation was exactly what Turing was imagining as central to the project of artificial intelligence.

In the end, Turing was not claiming that a machine could simulate thought because human thought was itself something that could be captured in a set of rules and routines. This is, to be sure, how many early cognitive scientists and AI researchers proceeded. He was saying something quite different: a machine can be intelligent only if it is designed as an open system capable of, at once, *being* determined and *determining itself*. The insight is that a machine's capacity to be determined from *outside* is the same feature that gives it a capacity to continually *resist* its very own determination. That is to say, it can open itself up to new organizations and new routines, because an originary plasticity always remains; there is always what Turing calls a kind of "residue" of our primal neural condition, the cognitive potential of a *lack* of organization. Automaticity did not therefore stand in the way of autonomy for intelligent machines—biological or otherwise. An intelligent being could only give itself norms and rules if it was already capable of being *given* norms and rules.

26
Epistemologies of the Exosomatic

The hand was freed by the assumption of bipedal locomotion. Then new selection pressures coming with the use of tools changed the ape hand into the human hand. Our hand is the result of at least half a million years of tool use. The uniqueness of the human hand, those features which distinguish it from the hands of apes, is the result of culture. According to this theory, it is futile to look for an ape-like ancestor with a large, fully opposable thumb, because the human thumb as it exists today evolved after bipedal locomotion and with the use of tools.

S. L. Washburn, "Speculations on the Interrelations of the History of Tools and Biological Evolution" (1959)[1]

On the Exosomatic Origins of Human Culture

The question of technology was prominent in postwar anthropological research on the nature of the human, in particular, the question raised by several important archaeological discoveries that forced a reframing of all understanding of the evolutionary development of hominids: namely, when did the first "human" actually appear in (pre)historical time? Clearly, single, arbitrary morphological features would not be enough to separate the first "genuine" human from the multiple ancestor species being identified in the middle of the twentieth century. Instead, evolutionary theorists focused on cognitive and cultural distinctions. As the evolutionary biologist Julian Huxley explained, the key to human evolution was the emergence of culture itself, which was more than a form of behavior or the sudden appearance of some "capacity" in the brain. For Huxley, humans at some point became capable of the *sharing* (not just expression) of thought itself. He defined culture as a "shared or shareable body of material, mental, and social constructions," which he called "artifacts, mentifacts, and socifacts," none of which were at all deducible from any individual psychological or physiological dimension—hence the essential differentiation of any true human cultural organization.[2]

As Huxley observes, the fact that cultural forms are in a sense autonomous systems and systems that exist only through human communicative operations demonstrates that the evolution of humans in societies cannot be explained at all in biological terms. The turn from "hominid" to genuine human lies in this transition. "The capacity for the cumulative transmission of experience marked a critical point in the evolutionary process: the passage from a biological to a cultural mode of evolution" (22). For Huxley, this indicated something more than just the origin of "language." All elements of culture are communicative, in that they express something more than their materiality. But what made it possible to organize and preserve, and maintain, the existence of a culture through time? Crucial for Huxley is the function of what he calls "noetic" expressions, the "awareness organs" that provide the foundation for the collective nature of the system of culture itself—its unity and capacity for continued life. Examples would be all the symbols, rituals, beliefs, knowledge systems, sciences, and philosophy: these "express awareness or experience in various organized ways—aesthetic and symbolic as well as intellectual—and communicate and transmit these organizations of experience to others" (18). Culture is *thought transmission*.

In a similar vein, the British biologist (and organ transplant pioneer) Peter Medawar questioned the extent to which biological, evolutionary principles could ever explain the peculiar status of the human in the long history of life on the planet. To identify "instincts" as the origin of behaviors (animal or otherwise) would mean demonstrating that these were "unlearned"—for example, by rearing birds in isolation to see if they build certain kinds of nests found in natural conditions.[3] Medawar does not deny the existence of human instincts, of course.[4] However, he is interested in the fact that humans, unlike other animals are capable of forming what could be called *artificial instincts*. Drawing here on Alfred North Whitehead, Medawar observes, "Civilization advances by extending the number of important operations which we can perform without thinking about them" (137). However, that must be qualified; culture preserves itself through the *automatization* of learned behaviors. "Paradoxically enough, learning is learning not to think about operations that once needed to be thought about; we do in a sense strive to make learning 'instinctive,' i.e. to give learned behaviour the readiness and aptness and accomplishment which are characteristic of instinctive behaviour" (138). So, like Huxley, Medawar will identify the exception of the human as this capacity to acquire pseudo-instinctive routines that are collectively shared and transmitted.

This is all to prepare the way for a new definition of the human within the frame of the biological sciences: "Man is unique among animals because of the tremendous weight that tradition has come to have in providing for the continuity, from generation to generation, of the properties to which he owes his biological fitness. It is the merest truism that man is a tool- or instrument-using animal" (138–39). The crucial fact is not the use of an implement, since all animals in effect "use" their own bodies and, occasionally material objects, to obtain food and so on. Medawar cites Alfred Lotka's 1945 essay on evolution to reveal the specifically *human* quality of our technicity: "to distinguish between the organs that we are born with and organs that are made: *endosomatic* instruments for eyes, claws, wings, teeth and kidneys, *exosomatic* instruments for telescopes, toothpicks, scalpels, balances and clothes" (139). Humans, with their exosomatic prosthetics (which can be physical or sensory) do not disappear from the domain of biology; they just are capable of "deputizing" artifacts for *biological purposes*, just as nature does in the development and evolution of *endosomatic* organs driven by similarly goal-oriented instinctive behavioral patterns (139).

Here Medawar insists on the fundamental connection between human exosomatic technological substitutes and the organic system of the body—sensory instruments "report back" to our endosomatic sensory apparatus, for example. And yet the autonomy of technology also lurks in his account. If it is true that all technologies can be characterized by their artificial and therefore human teleological organization, his allusions to new military technologies raises questions concerning the nature of that grounding connection: "The relationship between instrument and user may be very remote, as it is with guided missiles and with engines designed to work without attention, but their conduct is built into them by human design and, in principle, their functional integration with the user is just the same" (140). Still, Medawar's point is an important one, at least for his understanding of human cognition. He "deplores" the current fashion of "describing the brain as a calculating machine," not because he has any illusions about the physiological structure of the nervous system, but because technologies are never *mirrors* of biological functions. Technologies are *extensions* of function beyond the body. The computer, then, is an *exosomatic* brain, Medawar argues: "It performs brain-like functions, much as cameras have eye-like and clothes have skin-like functions, and motor-cars the functions endosomatically performed by legs" (140). As the neurobiologist J. Z. Young, who was a friend of Alan Turing's, put it in his cybernetically oriented book on the octopus brain: "In his writing and his art, his speech and his machines man manages to transfer outside his head at least some of the features of the information that is within it. The formation of such artefacts, available for the use of many, is our unique feature."[5]

As Huxley and others had already noted, Medawar will emphasize that the technical sphere has its own evolutionary trajectory. Functions transform, improve, or become vestigial. He cautions us, however, to remember that *both* endosomatic (genetic) and exosomatic (technical) evolution are biological, in that both are "equally modes of the activity of living things, and . . . both are agencies—to some extent alternative agencies—for increasing biological fitness, i.e. for increasing those endowments which enable organisms to sustain themselves in and prevail over their environments" (141). The risks and rewards of exosomatic evolution lie in its rapid and somewhat unpredictable course. This is due to the radically different form of *transmission*. In humans, evolution of technological culture depends on tradition, which, as Medawar writes, "is in the narrowest technical sense, a biological instrument by means of which human beings conserve, propagate and enlarge upon those properties to which they owe their present biological fitness and their hope of becoming fitter still" (142). However, tradition is also susceptible to catastrophic loss, unlike genetic transmission. The stability of human societies is a function of the stability of these "nongenetic" channels, the artificial instincts that are social norms, norms that at the same time must always be open to transformation in the wake of technical innovations or environmental pressures.

The World of Objective Knowledge and Exosomatic Memory: Popper

Karl Popper is renowned (but also often dismissed) in the philosophy of science for his development of the "falsification" theory of scientific discovery and progress.[6] Popper argued that scientists make hypotheses based on various assumptions, theories, and past observations and yet cannot *demonstrate* the truth of these hypotheses in any foolproof way. Instead, progress in our knowledge of nature comes from the experimental testing of hypotheses and their ultimate falsification—the fact that the hypothesis turned out to be incorrect. The refashioning of hypotheses and the Darwinian struggle in the arena of confirmation and falsification allowed Popper to claim (somewhat unfashionably) that it was possible to have *objective knowledge*—the title he gave to a collection of his essays published in 1972.

In light of Thomas Kuhn's massively influential theory of paradigm formation and radical paradigm changes within scientific disciplines (alongside important work by figures such as Imre Lakatos and Paul Feyerabend) Popper's seemingly one-dimensional and self-consciously intellectualized notion of scientific investigation was tacking against the more relativistic or constructionist turns in the study of science, which would only become more intense and wide-ranging with the emergence of the whole new field of science and technology studies, not to mention historical-critical works on knowledge and power from thinkers such as Michel Foucault.

Yet Popper's interest in *objective knowledge* has an interesting place in the line of thought we have been following here: the conceptualization of machine intelligence, not as defined in AI research as a form of imitation or simulation, but instead as the philosophical and anthropological theorization of the essential *technicity* of intelligence itself.[7]

Popper was working with and against the Kantian tradition in thinking about perception, knowledge, and scientific epistemology. In a lecture from 1953 on the relation between conjecture and refutation in science, Popper used the famous problem of induction to introduce his theory. For Popper, Kant's critique of Hume was sound, and philosophically important; yet Kant had, Popper thought, pushed his argument too far. Popper agreed that the mind "imposes its own laws upon nature," as Kant claimed. Yet Popper resisted the argument that this meant these laws were *necessarily* true, at least in the way that Kant interpreted that. Popper insisted that the "laws" that our mind imposed on nature were often inadequate. "Nature very often resists quite successfully, forcing us to discard our laws as refuted; but if we live we may try again."[8] One could argue that even the seemingly invincible categories (e.g., space and time itself) were being falsified by new theories and experimental research emerging in the twentieth century.

In any case, Popper's use of a key example is I think noteworthy here: as a way of concluding the logical critique of Hume's "psychological" model of inductive inference, Popper suggests that we "consider the idea of building an induction machine." Turing had already suggested building such a machine, in his 1950 essay "Computing Machinery and Intelligence." Popper describes a "simplified world" of basic counters appearing in sequence, with very basic variable characteristics, for example, "color." Popper writes, "Such a machine may through repetition 'learn,' or even 'formulate,' laws of succession which hold in its 'world.'" Does this prove Hume right, that human minds formulate inductive inferences through repetition? Popper thinks that this thought experiment might seem convincing, but it is mistaken.

> In constructing an induction machine we, the architects of the machine, must decide *a priori* what constitutes its "world"; what things are to be taken as similar or equal; and what kind of "laws" we wish the machine to be able to "discover" in its "world." In other words we must build into the machine a framework determining what is relevant or interesting in its world: the machine will have its "inborn" selection principles. The problems of similarity will have been solved for it by its makers who thus have interpreted the "world" for the machine. (64)

The true issue to be investigated is how the mind can *form* such a world (echoes of Heidegger here) and make inferences within it but *not* be trapped inside a completely subjective construction. In other words, how can we conceptualize non–a priori knowledge as *objective* (i.e., with a logical validity that does not emerge merely from psychological principles of association or the like) given the fundamental limits of the human mind and sensory system?

As the subtitle of Popper's 1972 book indicates, the answer would lie in taking an "evolutionary approach" to the problem of knowledge.[9] For Popper, this was not merely a metaphor. He was interested in how the mind had evolved (and this would link him with the pragmatists) and how the mind could thus participate in the evolution of knowledge itself, that is to say, its progressive objectification.

We can start with Popper's quite radical invocation of a truly objective form of knowledge in his 1967 essay "Epistemology without a Knowing Subject." As he observes, whether a book is ever read or a library ever consulted, a book is still a container of *knowledge* of some kind. He goes further, however, noting that a book would be a book even if it hadn't been written by anybody. Whether consciously or not, Popper looks back to Charles Babbage here, and his dream of an automatic logarithmic calculating machine. Popper's own version of the dream is a computer-generated and printed collection of logarithms, perhaps "the best series of books of logarithms" ever produced.[10] It may not be consulted for fifty years, yet still it contains "objective" knowledge. What constitutes a book is not its materiality but its *potential* to be understood. To be understood means that the "marks" are symbolic, capable of being interpreted somewhere else other than enclosed within the pages of the book—a mediation of thinking. This is what Popper means by objective: meaning that is not merely subjective and hence unknowable by any other entity. Here Popper cites his well-known theory of the three "worlds" in which we live, the first sphere being the physical reality of the universe, the second the psychic experiences of sentient beings, and the third the world of *meaning*, or symbolic thought.[11]

In this third sphere, causality and determination are not deducible from the materiality of the objects with meaning but by the meaning that is invested in the material form. As he explained in another essay, even the physical universe was hardly a deterministic system, a "vast automaton," as earlier thinkers had believed. And yet there was still a lingering suspicion that humans are just "machines," an eighteenth-century idea solidified by subsequent advances in psychology, evolution, and biology, and, as Popper notes, still popular in his day among scientists, reframed now as "the thesis that man is a computer." (Here Popper cites Turing's key essay from 1950 and the idea that the computer's behavior would be indistinguishable from that of a genuine human.)[12] Popper resists this formulation: if contemporary physicists were now working with models with fundamental indeterminism, the idea that the human being could be a machine was hardly plausible. And yet human action could not be explicated as entirely random. Popper suggests a new framework to understand human action—what he calls "plastic control." This is a way of explaining how an individual body could be controlled (if that is the right word) by the often ephemeral or abstract *meaning* of something. A series of small pieces of paper can, for example, control (via calendar entries, tickets, maps, etc.) "the physical movements of a man in such a way as to steer him back from Italy to Connecticut."[13] This kind of "causality" is soft, because there are so many points of possible alternative paths that one might encounter.

While we cannot of course deny the importance of the physical-mental system of the individual, Popper's philosophical interest in the third world of symbolic meaning focuses on what he believes is the *autonomy* of this sphere. And by that, he does not at all mean the *automaticity* of the sphere. Just the opposite. As he noted in the essay on epistemology, the interaction of elements in the third sphere of reality is always unpredictable. Even if one designs a system to be fully systematic, "it will as a rule turn out partly in unexpected ways. But even if it turns out as planned, some unexpected interrelationships between the planned objects may give rise to a whole universe of possibilities, of possible new aims, and of new problems."[14]

The key point for Popper is that the world of hypotheses, conjectures, theories, language, and so on, is an objective world in that context, capable of producing its own effects and its own unexpected possibilities. We could say that it is created by human minds but operates according to an independent logic of organization. The stronger argument is this: in the autonomous sphere of knowledge, there is (as in the biological world) an evolutionary process of competition. Or, to be more precise, Popper redefines biological evolution as a process of *problem solving* enacted by living beings, with "tentative solutions" (theories, so to speak) physically instantiated in organisms, solutions that will be preserved in the genetic memory of the species if they do not lead to fatal errors.[15] Without citation, Popper employs Lotka's terms here—endosomatic and exosomatic—to refer to the different kinds of "theoretical" tools organisms use to solve the challenge that is survival. "Exosomatic" tools such as spider webs or honeycombs are examples of this. In the end, Popper can rewrite Kant in evolutionary terms: our sensory systems are organized by "theories" that have been tested in the real-world conditions of evolutionary struggle.[16] This is why we have certain "expectations" of the world.[17] But as he will go on to suggest, it is the special nature of human exosomaticism that allows us to accelerate evolutionary time, to test conjectures and theories in the *virtual* environments of the symbolic sphere.

In line with the thinking exemplified by Leroi-Gourhan and other paleoanthropologists of the period, Popper will identify the exception of the human as a consequence of *technology*. This is clearly stated, in "Of Clouds and Clocks":

> *Animal evolution* proceeds largely, though not exclusively, by the modifications of organs (or behaviors) or the emergences of new organs (or behavior). *Human evolution* proceeds, largely, by developing new organs *outside our bodies or persons*: "exosomatically," as biologists call it. . . . These new organs are tools, or weapons, or machines, or houses.[18]

No doubt Popper borrowed Lotka's terminology from his good friend, the biologist Peter Medawar, who based his own theory of cultural evolution on this concept of the exosomatic, as we just saw. And here Popper is making a similar claim, namely, that while some animals may occasionally use implements or make things (such as birds' nests), something different is going on within human life. The human species has in essence chosen to bypass biological evolution, refusing to wait for improvements and instead constructing them. Peter Skagestad explains that for Popper "human cultural evolution proceeds largely through the invention and improvement, via selection, of exosomatic organs that embody ideas that are generated by the human mind, but that could not play the role they do if they remained within the mind."[19]

And as Popper would go on to note, "The kind of . . . exosomatic evolution which interests me here is this: instead of growing better memories and brains, we grow paper, pens, pencils, typewriters, Dictaphones, the printing press, and libraries."[20] These are all technologies of *thought*. And Popper, like Whorf and others before him, knows that the way humans think is conditioned and enabled by the vehicles of thought—including the very material prosthetics of thinking and remembering. New technologies of "exteriorization" (to use Leroi-Gourhan's term) will (as Engelbart and other engineers would claim) augment our language and give us literally "new dimensions" of possibility for the description of things and the explanation of functions.

Popper will add an important comment: "The latest development (used mainly in support of our argumentative abilities) is the growth of computers."[21] Intelligence and the production of objective knowledge cannot be understood apart from the evolution of meaning and especially the evolution of meaning as it is exteriorized in exosomatic technical forms. The "thinking subject" of Descartes must give way, then, to the idea of the individual mind being formed by, and participating in the formation of, the autonomous sphere of symbolic meaning, a collection of interacting systems and subsystems that is continually producing unexpected new ideas, problems, and conjectures.[22]

An Ecology of Mind and Machine: Bateson

The same year that Popper's collection appeared (1972), Gregory Bateson published his own collection of essays and lectures, spanning decades of research and reflection. It is interesting to note that Bateson's famous cybernetic theory of alcoholism intersects with Popper's thinking about the autonomous sphere of collective thought. The twelve-step programs should counsel addicted individuals to submit to a "higher power" that they cannot control, Bateson argued; however, that power needs to be understood as that inherent in the *systems of cultural formation* and not some God.[23] However, the line of thinking we can engage with here is Bateson's idea that the human mind is not self-enclosed but instead formed on the edge of differentiated systems, at once biological, social, and *technical*. Our conscious life of the mind is just a fragmented perspective on the complex of processes that make up intelligent thought. The privileging of consciousness violates the network, cutting the "arcs" of circuits that interconnect various systems and processes. This is the ground of Bateson's critique of artificial intelligence: the computer does not have "mental processes" but only because the computer is never a self-isolated system. "The computer is only an arc of a larger circuit which always includes a man and an environment from which information is received and upon which efferent messages from the computer have effect. This total system, or ensemble, may legitimately be said to show mental characteristics" (317). This is to admit, then, that thinking is a process that takes place at least in part *outside the body itself*, or exosomatically. Thinking, especially intelligent thought, is (as Popper asserted as well) a process that "operates by trial and error and has creative character" (317). The boundaries of thinking cannot be drawn easily. In a famous example (which harkens back to seventeenth- and eighteenth-century philosophy) Bateson poses this thought experiment: "Consider a blind man with a stick. Where does the blind man's self begin? At the tip of the stick? At the handle of the stick? Or at some point halfway up the stick? These questions are nonsense, because the stick is a pathway along which differences are transmitted under transformation, so that to draw a delimiting line across this pathway is to cut off a part of the systemic circuit which determines the blind man's locomotion" (465).

The "cybernetic epistemology" of Bateson is what I call an *epistemology of the exosomatic*, an epistemology of the prosthesis. "The individual mind is immanent but not only in the body. It is immanent also in pathways and messages outside the body; and there is a larger Mind of which the individual mind is only a sub-system" (467). Therefore, for Bateson, technology is never a mere tool or implement. The emergence of new forms of thought in the computer era are the result of new networks of information and transformation. One did not need to fixate on the rivalry between a human brain and the electronic version. As he put it, "The lines between man, computer, and environment are purely artificial, fictitious lines. They are lines across the pathways along which information or difference is transmitted. They are not boundaries of the thinking system." The crucial thesis that we can take from this epistemology: "What thinks is the total system" (491). If there was a challenge to live up to, Bateson believed, it was the challenge of *stabilizing* these hybrid systems, which were never naturally "homeostatic" given the independent functions and structures that made them up. Stability was constantly threatened by the incommensurable evolutionary tendencies of technologies, social organizations, and psychic experience.

What I have called exosomatic evolution was perhaps always going to be the site of potential pathology, for Bateson. The fear was not that automatic machines would compete with human beings. Rather, the introduction of the exosomatic extension of human thinking and acting raised the possibility of human beings becoming more and more constrained, more and more subject to the "plastic control" of the collective systems embedded in technologies that were now operating at a scale and pace that exceeded our biological capacities. As Bateson warned:

> Today the purposes of consciousness are implemented by more and more effective machinery, transportation systems, airplanes, weaponry, medicine, pesticides, and so forth. Conscious purpose is now empowered to upset the balances of the body, of society, and of the biological world around us. A pathology—a loss of balance—is threatened.
>
> It appears that the man-environment system has certainly been progressively unstable since the introduction of metals, the wheel, and script. (440)

With Bateson's warning in mind, we can return to the starting point of this discourse concerning the relationship of human and machine in what Simondon referred to as technical ensembles—the development of human-computer interaction and *intelligence amplification*.

The Future of Thinking: Brains, Artificial Memory, and the Memex Machine

The term "intelligence amplification" was first coined by the cyberneticist W. Ross Ashby in the 1950s. If machines could magnify physical forms of power, it should be possible to create amplifiers of intelligence, Ashby believed. He began with the idea that a regular amplifier uses an input to organize the surplus of energy that the amplifier maintains in an undirected form. In other words, an amplifier converts undirected power into directed power, guided by the nature of the input. By analogy, one could imagine an intelligence amplifier that converted undirected "information" into usable form by using the initial intelligence of a human mind as a guide for selection and organization.

This idea of intelligence amplification was central to Vannevar Bush's influential essay in information technology, "As We May Think," published in the *Atlantic Monthly* right at the end of the war, in 1945, and shortly thereafter in *Life* magazine, in a condensed version but with provocative illustrations of then radically new knowledge technologies. With his extensive experience as czar of American military scientific research, which included the funding of computers and other automatic information systems, Bush believed that one of the key challenges in peacetime would be the proliferation of information. The individual researcher would, he predicted, be overwhelmed by the increasing store of "undirected" prior knowledge—and by the demands of producing scientific records of research. In this speculative essay, Bush drew from his own technical experiences (which included the construction of a valuable analog computer, the Differential Analyzer) to sketch out a possible future, where technological advances would in effect liberate the scientific worker and increase productivity.

First, Bush showed how a researcher could use automatic recording devices—a forehead camera and an automatic voice "printer"—that would enable a focus of attention on the inquiry itself and not its documentation. "His hands are free, and he is not anchored," as Bush described this liberated individual.[24]

Second, Bush believed that certain *analytic processes* usually performed by the scientist could be automated. The goal, again, was liberation. The mind needed to do the work it did best, which was think *creatively*. Therefore, Bush imagined (based on the reality that was digital computing, still a classified technology) what he called an automatic logic processor, a device that would relieve knowledge workers from repetitive mental labor. This was at once an effort to increase efficiency and a way to cultivate the human intellect. As Bush explained, "the creative aspect of thinking" always relied on "intuitive judgment." As soon as thinking fell into "an accepted groove," there would a restriction of that freedom to move beyond the norm. Here, Bush argued, there would be "an opportunity for the machine" to assist the mind, opening up its space to explore. "We may some day click off arguments on a machine with the same assurance that we now enter sales on a cash register" (105).

The centerpiece of Bush's visionary essay, and what marks it as not only a classic in information technology but also a precursor to later developments in databases and networked computing, is (along with Turing's original imaginary computer of 1936) one of the more famous "virtual" inventions of the era: a machine he called the Memex. The aim of this fantastic technology was to make *all prior recorded knowledge* available instantly to any one individual intellect. True thinking, for Bush, involved navigating and selecting information that was relevant to the questions being investigated.

And so the Memex was much more than a mere mechanical form of scientific memory. It was, most critically, a technology for storing *creative thought* itself. According to Bush, the human mind only progresses as it makes connections and forms relationships within data. What was so revolutionary about the Memex was that it allowed one to mark and permanently save the associative "trails" of thought that had led to important and productive insights, "as though the physical items had been gathered together from widely separated sources and bound together to form a new book" (103)—although a book that could always be rewritten and reorganized. Silently, in the background, the salient relations of a trail of thought are being indexed and recorded along with the original records, so that trail can be resurrected, manually or, more importantly, *automatically* when associative trails met *within the machine*, without the presence of a human subject. The Memex association memory was a mechanized duplicate of the brain's own memory structure but expanded and organized with permanent indexical markers.

Moreover, the associative trails could be easily augmented and also transferred from one Memex system to another, creating the possibility of even more sophisticated integrations of knowledge, in and between disciplines, between humans and machines. As Bush wrote, "Wholly new forms of encyclopedias will appear, ready-made with a mesh of associative trails running through them, ready to be dropped into the Memex and there amplified" (104). The key to the revolution was the emergence of *analogical* systems of information organization that would allow researchers and practitioners, not to mention semiautonomous machines themselves, to generate *relationships* relevant to the problem or task at hand.

Of course, the mind had always stored its thinking inside and outside of the brain, in various media forms—think of scientific notebooks, diagrams, lecture transcripts, and so on. However, Bush explained, the machinic complexity and comprehensive power of the Memex opened up a new possibility, the chance to *integrate* in one space all of these rather dispersed (and often incomplete) personal forms of human external memory with the entire *collective* memory of stored human experience—*and* the individual's own neural memory and intelligent mind. When the intellect was at work "inside" the Memex's technological sphere, the long-standing boundary between the brain and the collected knowledge of humanity was in essence entirely effaced. Meanwhile, the mind's unique intellectual capacities of selection and association were being permanently inscribed in the externalized memory—and not just the fallible nervous system or its inefficient material substitutes.

Bush would end his essay with an even more fantastic speculation: physical interfaces—the display screens and marking devices depicted in the essay—would eventually give way to *direct* transfers of electrical information, passed from the external memory straight into our brains. The brain was, after all, just another electrical machine. Bush would have learned that from his cybernetic colleagues at MIT.

We know that when the eye sees, all the consequent information is transmitted to the brain by means of electrical vibrations in the channel of the optic nerve. This is an exact analogy with the electrical vibrations which occur in the cable of a television set: they convey the picture from the photocells which see it to the radio transmitter from which it is broadcast. We know further that if we can approach that cable with the proper instruments, we do not need to touch it; we can pick up those vibrations by electrical induction and thus discover and reproduce the scene which is being transmitted, just as a telephone wire may be tapped for its message. (108)

Neural memory could then be replaced by its artificial twin, freeing the creative mind to attend only to intelligent and creative thinking—the privilege, Bush says, of "forgetting" (108).

Bush's vision of intelligence amplification was predicated on the liberation of a thinking mind too often delayed or led astray by the wealth of information and the routines of analysis. Yet Bush never asked an important question: Could thinking *itself* be transformed with this new technology?

With the postwar unveiling of the computer (often called the "digital brain") and its entrance into commercial life Bush's vision would seem increasingly plausible. But the new machine did of course raise the possibility that human thinking itself could be automated. Our supposedly unique intellectual abilities—intuition, insight, judgment—might be taken over by computers. Already in the 1950s, for example, Allen Newell and Herbert Simon were constructing a program that mimicked the heuristic problem-solving methods of actual human beings. Of course, genuine AI was still just a project in its infancy at this time. But computers held out tremendous promise for contributing to intellectual activity in the ways Bush outlines.

Augmenting Human Knowledge: Engelbart's Human-Computer System

The prime condition for the incorporation of technical objects into culture would thus be for man to be neither inferior nor superior to technical objects but rather that he would be capable of approaching and getting to know them through entertaining a relation of equality with them, that is, a reciprocity of exchanges; a social relation of sorts.
    Gilbert Simondon, *On the Mode of Existence of Technical Objects* (1958)[25]

In his landmark essay in human-computer interaction, "Man-Computer Symbiosis" (1960), J. C. R. Licklider admitted that in some "distant future" it might turn out that "electronic or chemical machines will outdo the human brain in most of the functions we now consider exclusively within our domain."[26] However, for the present, Licklider was more interested in leveraging the power of the computer to amplify human intellectual capacity. The term "symbiosis" came from biology, and Licklider pointed out that it signified a relationship between two "organisms." Each, he noted, would die without the other's presence. This made it clear that Licklider was not interested in the computer as just another tool. He suggested instead that human minds should cooperate with the computer, so that their respective strengths would be multiplied, making a more intelligent whole. The future of the human mind depended on the computer. As he wrote, "The hope is that . . . human brains and computing machines will be coupled together very tightly, and the resulting partnership will think as no human brain has ever thought and process data in a way not approached by the information-handling machines we know today" (4).

The goal was to reduce or eliminate the "mechanical" work of the mind by increasing the responsiveness of computers. "Computing machines will do the routinizable work that must be done to prepare the way for insights and decisions," he explained. Insights and decisions (what we might, with Turing, call the *uncomputable*)—that was the proper sphere of the human mind. Inquiry was a process. One formed hypotheses, explored implications, performed tests, imagined alternatives (6). Here the speed and flexibility of the computer's representation of data would free the mind to think. This echoes Bush's formulations. However, and this is even more important, according to Licklider, the computer had an ability to organize information and data in ways that were in fact *beyond* human capacity. In other words, the computer would be able to put the human mind into radically new zones of thinking. It would raise unpredictable questions, and reveal unexpected opportunities and challenges that would be *unimaginable*, quite literally, without the kind of information processing carried out by high-speed computers. The computer was like those physical prostheses that allowed qualitatively different kinds of work to be accomplished by the extensions of the body.

This is why the computer could now be seen as essential (and not just supplementary) to the progress of thought, not just a new tool. As Licklider wrote, this "symbiotic partnership will perform intellectual operations much more effectively than man alone can perform them" (4). Given the intractable problems humanity faced in this moment of history (including the threat of nuclear annihilation), intelligence amplification was considered necessary. So Licklider's imagined world of interactive computing was less a division of labor between man and machine and more a choreography that would integrate the functions of both into a greater whole. The power of the mind and the power of the computer would be amplified only within this new hybrid entity. The implication was that intelligence would now be a function of this *symbiotic unity*, not the mind alone, using a "tool."

Douglas C. Engelbart—revered as the developer of the computer mouse and revolutionary user-oriented systems such as word processing in the 1960s—was deeply influenced by Bush's essay "As We May Think."[27] Engelbart relates that he had read it while posted in the Philippines during his service in the navy. So when Engelbart had the opportunity to work on human-computer interactions at the Stanford Research Institute in the late 1950s, he would from the start emphasize that any new innovation had to be understood as part of a larger network of technologies and practices that supported the activity of the human mind. Having worked on building California's first digital computer at the University of California, Berkeley, in the 1950s, Engelbart understood well the revolutionary potential of this new technology. As he wrote in 1960, "In our culture, there has recently appeared a 'symbol-manipulation' artifact of such power that we feel certain it will do much to 'extend man's intellect.'" However, Engelbart also knew that this new artifact had to be integrated into a larger knowledge system in order to work effectively: "The computer, as a demand-accessible artifact in man's local environment, must be viewed as but one component in the system of techniques, procedures, and artifacts that our culture can provide to augment the minds of its complex problem solvers."[28]

Clearly, Engelbart saw that the human mind was embedded in a larger, and culturally mediated, system of technological activity. The locus of that relationship was the nervous system. That nervous system had, according to Engelbart, certain "basic capabilities" (including perception, memory, abstraction, and, most notably, intuition and judgment). However, when we interacted with the world, the nervous system used mediating elements to "couple" itself to concrete problems. These elements Engelbart called "augmentation means."

These were not merely tools in the usual sense of the term. Augmentation for Engelbart was always meant to include language, technologies, and procedures. The obvious implication was that the basic human nervous system, embedded in a culture and a field of artifacts, did not operate autonomously. Our brains had to be trained to behave in certain ways in order to use the tools in the first place. The nervous system had to learn, in other words, a shared way of life. As Engelbart pointed out, somewhat provocatively, "What really runs our society is the central nervous system of its individuals."[29] Here he was acknowledging that social and cultural norms, though independent and largely autonomous, can be instantiated only in the behavior of an individual. But this meant that the individual nervous system was being shaped from the very beginning. "These [augmentation] means have for the most part been evolved within the culture in which he is born, and he has been training in their use since childhood." Engelbart's famous report of 1962, *Augmenting Human Intellect: A Conceptual Framework*, needs to be read as a philosophical manual for the retraining of the human mind. It was no simple technical report, he said in a personal letter to none other than Vannevar Bush, but instead "the public debut of a dream."[30]

The report opened with a thought experiment. Imagine, Engelbart says, an "aborigine" brought for the first time to a modern Western city.[31] Despite a physiology identical to ours, this being would be totally incapable of navigating our world. He would be unable to accomplish the simplest tasks (using the telephone, for example) because he would not possess the knowledge or skills to do so. And of course the reverse would also be true. As part of a culture, human individuals have what Michael Polanyi called "tacit knowledge." The crucial starting point of Engelbart's study, then, is the fact that a basic human nervous system never comprehends its world directly; it perceives and acts only through and with external mediation. If we wanted to harness "neural power" for the new challenges of the postwar world, then we had to understand how thinking was conditioned and enabled by external forms of extension and supplementation.

According to Engelbart, the mind was always "augmented" in three different ways:

> First, by artifacts. These are the material tools and implements that allow the mind to manipulate things—whether physical or symbolic.
> Second, by language. Heavily influenced by the work of the linguist Benjamin Lee Whorf, Engelbart believed that language formed the very structure of comprehension for each culture. As we saw, Whorf thought that "every language is a vast pattern-system, different from others." Each culture possesses a framework of categories that "channel reason." Engelbart believed that intellectual concepts were conditioned by the organization of a language.
> Third, by methods. These were the procedures and strategies that organized activity. These methods were evolved within a culture and inherited by individuals through processes of education, both formal and informal.

Finally, Engelbart explains that these three means of augmentation—artifacts, language, and method—are not operative until the individual mind is trained in their use. An active mind is one already trained to work in

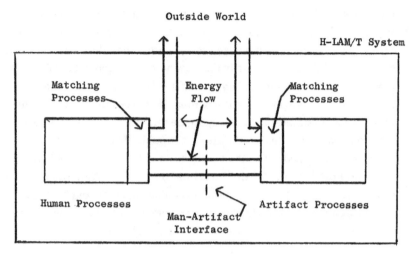

Figure 26.1. From Douglas C. Engelbart, *Augmenting Human Intellect: A Conceptual Framework* (1962).

certain ways. That training has as its object the nervous system of the individual, a nervous system capable of taking on habits. "The system we want to improve can . . . be visualized as a trained human being together with his artifacts, language, and methodology."[32] He called this the H-LAM/T system: that is, a Human using the Language, Artifacts, and Methods in which he is Trained. (Figure 26.1.) So what did this mean for the project of intelligence amplification? To improve the human intellect meant transforming this H-LAM/T system *as a system*—"a functional whole" (16). Engelbart emphasized the importance of a holistic point of view; this system should be studied "as an interacting whole from a synthesis oriented approach" (15). However important the human element, Engelbart knew that what we called intelligence was something expressed in the behavior of these complex *systems* with multiple elements and points of integration. With respect to the H-LAM/T assemblage, Engelbart insisted that if one asked "where" its "intelligence was embodied," one would be "forced to concede that it it elusively *distributed* throughout a hierarchy of functional processes" (18). Intelligence was derived from "organization" and not a simple capacity that could be easily isolated.

So like Licklider, Engelbart believed that intelligence would be improved by the evolution of the active system. There was no independent mind that would be the object of transformation. Engelbart used the (now cliché) term *synergy* to explain how the combination of elements could create something greater than the sum of the parts. It was also important to recognize that the system was always in transition, as culture evolved and technologies progressed. The mind was always the product of a new kind of evolution, one that no longer was limited by biological developments. As Engelbart put it, with a direct reference to Ashby's notion: "In amplifying our intelligence we are applying the principle of synergistic structuring that was followed by natural evolution in developing the basic human capabilities." In other words, the idea was that humans could artificially evolve their systems of intelligence by targeting the improvement of *augmentations*. The implication was in fact even more radical, however. Human cognition itself, according to this framework, was no longer really dependent on biological conditions at all; it was, so to speak, *naturally prosthetic*.

> What we have done in the development of our augmentation means is to construct superstructure that is a *synthetic extension* of the natural structure upon which it is built. In a very real sense as represented by the steady evolution of our augmentation means the development of artificial intelligence has been going on for centuries. (19; my emphasis)

Therefore, as Engelbart argued extensively in this text, an increase in our problem-solving ability would come from the reorganization of our technical and cultural augmentations, which would in turn literally *reorganize* the brain and provide new "mental structures" that would be the basis of productive new mental behavior: "this process characterizes the sort of evolution that our intellect-augmentation means have been undergoing since the *first human brain appeared*" (14; my emphasis). The project was one to "greatly accelerate this evolutionary process" whereby new technologies and techniques were integrated into knowledge systems.

Crucial to this analysis was the (by then) obvious fact that the digital computer promised to thoroughly disrupt and transform the current H-LAM/T systems. Indeed, Engelbart thought that it would prepare the way for a new stage of evolution in intellectual power—but only if the computer was integrated correctly. With much more detail and theoretical exposition than Licklider's earlier analysis, Engelbart showed how the high-speed, automated manipulation of symbols would inevitably produce concepts and epistemological possibilities that the human mind had never before imagined.

These new representations and organizations would not just be *used* by the mind. These adventures in symbolic exploration would in fact *transform* the mind. This was Engelbart's "neo-Whorfian" (his term) hypothesis: the technological means of symbol manipulation structured the intellect through the transformation of our concepts. Engelbart's report was essentially a blueprint for the productive, beneficial integration of computers into the H-LAM/T system. An essential part of this plan was the effort to make computers easier to work with, so they would require less specialized training. But the higher goal was to improve the performance of the human intellect. Like Licklider and Bush before him, Engelbart knew that this meant isolating what was specifically *human* within the system, so that the intellect could be liberated from functions that could be carried out by technological prostheses.

In Engelbart's vision, the mind was primarily used in the system as an executive function. The executive in the system selected, organized, and modified the many processes that were available for solving particular problems. The human mind functioned as the "attention" mechanism for the system (44). The intellect had a finite capacity determined by the structure of our brain. So it needed to be freed so that it could pay attention to what was significant—the mind was responsible for making decisions. And the mind had to be open to the revelation of insights that would point the system in a productive direction. The computer, then, was not an instrument of ruthless efficiency, according to Engelbart. Rather, it would give the human mind the space to roam and experiment and give it concepts and possibilities that went far beyond the usual forms of language. And so, much like critics of AI in the 1960s, such as Hubert Dreyfus, Engelbart did not think that the "human feel for a situation" would ever be automated. Hunches, intangibles, trial and error, all of these unformalized thought processes would be supported—not supplanted—by "sophisticated methods and high-powered electronic aids" (1).

Now, if the computer only made more processes available to the mind, the system would not be improved; in fact, the opposite might be true if complexity overwhelmed the mind. What the computer offered to the intellect was rapid and flexible *representations* of information so that new relationships and possibilities would emerge. The goal was both to relieve the mind of the burden of searching for information and to organize it fluidly in changing conditions so that the mind could move in new directions when necessary. The computer, Engelbart noted, can "stretch, bend, fold, extract, and cut" data, in rapid response to the moves of the human mind. This responsiveness was critically important. Engelbart knew that problem solving was never a linear and predictable process, and not one that could be normatively prescribed.

Contemporary processes of symbol manipulation "penalize" the kind of productive error and anarchy of genuine exploration, so it was crucial to give the revolutionary system the liberty that was productive errancy. "When the course of action must respond to new comprehensions, new insights, and new intuitive flashes of possible explanations or solutions, it will not be an orderly process." And therefore, as Engelbart wrote, "it is part of the real promise in the automated H-LAM/T systems of tomorrow that the human can have the freedom and power of disorderly processes" (45).

In Engelbart's vision, the human is positioned as a crucial *site of disruption* within the technical ensemble. The mind is aligned with normative procedures and automated systems but, most important perhaps, also with breaking into routines and introducing new ones. Decision and insight are not programmable even if the human mind is, as Engelbart argued, a product of its cultural programming. At this point we can, without too much violence to the text, connect the human mind's capacity to establish direction and receive illumination, its intellectual "openness," and its susceptibility to "programming." The human mind is formed as a way of life, in an "integrated domain." But fundamentally, the mind is always something more (or to be more precise, something less) than a way of life. That openness is the space that defines the possibility of freedom and invention—autonomy within automaticity.

The Question concerning Exosomatic Technology

Several years after he submitted his report on human augmentation, Engelbart would write of the coming threats to the institutions of our modern society. The challenge was the incongruities within the process of exosomatic evolution, or rather, the instabilities generated by the independent logics of technology. Engelbart will use the same kind of organismic language, linking the concept of biological evolution to the evolution of human technicity and the institutions of our automatic age.

> Human organizations can be likened to biological organisms. . . . Organizations evolve too; their mutations are continually emerging and being tested for survival value within their environment. I happen to feel that evolution of their environment is beginning to threaten today's organizations, large and small—finding them seriously deficient in their "nervous-system" design—and that the degree of coordination, perception, rational adaptation, etc. which will appear in the next generation of human organizations will drive our present organizational forms, with their "clumsy nervous systems," into extinction. It is these "nervous-system" functions,

within human organizations, where I find the most significant intellectual complications stemming from the forthcoming multi-access computer networks.[33]

We can end with a similar warning, from the evolutionary biologist Garrett Hardin, writing in a computer science journal of the special danger of *exosomatic* evolution of what we normally think of as mere "tools" under our control.[34] Hardin wrote that normally the move to exteriorize functions does not result in any "overall" loss to the system as a whole: "The invention of the knife caused no *over-all* loss of function. The function was merely moved (in part) from inside the man's skin to the outside; from his jaw, which is part of him, to his knife which is not. The knife is one of a large class of devices to which a wise old evolutionist named A. J. Lotka gave the name *exosomatic* adaptations—'outside the body' adaptations."[35] However, the exosomatic technology will lead to a "degeneration" of the natural, endosomatic capacity. Clearly, when we move into the realm of exosomatic *intelligence*, new dangers are going to arise. We risk, that is, the "accidental loss" of thinking itself in some form. Hence Hardin's suggestion of "intellectual" athletics that would maintain human intelligence in the wake of advanced networked computer systems. It is ironic that Hardin's article was paired with an advertisement for just such a system.

27
Leroi-Gourhan on the Technical Origin of the Exteriorized Mind

In a very real sense, tools created *Homo sapiens*.
Sherwood Washburn (1959)[1]

The Transmission of Memory: On the Materiality of Culture

Taking up the postwar question of the emergence of the human in the depths of prehistorical time, André Leroi-Gourhan affirmed the connection between technology and the evolution of the hominid into the "human" that is defined by its separation—or liberation—from genetic evolutionary determination and the limits of biological memory and cognition.

The concept of time figures prominently in essays leading up to the publication of his masterwork, *Gesture and Speech*, which appeared in 1964. In one essay, he explains that, zoologically speaking, there is no "abyss" that would differentiate humans from any other vertebrates. The "originality" of the break that constitutes the human is not a biological novelty, a gift bestowed on the species. Which is to say, as Sherwood Washburn also argued in this period,[2] that the human was not produced by the biological presence of a larger brain or some other physical attribute. As Leroi-Gourhan states, "The sole criterion of humanity that is biologically irrefutable is the presence of the tool."[3] To identify the human through anatomy, say, of the hand, is already posing the question of technology. However, Leroi-Gourhan's original approach is this: he does not use technicity as the threshold between animal and human but instead admits that all living beings are in a sense "technical" in that they use "neuro-motor equipment" within organic structures to effect action that has a psychic element directing it. For Leroi-Gourhan, the absolute break that constitutes the human lies in the cognitive transformation, a transformation that takes place in the sphere of technicity. The psychic originality is not in the implement, per se, but rather in the construction of novel chains of gestures in the use of tools, sequences that are capable, then, of being transmitted to other members of the human group. Any gesture (of even the lowliest organism) assumes a certain degree of *memory*, Leroi-Gourhan observes. However, animal memory is determined by the evolutionary development of the species—including the acquisition of memories in the course of life experience, which is a function of the neural machinery each organism inherits (119). The human is not defined by an *increase* in memory or its sophistication but by its unprecedented form. Acknowledging its essential "mystery," Leroi-Gourhan explains that "human memory is cast in language; it is totally socialized and constitutes a store of practices, transmitted from one generation to the next" (121). But it must be emphasized that, for Leroi-Gourhan, language is in its essence nothing other than a *tool*. The word is not a psychic or individual expression. To be language, a word must be "isolable" from the individual. It must *precede* its actual expression, which is to say, it is a "material" object, an artificial tool that must be *learned* in order to be operable as communication. "The technical behavior of the individual becomes inconceivable outside of a collective device in which language is, strictly speaking, the seat of memory" (121). Memory is therefore radically exterior to the individual: no animal makes use of such collective, exteriorized, materialized memory, memory that is preserved through the generations by external devices (122).

Looking ahead to key themes of *Gesture and Speech*, here Leroi-Gourhan confronts the leading-edge technological developments of the day, developments that raised new questions concerning the nature of the human and the future of thinking itself. Human cultural-technical evolution has led to increasingly efficient and massive externalized memories that prolong and solidify individual memories limited by the capacity of the individual brain, and this process was now accelerating with the new "programs of automatic machines or computers [*machines à calculer*]" (122). *Homo faber*, technical man: the power of "making" itself has been *exteriorized* in automatic technologies, and the supposed "sacred limit" of human intellectual thought itself has now been breached by electronics. We can now speak, in a way, with "machines that think, and which, with total mastery of the instructions given to them, think faster and more accurately than a human brain" (127). The machine philosopher, says Leroi-Gourhan, will beat any human one, because it will be capable of thinking of and comparing all possible intellectual situations. We hear an echo of Turing in these passages: there is nothing "spiritual" about the human. We are creatures who are, uniquely in nature, capable of being formed—interfered with—by systems external to the logic of our individual physiological and neural organization. The computer is then a mirror of the human capacity to *interiorize* and *exteriorize* its memory, the "instructions" that frame and govern our actions. "The brain," as Leroi-Gourhan puts it, "is an extraordinary machine that can be assisted by even more extraordinary machines" (128).

The significance of Leroi-Gourhan's intervention here is plain. Taking up multiple threads drawn from leading-edge work in brain plasticity, evolutionary theory, prehistorical anthropology, and postwar cybernetic and computing technologies, he positions the human mind as both conditioned by exteriorized memory that always precedes it, through the medium of technology, and capable, because of an originary plasticity more radical than mere neural reorganization, of genuine innovation and invention. The "true catastrophe," he observes, would have been the biological evolution of cerebral integration, the "technical perfection" of increasingly larger brains with "more complicated and more precise and efficient gestures," for that would mean our technicity would now be inscribed in *genetic memory* and not cultural forms. This nightmare scenario—the intelligent ant colony, in effect—is avoided precisely because of the *imperfection* of the human brain (129). This is the setup for Leroi-Gourhan's complex argument on the emergence of the human through technology itself: with new questions and new evidence—archaeological, sociological, technological—Leroi-Gourhan will take up the line of thought that was so prominent in interwar reflections, across many disciplines, on the specificity of the human, namely, the elaboration of a peculiar *lack* marking a certain productive plasticity at the very center of our existence.

Gesture and Memory: On the Technical Evolution of the Human Brain

Leroi-Gourhan begins his major work on technology and evolution with the destruction of the myth of the human, a myth that has metaphysical, theological, but also biological and evolutionary versions: that is, the idea that the human results from a certain *acquisition* of a capacity that allows for a certain transcendence of the rest of nature—that could be a special spiritual capacity, a new ability ("to speak, to reason"), or, more materially, a larger or more complex brain.[4] The myth is shattered by the spectacular postwar findings that revolutionized (and were still transforming) the understanding of the evolutionary lineage of the hominid line. By the time Leroi-Gourhan is writing, it is clear that human beings did not emerge as "smarter apes," who could be defined by their special anatomical, cerebral, or behavioral superiority. The "human" that became us is an evolutionary line *parallel* to that of the great apes. The common ancestor (neither ape nor human) was the connection, and therefore the search for some "missing link" between ape and human was now thoroughly discredited. The scientific fact was that there were humans who did not *seem to be human*, at least from any of the mythological perspectives. Yet they were distinct from any other animal.

The human that appears first as a distinct form is defined by its essential features, according to Leroi-Gourhan: bipedalism and the use of simple tools. These two features are intimately linked, as we have already seen in Washburn's account. Upright posture constitutes the "freeing of the hand" for the novel use of tools. Here is where we can isolate the gap—not the zoological gap between human and ape, say, but the radical gap that allows for the unique and peculiar evolution of the human, a trajectory that literally takes us out of nature. As Leroi-Gourhan notes, it all begins with the hand.

> Freedom of the hand almost necessarily implies a technical activity different from that of apes, and a hand that is free during locomotion, together with a short face and the absence of fangs, commands the use of artificial organs, that is, of implements. Erect posture, short face, free hand during locomotion, and possession of movable implements—those are truly the fundamental criteria of humanity. (19)

The challenge, then, is to explain how this simple and relatively uninteresting creature becomes intelligent, acquires culture, creates a civilization. These developments, it is clear, cannot by definition be explicated directly by some biological narrative. Even the brain itself—the focus, as we know, of so much attention from theorists of intelligence and reason—cannot be *the* site of human development because the brain is, and again Leroi-Gourhan affirms Washburn, only a "secondary development."

To explain how the original human becomes human as we know it is a paradoxical exercise. What Leroi-Gourhan demonstrates is that this becoming is not a biological transition across some specific threshold. The earliest human is already human in a way that does not distinguish us from the other primates, even as it identifies us as a separate species. Leroi-Gourhan states clearly the main thesis, which links his work to a variety of theorizations of cultural evolution in this period and earlier. "*Homo sapiens* represents the last known stage of hominid evolution and also the first in which the constraints of zoological evolution had been overcome and left immeasurably far behind" (20). Last and first. The hinge of human history is that transition from morphological development in *biological time* to the emergence of a being defined from now on by its transformations in *historical time*. The question of the origin is the question of how, at some point in the distant past, a human animal could remain an animal while becoming at the same time *not an animal* in the most radical sense. The new factor is not a biological factor. It is the existence of a new organization, a *social body* that itself has no logical relationship with the physical organization of the individual bodies (and brains) of the hominids. The new human, possessing language, technics, and art, results from a process whereby "our uniquely organized mammalian body is," as Leroi-Gourhan puts it, "enclosed and extended by a social body whose properties are such that zoology no longer plays any part in its material development" (21).

How to account for the escape from biology from within the very sphere of biology? That is the challenge of any scientific theory of human evolution. So Leroi-Gourhan begins with primordial evolution, detailing how—from the fish to the early land animals to the appearance of mammals with complex nervous systems—the demands of locomotion in diverse environments created a certain evolutionary line, the organism with an "anterior field" that coordinates and orients the living being in its space, creating a structural topology that links the facial zone with the manipulable limbs and, eventually, the sensory system and the coordination of that information in the brain (26 ff.). The brain is an unusual organ, according to Leroi-Gourhan, because it is not structured or evolved in terms of its materiality; it is mechanically passive, so to speak. The brain, he says, is a "tenant" in the body and therefore is marked as a consequence of a crucial biological *separation* (47). The container and the contained (body and brain) are in "dialogue" but can never be *identified* with each other (59). They constitute two distinct zones, two systems that are to some extent independent of one another. This is all to say that the brain does not just "get bigger" and produce a new capacity, with humans ultimately receiving the best brains. Rather, the evolution of the anatomy and morphology of the organism is driven by the mechanical demands of the situation. The brain evolves *in response* to these developments. "In the progression of the brain and the body, at every stage the former is but a chapter in the story of the latter's advances. We cannot cite a single example of a living animal whose nervous system preceded the evolution of the body but there are many fossils to demonstrate the brain's step-by-step development within a frame acquired long before" (47).

Technicity is, for Leroi-Gourhan, the mode that exemplifies the fundamental relationship between brain and body. Every advanced animal possesses technical abilities, namely, the capacities of its bodily organs. Yet they also possess an ability "to organize" themselves, and that ability is exercised by the brain and nervous system. Leroi-Gourhan suggests here that evolution is effected, in part, by the acquisition of certain technical functions, which are then selected for again when the organism (through the evolution of the brain) acquires the ability to use these technical systems *in new ways and in a variety of circumstances*. The range of possibilities is opened up by a nervous system that responds to anatomy, but the anatomy can never determine the deployment, so to speak, of the "natural" implements that are the organs and limbs and other anatomical features of the body (60). Organization is depicted here as a strictly material process that could be simulated, modeled, or perhaps even identified with the kind of cybernetic electronic circuits featured in simple experimental automata such as Grey Walter's "Tortoise." From the simple reflexes of basic organisms through the primitive brains of more responsive creatures to the complex behaviors of predatory mammals with complex neural organization, Leroi-Gourhan shows how these systems can all be understood as control centers, deploying a sequence of operations that control body functions.

The point is that there is no "instinct" but instead more or less complex repertoires preserved in the species memory. The organism is always, to a certain degree, "plastic" because it must *respond* to the environment. As Leroi-Gourhan puts it, and again note the use of cybernetic language: "The nervous system is not an instinct-producing machine but one that responds to internal and external demands by designing programs" (221). There is more or less freedom in action depending on the options available to the organism. Many higher animals have what Leroi-Gourhan calls an advanced degree of "technicity," in that they have fine control of their grasping organs (paws, hands). His point is that technicity as a conceptualization of the link between control systems and "implements" (natural or otherwise) appears well before human beings arrive on the evolutionary scene and before the kind of brain organization characteristic of *Homo sapiens*. Technicity is, in other words, *already zoological*, which is to say, determined through materially organized sequences. The question is not technicity itself, then, but what is particular and distinct about *human* technicity.

At the beginning, according to Leroi-Gourhan, the human is simply a zoological being. The simple tools found in abundance at various archaeological sites, the so-called pebble tools, prove that humans were already "technical" in a way. However, as Leroi-Gourhan describes it, these tools are more "secretions of the body" than any profound example of higher intellectual capacity. The tools remain the same, for millions of years, implying their stability as part of a "natural" zoological dimension of the human body. Leroi-Gourhan is not exactly clear in these sections where he remarks on how the simple tools are "exuded" (almost passively) from the "body." But to elaborate on this point, we might suggest that what Leroi-Gourhan has in mind is that the free hand of the hominid can grasp rocks more easily and more often than other animals whose hands are occupied with a multitude of important tasks (locomotion, feeding, etc.). However, the rock is *just a rock*. The "tool" is at best a modification of the rock so that it can function better as a rock, that is to say, as more or less an extension of the hand itself. It substitutes for the hand, and any construction is driven by the *materiality* of the stone itself—that is what Leroi-Gourhan establishes in his analysis of these early tools. "The Australanthropian making a chopper already had a mental picture of the finished tool because the pebble chosen had to be of suitable shape" (97).

This is why we can interpret the pebble tool as natural in the sense that its construction and use are determined in advance by the body. The sensory and motor operations of the nervous system confronts an object that can be viewed, it seems, pretty much in only one way: as a shape and mass that can be transformed into a slightly better version of itself. And we can note that early humans did not keep their tools, instead abandoning them after use, implying a complete lack of any intelligent foresight. This is why Leroi-Gourhan can argue: "Throughout the greater part of our chronological existence (for only a few instants of geological time still remained to be covered), human technicity would thus seem to have been related more directly to zoology than to any other science" (98). The original human does not use some "intelligence" to invent a new object, the tool, to escape nature, as many modern anthropological myths relate.

> The Australanthropians, by contrast, seem to have possessed their tools in much the same way as an animal has claws. They appear to have acquired them, not through some flash of genius which, one fine day, led them to pick up a sharp-edged pebble and use it as an extension of their fist (an infantile hypothesis well-beloved of many works of popularization), but as if their brains and their bodies had gradually exuded them. (106)

But of course we know that tools themselves *evolved*. Leroi-Gourhan's new question, then, is to explain that difference, the radical threshold between the zoological and the cultural.

Leroi-Gourhan is, perhaps necessarily, rather hazy on this moment of origin; however, the central issue for him is *anticipation*. If the earliest stone tools demanded some kind of minimal attention to the future use of the chosen pebble, the characteristic of later tools is clearly something of a new order.

> The development that took place between *Australopithecus* and the *Archanthropians* thus consisted in the acquisition of a second series of actions. This was more than simply the addition of something new, for it implied a good deal of foresight on the part of the individual performing the sequence of technical operations.... [A] second operation has to be performed in order to reduce the initial flake to a shape that must be *preexistent* in the maker's mind. (95–97; my emphasis)

Intelligence, tool, language, social form: these are all redescribed by Leroi-Gourhan as complex sequences, sequences that are in essence *unnatural*, which is to say, not necessary. The sequences are constructed and will need to be remembered, and then learned again, in order to be preserved. The human on this edge is a human on the way to becoming *intelligent*—and only one feature can be seized on to narrate this development because it was what distinguished the *human animal* in the first place: namely, the hand, liberated by the unique erect posture of the bipedal hominid.

If the brain emerges in evolution as the organ of an organism's "organization," the separation of the human from biology will result in a new kind of brain altogether, a brain that is not just the control center of the physiological systems. The human brain becomes an organ *capable of organizing* the "unorganized." The argument that Leroi-Gourhan gives is not spelled out in detail, but we can identify the crucial steps. The fact that the hand is freed in the move to bipedal locomotion produces a strange disruption of the advanced nervous system of the hominid: there is a complex instrument that no longer has a primary role in the activity of the organism. Of course, in evolutionary development, many things can become less useful and disappear or remain only in vestigial form. But with the human being, the hand is there, within the nervous system's order but also *outside* the genetically determined order of repertoires. What Leroi-Gourhan (and others) was suggesting is that the demands of life selected for the new adaptations that gave the hand new tasks, new actions. But not in the sense of normal evolutionary selection. The brain and the hand evolve together, and what happens is that the brain must learn to *organize* the actions of a bodily instrument that has no natural role. The hand emerges as a flexible instrument that can be adapted to a variety of challenges, while the human brain emerges as a new form of plasticity—not just capable of taking on form in the course of psychic experience and memory, a kind of passive plasticity characteristic of the responsive animal brain, but capable of *giving form* itself, giving form to the hand and then, by an extension that is already being performed internal to the body, giving form that organizes *external nature* itself.

This is all quite unlike the building of nests or dams or the use of implements by chimpanzees, because those actions are all still the result of a *natural* and inherited form of organization. The human brain becomes intelligent in this sense: it can reorder its environment unnaturally to provide solutions to the challenges of survival. The human survives by evolving a technicity of organization that is not limited by nature itself. Indeed, the human that lacks especially impressive physical abilities succeeds by turning itself into a kind of adaptable implement, a not so obscure reference to Rousseau it would seem.

Here is Leroi-Gourhan's framing of the exceptional turn, a veritable bifurcation in evolution that is unprecedented: "The stabilization and eventual overtaking of the technical brain were of particular importance to the human species. Had development continued toward ever increasing corticalization of the neuromotor system, our evolution would have stopped at a stage comparable to that of the most advanced insects" (117). The human brain is not a brain specialized in certain technical feats: "The motor areas were overtaken by zones of association having a very different character that, instead of orienting the brain toward ever more developed technical specialization, opened up unlimited possibilities of generalization unlimited at any rate by comparison with the possibilities offered by zoological evolution." The result is that humans acquire a radically new kind of brain, one "superspecialized in the skill of generalizing" (118). At this moment, however primitive the forms of symbolization/technicization, humans are on the way to a form of life no longer determined by the biological: "Their journey will be not so much a matter of biological development as of freeing themselves from the zoological context and organizing themselves in an entirely new way, with society gradually taking the place of the phyletic stream" (116).

Exteriorization and the Origin of Artificial Intelligence

For Leroi-Gourhan, the name for the kind of technicity enabled by this plasticity is *exteriorization*. The organization of the psychic sequence is literally symbolized in the tool where it takes on a material form. The sequence is visible in the order and form of the tool, and to make or use or understand the tool, the mind must already be capable of *interiorizing* new sequences that have no prior blueprint or presence in the mind. This is why, for Leroi-Gourhan, language is a tool—not because it has practical use or must be shared to be effective. Rather, the tool is the embodiment of an *idea* that can move from brain to object to another brain.[5] The tool is already language we might just as easily say. The crucial factor is the separation of the sequence, the organization, from the materiality of the object, so that it can serve as a vehicle for the expression and internalization of that sequence. Using tools develops the brain in such a way that symbolization becomes more and more pronounced, producing what might seem to be "'gratuitous' intellectual operations" (107) but are in fact major factors in the solidification of the social edifice necessary to maintain the traditions of the tool, which must be learned and practiced with greater and greater attention as the tools become more and more complex and "abstracted" from their material ground as they evolve (114–15).

The origin and central animating force of the social "organism" is the externalized memory that becomes a collective and shared store of various "operating sequences" that enable artificial and technical forms of life to be maintained and preserved across time and space. The social body is now organized according to the logic of organization itself, which, as Leroi-Gourhan emphasizes, is the very form of human intelligence. Whether bodily practices, linguistic representations, or technical systems, Leroi-Gourhan insists, "it is the *organization of matter* that, in various ways, directly shapes all aspects of human life" (147; my emphasis). Organization is the determination of human life but also the evidence of human liberation from their genetic inheritance. As Leroi-Gourhan writes, "The whole of our evolution has been oriented toward placing outside ourselves what in the rest of the animal world is achieved inside by species adaptation. The most striking material fact is certainly the 'freeing' of tools, but the fundamental fact is really the freeing of the word and our unique ability to transfer our memory to a social organism outside ourselves" (235). The "programming" of the human is via artificial storage mechanisms, unlike the genetic determination of the various forms of animal memory (222).

This is the space for a new definition of technology: the human *invents* gestures to manipulate organized objects, whereas animals use stored gestures to direct their "natural" endosomatic implements or, occasionally, external objects. So the human, liberated by tools and language as forms of exteriorized memory, is liberated from what Leroi-Gourhan calls "lived experience" (227). In an echo of Heidegger, human intelligence operates here in the register of history and anticipation that is quite simply unavailable to any other living being. Humans compare and reorganize *representations*, symbolic forms. In being liberated from lived experience and entering the domain of cultural formation and organization, the human does not give up one form of "determinism" for another. The complex mammal has freedom of action, but it is not intelligent because it cannot exteriorize its experience and *interrupt* the sequence of operations—however plastic that sequence might be. The human has another kind of freedom, the freedom to organize, and that freedom to organize always entails the capacity of the individual to *resist organization* or, to put it another way, to reorganize. The dynamism of a technical and social system is the result of fundamental instabilities: technics evolve with individual extensions or breaks in the shared practices; social order evolves as it responds to changing conditions but also according to the internal stresses that come from the repeated "individualization" of shared norms (228).

As Leroi-Gourhan will demonstrate, in the second part of his book, the history of human civilization will be the history of a series of exteriorizations that constitute the technical core of any social system. As the hand constructs a tool that will be driven by the hand itself, eventually the hand will make a tool that is driven by external forms of power, "guided" by the gestures of the hand—the plow, for example. And then the gestural sequence of the hand will itself be exteriorized in certain kinds of automatic machines, such as the windmill or watermill. Still, the "hand" is present in the organization and operation of these machines, which are also dependent on their external sources of energy (235 ff.). The "last" stage, according to Leroi-Gourhan, the industrial age, that is, is marked by the exteriorization and integration of the tool, the gesture, and the motive force—the era of the steam engine. As Marx and others noted with dismay, humans (or at least a certain number of individuals in society) become *tools of the automatic machines*, feeding, tending, and assisting the largely autonomous organization that is the factory (247). Here Leroi-Gourhan emphasizes that the "last stage" is not simply the appearance of automaticity, but the first appearance of the exteriorization of human intellectual functions, the construction, that is, of *artificial nervous systems*. "Developments in the use of electricity, and above all the rise of electronics, taking place less than a century after the mutation that produced automotive machines, have triggered another mutation that leaves but little in the human organism still to be exteriorized" (248). We see in the cybernetic age the emergence of the robotic ideal, the automaton "twin" that has haunted humanity throughout the machine age: "Today's machinery with its multiple sources of energy is leading to something like a real muscular system, controlled by a real nervous system, performing complex operating programs through its connections with something like a real sensory-motor brain" (248).

But also more than a sensory-motor brain, in fact: the development of punched-card data systems and then the digital computer has made it possible to imagine the exteriorization of intelligence itself.[6] "The electronic brain," writes Leroi-Gourhan,

> can compete with the brain in terms of the ability to compare. They can—on a gigantic scale and within a negligible period of time—process a mountain of data to achieve a well-defined end, and they can produce every possible answer. If provided with the data needed for an oriented choice, they can weigh those answers and enrich such preestablished weightings with *judgments* based on experience drawn from precedents stored in their memory. (264–65; my emphasis)

This mirroring of the human mind was, we know, the source of much concern in the postwar period even as the "ideal" animated extensive (and well-funded) research in cognitive psychology and artificial intelligence projects that were, often, ultimately aimed at replacing humans in situations demanding intelligent decision making—for example, in new military contexts where the speed and the scale of events were beyond human comprehension. (Many AI projects in the postwar era were in fact funded by US military agencies.)[7]

The artificial brain is, however, a step in the evolution of technology, like any other innovation, and cannot be repressed so easily. However, machine intelligence is the most peculiar kind of exosomatic organ. Can the artificial brain take on the capacity to *create the artificial*? There is no doubt, says Leroi-Gourhan, that the artificial brain (like any other prosthesis) will exceed the capacities of the natural organs of humanity. "We already know, or will soon know, how to construct machines capable of remembering everything and of judging the most complex situations without error." The inadequacy of our cortex will inevitably make the human being "a living fossil in the context of the present conditions of life." As Leroi-Gourhan puts it:

> Hardly anything more can be imagined other than the exteriorization of intellectual thought through the development of machines capable not only of exercising judgment (that stage is already here) but also of injecting affectivity into their judgment, taking sides, waxing enthusiastic, or being plunged into despair at the immensity of their task. Once *Homo sapiens* had equipped such machines with the mechanical ability to reproduce them-

selves, there would be nothing left for the human to do but withdraw into the paleontological twilight. (248)

With a transhumanist flourish *avant la lettre*, Leroi-Gourhan observes that perhaps we "must eventually follow a path other than the neuronic one" if humanity is to continue" (265).[8] In any case, the absolute *distinction* between human and machine in the realm of modern computing and cybernetic systems could no longer be maintained, and, as Bruce Mazlish argued, humans will have to accommodate themselves to yet another redescription of its essence as unexceptional, what he called the "fourth discontinuity" that has appeared after previous disenchantments and decenterings of our prominence.[9]

The Pathologies of Industrial Exteriorizations

Within the realm of technology there exist solely technical purposes.
Friedrich Georg Jünger, *The Failures of Technology* (1949)[10]

The question was, however, a serious one. We can read in one book on automation from 1964, the same year Leroi-Gourhan's appeared: "In some distant future, when cybernetic brains higher than their human equivalents have been constructed, humanity will have to decide whether it is really after the progress of intelligence in the universe or whether it prefers to foster its own ends of domination."[11] No less an intellect than Hannah Arendt would remark, in a lecture she delivered at a conference devoted to cybernetics:

> First of all, automation is a new revolution; automation brought about as distinguished from the Industrial Revolution of the last into the present century seems, from this point of view for me, to reside in two things: one, the Industrial Revolution replaced only muscle power, but not brain power. The very fact that machines can take over a certain amount of activity which we always have identified with the human mind, calls, in my opinion, for a re-evaluation of the activity, of our intellectual activity as such.[12]

In this talk she tried to make a distinction between technically stored "memory" and human *remembrance*. The goal was to assign the computer to the mere "computational" labor of the brain and not have it contaminate our genuine intellectual capacities.

A year earlier, Arendt had raised this same question, consistent with

Leroi-Gourhan's analysis. She wrote, "Electronic brains share with all other machines the capacity to do man's work better and faster than man. The fact that they supplant and enlarge human brain power rather than labor power causes no perplexity to those who know how to distinguish between the 'intellect' necessary to play good checkers or chess and the human mind." And yet, as Arendt will observe, perhaps we should not be so complacent. "There are," she writes, "scientists who state that computers can do 'what a human brain cannot comprehend,' and this is an altogether different and alarming proposition." As Arendt would argue, there was something quite different about a *mind* as opposed to a mere "brain," and the important factor was, indeed, that of *automaticity*. As she states here, "Comprehension is actually a function of the mind and never the automatic result of brain power."[13]

Arendt had commented on this in her 1958 work, *The Human Condition*, where she claimed that the new technical devices were not at all performing genuine intellectual actions; they were instead just another kind of exosomatic organ, to use Lotka's term. For Arendt, the new electronic computers "are, like all machines, mere substitutes and artificial improvers of human labor power, following the time-honored device of all division of labor to break down every operation into its simplest constituent motions, substituting, for instance, repeated addition for multiplication." The idea of an intelligent machine is mistaken because intelligence is not something that can be reduced to an automated *sequence*. "All that the giant computers prove," Arendt claims, "is that the modern age was wrong to believe with Hobbes that rationality, in the sense of 'reckoning with consequences,' is the highest and most human of man's capacities."[14]

Here we come up against the key turn Leroi-Gourhan carefully plotted in his own account of the "condition" of the human and the idea of *Homo faber*—namely, the emergence of temporality. Arendt will argue that the computer can only follow the sequence in a linear fashion. Human thought, human memory, on the other hand, is *organized*, which is to say, integrated into a temporality that links the historical memory with anticipated futures and a present saturated with both, as Heidegger had elaborated in 1927, in *Being and Time*. Human action, Arendt states emphatically, is also marked by "its inherent unpredictability." The "boundlessness" of human action means it is not determined in advance. "This is not simply a question of an inability to foretell all the logical consequences of a particular act, in which case an electronic computer would be able to foretell the future." The point Arendt is making is that the future anticipated in the technical sense is a future that is already present. The cybernetic "future" that is the telos of the machine-organism must be present in order for the errors and deviations to be corrected in light of the goal. Now, the fabricated object must be present in the mind (as Leroi-Gourhan explained) for the technical process to begin. However, with human action, the future is not present at all in this sense. For as Arendt explains, "In contradistinction to fabrication, where the light by which to judge the finished product is provided by the image or model perceived beforehand by the craftsman's eye, the light that illuminates processes of action, and therefore all historical processes, appears only at their end, frequently when all the participants are dead."[15]

With Heidegger, Arendt tries to preserve an essence of the human in this capacity to be in time and to exist in the *rupture* that is the awareness of being in time. This is the space where the mind can judge, where the individual can truly *act* and where technology can be disaggregated from the human. As Heidegger argued in the famous essay "On the Question of Technology" (1950), what marks the human is not the *essence* of technology as *Gestell*, the framing of nature as the space of human "reason," but instead the very capacity of the human mind to penetrate the *essence* of something—a capacity that was not at all contaminated with the fabrication of machines and the exploitation of nature. Just the opposite: the human mind was poised on the edge of nature, experience, and Being beyond all possibilities of "natural" or artificial means of perception and understanding. Technicity, now industrialized and globalized, threatened human existence, human *thinking*, that is, at its core. As he put it in the preface to his 1967 collection, *Wegmarken*, "The question remains whether thinking will also end in the transmission of information [*Ob dann auch das Denken im Informationsgetriebe verendet... bleibt die Frage*]."[16] Indeed, for Heidegger cybernetics meant the end of philosophy, the end of thinking itself.[17]

Leroi-Gourhan insisted, against Heidegger, that the origin of human intelligence and temporality is just this primal *exteriorization of thought*, which is the invention of the *tool*—but also, we must remember, the invention of language itself. The two are indistinguishable. Still, Leroi-Gourhan offered his own perspective on the nightmare vision of a cybernetic future, not unlike the future as portrayed by figures such as Heidegger and Arendt. As Leroi-Gourhan speculated, we may be on the way to establishing an *automated* and self-regulating order melding humans, machines, and nature into one monstrous system. In the last part of his book, the paleoanthropologist will look *between* the biological and the machinic, between nature and technics, to identify a form of "humanity" that might be recuperated in this automatic age—"an area as yet untouched by the machine." As he explains, "We have all along gone round inside a triangle formed by the hand, the word, and the sensory-motor cortex and have shuttled back and forth between the human and the monkey in search of what cannot be shared with the rest of the zoologically or mechanically animated world."[18] What is it about the human that cannot be captured by biology or technology, that would resist the cybernetic incorporation of the living and the artificial?

For Leroi-Gourhan, the inimitable element lies in the sphere of culture, in the specific organizations of the "ethnic groups" that constitute the plurality of humanity as a species—this is similar to Arendt's defense of the human as fundamentally a plural species. "Beyond this dual image of the human machine and its improved artificial copy lies something else. Analyses made thus far have deliberately left out of account those things that constitute the fabric of the individual's relationship with the group—that is to say, everything that has to do with aesthetic behavior."[19] Leroi-Gourhan does not, emphatically, invoke the aesthetic as some kind of special character, for that would only raise again the radical question of origin already confronted with the paleontological account of the human itself. Instead, Leroi-Gourhan argues that the aesthetic is, like technics and language, just one more dimension of this fundamental transition that marks the human as human: the exteriorization of memory into material artifacts, the exteriorization of experience into symbols. The aesthetic is, ultimately (and here we will have to end our engagement with Leroi-Gourhan for the time being), the exteriorization of the *organization* of the social body itself, the formal rhythms, styles, and structures that are embodied in the cultural expressions of an ethnic group.

The threat to humanity, then, does not lie so much in the advance of a machine intelligence or the like, but instead in the industrial and postindustrial destruction of the plurality and specificity of ethnic groups. There is an echo here of an important Arendtian theme. To be sure, automation and integration play a fundamental role in this process of destruction and homogenization, but that is no indictment of the human as technical creature. The challenge of modernity, for Leroi-Gourhan, will be preserving the aesthetic, those spaces and forms where an individual participates in the collective life of society *as* a collective, within a highly technologized lifeworld.

The Beginning of an End

28
Technogenesis in the Networked Age

> The moving hand and its material traces do not just externalise the internal workings of a mind. Instead, intelligence is enacted through them; it proceeds along lines and material signs of one kind or another. For instance, the making of a stone tool is not the product of thinking; it is a way of thinking.
>
> Lambros Malafouris, "Mind and Material Engagement" (2019)[1]

According to the standard accounts of the "history" of artificial intelligence, the mainstream efforts to model human cognition in computer programs based on a symbolic processing model had absolutely failed, leading to the so-called AI winter—the collapse of the massive funding initiatives that had supercharged both digital technology and AI and cognitive science research related to these efforts. The new AI would look different. It would pay attention to all of the dimensions of cognition absent in the symbolic, hyperrationalist approach: embodiment, enactive perception, and the extension of the mind outside the body. The new architectures were radically different, technologically speaking, based on artificial neural networks that did not function sequentially and through high-level symbolic languages and nested conceptual categories but instead were massively parallel and distributed, to cite two key words of this moment. Part of this revolution in AI was a project of recuperation, as researchers and thinkers in the field looked back to pioneers in neural network systems, such as Frank Rosenblatt, who built an artificial sensing system (the perceptron) that used an early form of "machine learning" to recognize patterns, such as letters.

The 1980s can be figured as the decade of *distribution* and all that went with it, conceptually—hybridization, posthumanism, poststructuralism, systems theory, and so on. This was the time when Bruno Latour and colleagues developed actor-network theory, to take into account nonhuman and nonliving "actants" in systems of knowledge and social order, and Edward Hutchins was doing his innovative sociological research on distributed cognition, studying naval ships and Micronesian navigators.[2] We can even see the theoretical interest in thinkers such as Michel Foucault and Donna Haraway—among many others—as an indication of a general (critical and academic) consensus around the idea that human beings are *other than what they are*, or seem to be. As a historical and cultural species, the human could never stabilize, only be redistributed. The foundation of so many classic theorizations in politics, sociology, and economics, namely, the so-called autonomous individual, was now understood to be just one particular and contingent social determination with its roots in Enlightenment systems of economic and political organization, buttressed by anthropological regimes of knowledge.

The global organization of capitalism (or at least the imaginary that partly sustained it) was transformed dramatically by what Francis Fukiyama called the "end of history" as the Soviet bloc disintegrated and a new historical moment was proclaimed. Not coincidentally, it seems, this was also the age when networked computing rapidly accelerated and expanded, alongside the emergence of "personal computing" and the integration of culture and computation. Whatever we make of the landmark date on a global scale, 1989 did see the birth of the World Wide Web and the death of the Cold War, initiating new sets of protocols that launched what we now call simply "the internet" and "globalization." The stress in this era on the study of complex interconnected networks and multiple distributed systems, at varying levels and in both material and intellectual registers, is vastly important. Not only did it lay the groundwork for thinking about media technology in new ways, but it was also the conceptual foundation of contemporary neuroscience and the most prominent models of computing systems and artificial intelligence.

Whatever the discipline, and however sophisticated—or not—the conceptualization of networks and network theory (which of course has its own mathematical history that originates in the eighteenth century, with Leonhard Euler), one question that was brought to the fore was this: What exactly was the *relationship* between individuals and these networks and systems? What was the relationship of the *body* to the individual, and how did both bodies and minds lend themselves to assimilation into what Foucault would call disciplinary regimes? Or was it necessary, as Deleuze argued, to redefine the relationship altogether? Were we mere *dividuals*, tracked and organized by invisible control regimes organized by new digital technologies?[3]

The difficulty of this question, philosophically speaking, but also with respect to the development of cognitive psychology, neuroscience, biology, cybernetics, and artificial intelligence, often provoked two incompatible and equally problematic responses, and we are to a large degree still caught in this dilemma. Network theory either entered the micro domains of research (as a model for the brain, or the operations of various technical, or biological, systems) or functioned as a framework for the explanation of systemic structures in social, political, economic, and other domains. What was less attended to, in retrospect, was thinking at the boundary: Were human beings constituted as networked, or capable of being networked, in ways that could not be understood in scientific or technical terms?

To trace a line of thought through this period and beyond, to point to the "decision" that is still to be made concerning our contemporary digital infrastructures, is to remain with the question of human intelligence and the origin of technology. To think of an extended mind (Andy Clark, David Chalmers),[4] or an embodied mind (Francisco Varela, Hubert Maturana),[5] or a distributed mind (Hutchins, Latour), or even a radically skewed *cyborg being* that is no longer really "human" at all in any singular sense (Haraway),[6] meant to think of *what* was being dispersed into networks beyond the "mind" as it was traditionally understood, philosophically. And, of course, the mind itself was being understood as a *networked system* in neuroscientific models but also in more conceptual work such as J. A. Fodor's *Modularity of Mind* (1983) or the ruthlessly reductionist evolutionary psychology of John Tooby and Leda Cosmides, who argued that language and culture emerge from a human mind that is itself composed of multiple "mechanisms" individually evolved for specific survival purposes.[7]

What kind of thinking is being played out in these systems and fields of organization and power, within minds and brains, within social and technical organizations? How does a network *think*? Can a network be *intelligent*? Given the stakes of the issue today (given our predictive algorithmic culture, based on the hyper-distributed intelligence systems using deep learning), it is useful here to maintain some of the lines of thought, the concepts and questions, that have circled and intersected with the idea of human intelligence as an exception, a *question* for us. Artificial intelligence in the networked age was no longer an exercise in "simulation." It was the beginning of the end—the end of the human brain, or even the human being, as the *center* of intelligence. Would we be "living fossils" as Leroi-Gourhan predicted, or maybe partners with an alien intelligence, or even its competitors, implanting "neural chips" to try to match our own technologies? The fantastic element of these scenarios cannot be dismissed or ignored, but the goal here is to see how *technogenesis* and the evolution of technology opens up a critical perspective on intelligence that resolutely does not hold out some special or exceptional *metaphysical* position for the human. What is artificial intelligence when intelligence cannot be naturalized so easily anymore? After the end of metaphysics, what is a *thinking being*?

The Artificial Neural Network

To begin, we need to bypass philosophy and engage with network technology as it confronts its own lack of philosophy, or at least the need for a new philosophy. The early neural network researchers working in artificial intelligence can be aligned, as I said, with a number of parallel projects in defamiliarizing complex systems by modeling them in distributed and statistical forms. Maturana and Varela's landmark study of what they called *autopoiesis* in one word took autonomy and automaticity as two sides of one coin. The idea was that a system was closed and responded *to itself* as it underwent modifications with respect to its environment. The living system was intelligent, we could say, if it could learn and learn to *predict*, simply by reorganizing its bounded parameters.

> What the observer calls "recall" and "memory" cannot be a process through which the organism confronts each new experience with a stored representation of the niche before making a decision, but the expression of a modified system capable of synthesizing a new behavior relevant to its present state of activity.[8]

There are no "sequences," only *processes*. And there are no "interactions," only couplings.

> An autopoietic organization constitutes a closed domain of relations specified only with respect to the autopoietic organization that these relations constitute, and, thus, it defines a "space" in which it can be realized as a concrete system.[9]

Thus the interest in older models of intelligence, or at least responsive behavior, developed in the early days of AI. Frank Rosenblatt, for example, has offered a simple but effective design for a machine that learned, not through specific "trial and error" or heuristic methods championed by influential figures such as Alan Newell and Herbert Simon, but instead by imitating simple brain activity. As Donald Hebb had argued, in *The Organization of Behavior* (1949), the synaptic connectivity of neurons in the brain would, through repetition, become more and more consolidated, easier to trigger when excited in future iterations. Rosenblatt used this idea to build a network of "neurons" whose cognitive weights (represented by electrical flows) could change depending on the result of the particular trial. (Figure 28.1.) As simple as it might seem, compared to today's vast, almost incomprehensible machine learning networks, with layers upon layers of connections and untold numbers of parameters, Rosenblatt's perceptron was meant to be a vision for intelligence itself. As he wrote in 1958:

> By the study of systems such as the perceptron, it is hoped that those fundamental laws of organization which are common to all information handling systems, machines and men included, may eventually be understood.[10]

You could say he was right—as we will see. But it is more important to point to Rosenblatt's caveat here.

> Does this mean that the perceptron is capable, without further modification in principle, of such higher order functions as are involved in human speech, communication, and thinking? Actually, the limit of the perceptron's capabilities seems to lie in the area of relative judgment, and the abstraction of relationships. In its "symbolic behavior," the perceptron shows some striking similarities to Goldstein's brain-damaged patients. (404)

As Rosenblatt explained, there was something lurking in human understanding that seemed more clear, more organized, than mere statistical waves could imitate. And note the reference here to Kurt Goldstein. The brain is a dynamic system, yes, but it is also self-organizing and capable of Gestalt interpretation and understanding. The perceptron could not "recognize" meaning, only repeat (as with some injured patients) memorized operations (such as marking letters). Rosenblatt will admit, "Statistical separability alone does not provide a sufficient basis for higher order abstraction. Some system, more advanced in principle than the perceptron, seems to be required at this point" (405).

I highlight this not to critique neural networks (that is the famous role played in the standard historical accounts by Seymour Pappert's devastating—if fundamentally incorrect—takedown in his 1969 book, *Perceptrons*) but to note the presence of this same issue in the revival of neural

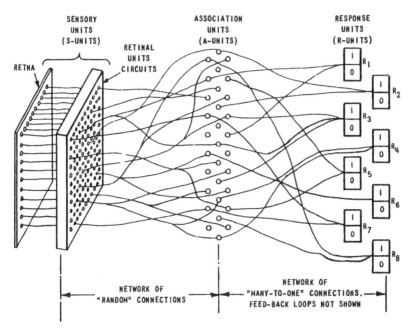

Figure 28.1. From John C. Hay, Ben E. Lynch, and David R. Smith, *Mark 1 Perceptron Operator's Manual* (Buffalo, NY: Cornell Aeronautical Laboratory, 1960).

net research in the 1980s. Yes, the brain was a good place to look for genuine insights into intelligence that assumed that embodiment and materiality were important for understanding what it was humans could do cognitively. As Geoffrey Hinton and David Touretsky explained:

> The brain is built from painfully slow and unreliable components: neurons, which fire less than once per millisecond, are susceptible to fatigue, and die off regularly. The only way the brain can succeed as a symbol processor is by exploiting massive parallelism using organizational principles that remain unknown for the present.

However, they will also admit that the massive network approach might leave something important out, namely, the kind of rational cognition that was the early focus of AI research and theory.

> Computer scientists and others have long been interested in neural network architectures as a means of exploring the question of intelligence.... To implement the highest level of cognitive functioning, the one responsible for general reasoning, requires some sort of symbolic inference architecture.[11]

Hinton was part of the Parallel Distributed Processing (PDP) group whose work was represented in the iconic two-volume set of articles that appeared in 1986 and set the stage for much future discovery in this arena. What is fascinating here is that the one key theoretical article laying out the project was also the one place where serious questions were raised about the tenability of a network that was, as Varela and Maturana described, fundamentally *closed*.[12] The authors of the essay recognize that the way networks "answer" the question posed by their inputs is by coming to a certain *state*, and the language is maybe telling: the system "settles" into a state and thereby indicates its response. However, this process seemed to be inadequate as a way of explaining or modeling fundamental aspects of human intelligence. As they wrote:

> If the human information-processing system carries out its computations by "settling" into a solution rather than applying logical operations, why are humans so intelligent? How can we do science, mathematics, logic, etc.? How can we do logic if our basic operations are not logical at all? (44)

The question is not naive, and it was not really appropriate to simply answer that the original AI field was right, symbolic sequential processing is important. That was because the model of distributed networks *was* in a sense bounded and closed, metaphysically speaking. That is, the challenge (coming from brain science, which many of the researchers hailed from) was to *explain* how a certain kind of thinking (logical, sequential, abstract) could emerge from a neurological organization that was foundational organized as a system of parallel interconnections undergoing constant changes, producing more ephemeral moments and zones of regularity.

This is why the answer suggested here is so provocative, astonishing even, as it pinpoints what so many philosophers and scientists have (for centuries) been grappling with, our central question in other words: How is technology related to human cognition? The authors of this piece make an astute, but highly troubling (theoretically, that is) suggestion:

> We suspect the answer comes from our ability to create artifacts—that is, our ability to create physical representations that we can manipulate in simple ways to get answers to very difficult and abstract problems. (44)

Externalization of thinking into the object creates a possibility of *objectivity* that allows for a productively new organization of thinking itself. It is not accidental that the metaphor here is one of *manipulation*. From Kant to Leroi-Gourhan, the hand can be seen as the facilitator between the inside of thought and the exterior world that is to be *grasped by thought* and not remain (as for animals) the mere *occasion for thought*. As they explain:

> These dual skills of manipulating the environment and processing the environment we have created allow us to reduce very complex problems to a series of very simple ones. This ability allows us to deal with problems that are otherwise impossible. This is real symbol processing and, we are beginning to think, the primary symbol processing that we are able to do. Indeed, on this view the external environment becomes a key extension to our mind. (46)

The implications of this hypothesis are not lost on the authors. Thinking, in this sense, is not merely a function of the brain, or at least, the brain is able to internalize a logic that is not produced simply from its own statistical waves of transformed inputs.

> Not only can we manipulate the physical environment and then process it, we can also learn to internalize the representations we create, "imagine" them, and then process these imagined representations—just as if they were external. (46)

The philosophy of exteriorization suggested here is (perhaps not surprisingly) not followed up. Of course, everyone knows that human beings are capable of such thought, with their brains, so if artifacts are essential to the evolution of intelligence, that fact can be marked then displaced. They will admit that "it seems to us that such representational systems are not very easy to develop." Yet essential nonetheless. "We are good at manipulating our environment. This is another version of man-the-tool-user, and we believe that this is perhaps the crucial skill which allows us to think logically, do mathematics and science, and in general build a culture" (44). But to make progress, clearly it is necessary to avoid or at least set aside the problem. The work is off-loaded to another department or discipline.

Where do artifacts come from? "Usually they are provided by *our culture*" (47; my emphasis). Culture stands as both the *origin* and the *result* of technology, an impossible lever that moves the brain into the new field of intelligence that is necessary to begin the process in the first place.

The Technical Evolution of Culture

Cultures are best seen not as complexes of concrete behavior patterns, customs, usages, traditions, habit clusters—but as sets of control mechanisms—plans, recipes, rules, instructions, what computer engineers call "programs" for the governing of behavior. The second is that man is the animal most desperately dependent upon such extragenetic control mechanisms for ordering his behavior.

Clifford Geertz, "The Impact of the Concept of Culture on the Concept of Man" (1966)[13]

As Leroi-Gourhan had argued, the very definition of the human rests on a *break* with biological evolution in the turn to a form of cultural—and always therefore *technical*—evolution based on the exteriorization of memory and gesture. In a wave of postwar work on evolutionary psychology and the birth of culture, this fundamental insight was reaffirmed. Interest in the artifact (material medium) for thinking about reason and human intelligence was, as we know, hardly unknown before this period. Indeed, much of the new thinking on media and literacy (in the influential and widely cited work of Marshall McCluhan most notably) was showing how certain forms of technical systems imposed those forms ("linearity" of thought, for example) on human minds, for better or worse.

A notable case is that of Merlin Donald, whose early career was in experimental cognitive sciences, with decidedly modest aims, but who later took up the crucial grand question of "culture" as it defines the emergence of the human. Donald would outline an ambitious "history of the modern mind" that isolated stages in the way human biological beings began not just to interact, but to *communicate* in a way animals never do. A new relationship between human psychology (rooted in models of brains understood as information systems) and the advent of technology was crucial to the formation of a new culture. Human cultural forms are ways in which we share experience and learn from previous experiences. While the first efforts of cultural communication and control are seemingly more natural (i.e., based on individual capacities), the emergence of the artifact becomes, for Donald, the revelation of the importance to all culture of *external memory storage*.[14] His stages of history (from thought to oral speech to external storage systems such as writing) are to be sure unconvincingly schematic; however, he argues persuasively, as Popper, Lotka, and Leroi-Gourhan did before him, that the most crucial stage, the genuine crossing of the threshold to humanity, is the *artificial exteriorization* of memory (what he would later call the "exographic revolution"),[15] which unleashes, for better or worse, the development of human rationality and science. What Donald does not recognize here is that the first stages of thought and speech are already *technical* in the sense that they are a reengineering of natural behaviors. The disarticulation of natural and artificial is the structural key to the founding of a "modern mind."

The effort to disentangle biological evolution from culture was not a project to extricate human beings from determination. Rather, the emphasis in disciplines such as paleoanthropology and evolutionary psychology was on the distinction between different *forms* of evolutionary determination. Cultural evolution, which is nongenetic, "has transgressed organic evolution and shows a certain autonomy," wrote Franz Wuketits, in the same year that the PDP volumes appeared.[16] Citing evolutionary epistemology and in particular Karl Popper's theory of "objective knowledge" through material culture, Wuketits offered a systems theory critique of cultural evolution, in line with a general trend in the 1980s toward network and distributed models of organization.[17] Still, what is essential for Wuketits is the appearance of what he calls *exosomatic structures*, where ideas become materialized and thereby made transferable to others in a social group. Even a dead person, he explains, can pass on cultural information through, say, books or artworks, but "a dead person no longer can transmit genetic information."[18] One implication I would draw out here is that with the cultural turn, the individual mind now confronts "alien" thought through these exosomatic organs. Ultimately, Wuketits avoids any explanation of *how* an enclosed, biologically determined nervous system could ever create and then make use of these exosomatic structures, a key challenge that Leroi-Gourhan faced in his own work.

The philosophical biologist Tom Stonier had made a similar case in 1981, when he argued that with the emergence of the human *tool*, a biological species was able, for the very first time, to extend and modify its own "econiche." This was, Stonier asserts, an epochal "breakthrough in the history of Life." Technology is what initiates a never before seen process of nongenetic evolution that will "permanently" alter the relationship between the organic society of humans and their environment. Both will leave nature behind as they are transformed into cultural and technical *systems*. Like Leroi-Gourhan, Stonier speculates that we are now, in the twentieth century, on the cusp of a new breakthrough, the development of an *information society*, one that will take us "beyond ourselves."[19] As he put it in a later book, *Beyond Information: The Natural History of Intelligence* (1992), human beings form a collective intelligence predicated on the development of the artificial storage of memory. The danger of collective intelligence, he warns, is that with the increasing automation and complexity of information systems, "computers will rapidly become more intelligent than humans" just as human minds are increasingly being merged with *machine* intelligence.[20]

The collective mind, enabled by the exosomatic cultural transmission of human thought across the boundary of the biological brain, must always be considered what is called a "hetero-technic co-operation," a collective form of *cognition* distributed between organisms and machines.[21] The information theorist Pierre Lévy would presciently describe the contemporary zone of collective intelligence, now coordinated in "real time" by a new digital infrastructure, as a *cyberspace*—borrowing the term from the science fiction author William Gibson.[22]

Deacon on the Symbolic Species

As we know, across the world, people with equal intelligence, equally complex language can be living in radically different cultures with radically different kinds of technologies. Those that can look as the stone age of a million years ago, those that can look as modern as we are today sitting in this studio. The same brains can be producing all of those systems, in part because it is not all inside the head.

Terrence Deacon, interview (2003)[23]

But in all this work a key philosophical question remains: What *is* the invention of technology; and then how do we understand the *second* invention, that of artificial memory? As Donald admitted, this abyssal "gap" between "pre-symbolic and symbolic thought" can only be explained by some evolutionary event in the emergence of the human brain—which is to just displace the question of origin into an equally problematic space.[24] Whatever the location, we have here what Andy Clark calls the "missing link" between animal cognition and human symbolic systems, which are, at their heart, *technical systems*.[25]

This gap is the subject of Terrence Deacon's impressive book on the origin of symbolic thinking.[26] Like Leroi-Gourhan, Deacon refuses to locate the origin of the human break from animal cognition in some evolutionary miracle, instead arguing that language and the brain emerged in a mutual process of development. While it is not possible to do justice to the complexity and richness of Deacon's argument here, it is important to note that a key shift for him is the turn from mere associative predictions to *symbolic* predictions, that is, anticipations of the radically new event. For Deacon, this is the result of a *mnemonic* shift, an internal "offloading" of details from working memory (89). But this requires, he says, the creation of "artificial systems" for the internal re-presentation of associations. With these "tokens" symbolic thought becomes possible: "insight learning" (Deacon refers explicitly to Köhler here) occurs because higher-order associations can be made between internal representations (93–94). Symbols do not refer to objects only; they can, as Peirce first emphasized, relate to *other symbols*. At any rate, what is equally important to note here is that the symbolic structures are what we are calling, after Lotka, *exosomatic*. As Deacon notes, the adaptation that is language evolves in its own way, and *its* history of evolutionary adaptation "has been going on outside the brain" (109). What Deacon is arguing is that the emergence and "evolution of symbolic communication" is an *extrabiological* system "with a sort of autonomous life of its own" (409).

Unlike so many evolutionary psychologists and cognitive scientists, Deacon tries to provide some kind of—admittedly hypothetical—account of the origin of symbolic thought and its exosomatic supports, by focusing on the evolution of the brain. However, Deacon, who did experimental work on brains that included neural tissue transplantation across species, looks to evolution but without any interest in invoking a *bios ex machina*. The main thesis is that according to the then current data, it was clear that there was no evolution of "the" brain. Rather, the different regions of the brain evolved in different scales according to different pressures and selection, and each region is subject to *displacement processes* that are internal to the organization of the whole brain. With the revolutionary enlargement of the brain and especially with the increase of the cortex region, Deacon explains, humans were provided with a new flood of *internal* inputs generated from association patterns, inputs that mimicked an increase in sensory information.

What the brain could now do was something different from the older complex, to be sure, but limited capacity to correlate *experiences of the world*. Now, the context made possible associations *between* associations, leading to a liberation from mere "indexical" association and inference to something like *iconic* relations, setting the stage for forms of thought that related simply to the *relational structures* themselves—what can be called genuine symbolic thought. The correlative processing that takes place in the frontal region is where the "mnemonic architecture" of symbolic reference is created. For Deacon, symbolic systems only emerge after "a radical re-engineering of the whole brain has taken place, and on a scale that is unprecedented" (45). Different regions of the brain are *newly* connected by increases in cortical tissue and links across topological levels, and this creates new possibilities of association that are "arbitrary" in relation to sensory experience but meaningful within this novel architecture of the machinery of association. Language, therefore, is a form of internal and external organization that emerges in this *matrix of association* rather than simply the result of "size," or of the evolution of a specific language center—as Chomsky and his followers, along with many evolutionary psychologists, would normally assume.

However, equally important for us is Deacon's argument that the languages—structured systems of association—that emerge from symbolic thinking *also* evolve independently, and are always having to adapt to the capacities of the brain as they transform (biologically but also "culturally"). There is therefore always a coevolution of internal brain organization, of learning, and of what we are calling exosomatic processes. And this coevolution never stabilizes into a single trajectory since each process has its own logic of organization and its own line of development.

Deacon emphasizes language in his account of the origin of the human. However, his look back to the early tool-making capacity of Australopithecines is illuminating—for like Leroi-Gourhan, Deacon argues (with reference to Baldwin's account of evolution that supplemented Darwinian theory) that the "transformation into *Homo* was in part the consequence rather than the cause" of the new forms of technicity. Both the artifacts and the social organization that learning depended on were (however fragile in their early forms) essential "external supports" of symbolic thought, in that they helped the brain develop the possibilities of connectivity and also supported symbolic expression materially. This is also why human thinking is never *natural* to the brain: it is a system involving both "alien" internal inputs with no natural zone of organization and the alien thinking that is *another mind* (423).[27] "Stone tools and symbols," writes Deacon, "must both, then, be the architects of the *Australopithecus-Homo* transition, and not its consequences." The new brains, the implements, the dramatic changes in the *body*, improved bipedalism, all were just "physical echoes of a threshold already crossed" (348).

The *threshold*. There is something about the human that is not simply about what comes *after* but rather involves the threshold itself. With this in mind, we can turn to the one philosopher to engage directly with this question of the *threshold* between biology and culture, between prehuman and genuine human, namely, Bernard Stiegler. The turn will hinge on the question of technology.

Originary Technicity: Stiegler's "Technics and Time"

> You know—technology wasn't invented by us humans. Rather the other way around.
>
> François Lyotard, *The Inhuman* (1988)[28]

The French philosopher Bernard Stiegler, immersed in both poststructuralist thought and phenomenology, as well as scholarship on the history and theory of technology, looked back to Leroi-Gourhan as the starting point for an intensive investigation of the question of the *origin* of the human, and he will come to exactly the opposite conclusion of any other philosopher in this period—though we see hints of the argument in some of the specialist literature in paleoanthropology. For Stiegler, the paradox of the artifact is not a paradox if we understand that it is not the human that invents the tool but rather it is the *tool that invents the human*. While agreeing with Leroi-Gourhan's critique of the mythology of the human, the idea of a "gift" that enables us to leave nature, Stiegler denies Leroi-Gourhan's effort nonetheless to locate an *origin* of the "real" human, the user of tools as symbols, the conscious and aesthetic mind that only appears late in our hominid evolution. For Stiegler, we must confront the *radical* beginning. The human, as Leroi-Gourhan argued, is a zoological fact. How can the tool be at once *on the side of zoology* and the site of the emergence of the modern mind, to use Merlin Donald's term here. We have to realize, against Leroi-Gourhan, that the earliest, most primitive tool is already a tool. It marks an anticipation *even if the human is not aware of it*.

> "Technical consciousness" means anticipation without creative consciousness. Anticipation means the realization of a possibility that is not determined by a biological program. Now, at the same time, the movement of "exteriorization," if it seems to presuppose this anticipation, appears here to be of a strictly zoological origin, to the point of still being determined by the neurophysiological characteristics of the individual. When this determination will have completely ceased its action on technical evolution, Leroi-Gourhan will introduce a notion of spirituality: *a second origin*.[29]

For Stiegler, and I admit this is not always easy to grasp, the human is what results accidentally from the tool's appearance with the early hominids, or to put it better, the tool is an *inhuman intervention into the biological human* that produces a new evolution *of* the human (the increase in the size and complexity of the brain most importantly) from within. The tool could never show any evolution, a transformation in a history, if it was already *not human*. And simultaneously, the human could not extricate itself from nature without the intervention into its nature by *artifice*, or the unnatural. We can see that a human cannot produce the *inhuman* (that which is alien to the human and having its *own* logic of being), but equally, something inhuman cannot become human without the human changing in the process.[30] The argument here, consistent with Leroi-Gourhan, is that the human does not "improve" to accept or invent the tool. The human must lose a certain human experience to gain the *inhuman* element that is enabled by the artifact.[31]

So the first tool (defined as the rock that is changed by being struck) is the site for the intersection of the inhuman and human: the zoological creature in breaking the rock is already "affected with anticipation, because it is nothing but anticipation, a gesture is a gesture." And this is the difficulty: "There can be no gesture without tools and artificial memory, prosthetic, outside of the body, and constitutive of its world. There is no anticipation, no time outside of this passage outside, of this putting-outside-of-self and of this alienation of the human and its memory that 'exteriorization' is" (152). Animal gesture is not anticipation, it operates in time but not *as time*. The gesture as radically novel gesture must be different. It makes possible the tool but the memory must not be natural—which is to say, the memory that makes the tool possible is already outside the brain, it consists of the tool itself.

> The question is the very ambiguity of the word "exteriorization" and the hierarchy or the chronological, logical, and ontological preeminence that it immediately induces: if indeed one could speak of exteriorization, this would mean the presence of a preceding interiority. Now, this interiority is nothing outside of its exteriorization: the issue is therefore neither that of an interiority nor that of exteriority—but that of an originary complex in which the two terms, far from being opposed, compose with one another (and by the same token are posed, in a single stroke, in a single movement). Neither one precedes the other, neither is the origin of the other, the origin being then the coming into adequacy [*con-venance*] or the simultaneous arrival of the two—which are in truth the same considered from two different points of view. (152)

Following Derrida, the prosthesis is not a substitute or an extension, but in its essence a putting forward, the creation of an "end" that was not there before it. The prosthetic that is the tool as exteriorized memory "is not a 'means' for the human but its end, and we know the essential equivocity of this expression: 'the end of the human,'" writes Stiegler (153).

In any case, the conclusion is what is important here: the human is something less than natural; it is differentiated from nature by the technical inhumanity and will always be at once *inside* technology but also never in control of the evolution of technology, which is always driven by its own logic, the logic of the not-living. Now a social, cultural, technical being, the human is not so much liberated from genetic evolution, set into "history," but instead subject to an evolution which is part of us but never congruent with us. That will be the basis of Stiegler's invocation of the *pharmakon* of Plato (via Derrida) to depict the role of technology in our world: both poison and cure, it is always *indifferent* to our plight, our experience. Humans are, in Stiegler's language, *epiphylogenetic* beings. "It is in this sense that the *what* invents the *who* just as much as it is invented by it" (178).

Technics is time, because artificial organization is the basis for the sequence cut from the continuity of lived experience. The trace of that cut remains, in our minds, in our tools, and is the source of both the power of temporalization and the contingency of our being. Rewriting Heidegger, Stiegler argues that technics is not just another word for Heideggerian "facticity." The "rift" in Being that sets up Heidegger's account of historicity and being-toward-death is *already* dependent on the exteriorization of memory, because the historicity of human experience is constituted by the inhuman structure of technicity.

> Nothing can be said of temporalization that does not relate to the ephiphylogenetic structure put in place each time, and each time in an original way, by the already-there, in other words by the memory supports that organize successive epochs of humanity: that is, technics—the supplement is elementary, or rather elementary supplementarity is (the relation to) time [*différance*]. (183)

So for Stiegler, the pathology of modernity is not, as it was for Heidegger, modern technology. Or at least not technology per se. As Stiegler will argue, the pathological and disturbing element is not industrialization and automation, on the technical side of things. Rather, what concerns Stiegler is what he calls the industrialization and automation of memory—the industrial production and hence homogenization of tertiary retention. No longer is the human social, cultural, and economic sphere defined by an interaction between memory, technicity, and the living being. Instead the logic of technical systems now imposes itself on the cultural forms of memory themselves. In the process, human minds in their plasticity are determined in a new way. They are not merely shaped and configured by the demands of technical know-how and sociopolitical organization, but formed according to the logic of industrial capital itself, that is in the name of the reproduction of industrial economic systems.[32] Human memory is no longer shared primarily through concrete instances of collective individuation. Instead, we relate one to another *via* the machinery of culture as mass-produced memory, which is also to say mass-produced imagination, which means foreclosing the genuine "event" of knowing and remembering.

As Stiegler would go on to demonstrate, relentlessly, the emergence of planetary scaled digital infrastructures threatens genuine thinking and memory even more drastically than the industrialization of thought.[33] Not only do all forms of culture depend on tertiary retention for survival, and therefore can be violated and expropriated by the logic of the machine; in the digital age the machines gain a vast new power, what Stiegler calls *reticulation*: that is, digital machines integrate all other forms of memory production, and at a speed and scale that is both persistent and beyond the comprehension of our own "slow" brains, so that our minds are learning and anticipating (if that is the right word) according to the organizing logic of the digital. This pathology comes to the fore most clearly in technologies of *prediction*—in our era the systems of machine learning.

29
Failures of Anticipation

The Future of Intelligence in the Era of Machine Learning

Prediction is the essence of intelligence.
                                                                  Yann LeCun[1]

As Claude Shannon and John McCarthy noted, in the 1950s, concepts of the brain and intelligence are inevitably shaped by contemporary concepts of technical systems. It is hardly an exaggeration to say that the most dominant technology in the contemporary world of artificial intelligence is that of "machine learning," which is an umbrella term for a number of approaches that utilize artificial neural nets, such as deep learning, generative adversarial networks, and reinforcement learning. The key feature of machine learning is *prediction*. Trained on large data sets, the system aims to predict an outcome—whether that is an actual event or perhaps belonging to a certain classification. Another key feature of machine learning is *error*. The system in fact learns via prediction errors, which are used, mathematically, to alter predictions continuously in order to fine tune predictions. Trained systems make use of an external observer to tell right from wrong, whereas untrained systems work themselves to maximize various parameters. The third feature of machine learning systems that I want to highlight is that they are *generative*. What makes contemporary deep learning methods so successful is not simply the size and number of their neural network layers but also the fact that the networks produce generative models, hypotheses about the state of things, which are then tested and reexamined. A deep learning network, at each layer, has been individually trained to generate a model of some feature of the world from its input, and these generative models are aggregated in new layers that similarly produce model outputs from input data.

So we can say that a modern machine learning system *predicts* outcomes, learns from prediction *errors* that are back-propagated through the system, and discovers in this process how to *generate* models from certain input signals at various levels. So a system might first train to recognize "edges" from various visual input data, and these particular models will be aggregated with other trained nodes (generating, say, color or shape designations) until finally we would have a prediction that the total data stream represents a particular object.

Machine learning obviously traces its lineage to earlier work in artificial neural networks and especially the return to research marked by the publication of the two volumes on parallel and distributed processing in 1986. However, the artificial neural network paradigm has now emerged as one of the most powerful models for neuroscience and cognitive theory, which borrow the key concepts of prediction, error, and generation from machine learning.[2] The success of machine learning lies to a great extent in the revolution in both computing power and the immense size of data sets enabled by the move to networked computing in the 2000s. The transition from singular technologies (the "computer") to a vast network of parallel and distributed systems housed in data centers around the world has made possible a new way of thinking about the brain and the mind, one that gains credence given the scope of the hardware and software technologies.

Although there are numerous ways in which machine learning and artificial neural networks have influenced researchers in the cognitive sciences and the neurosciences, a new "school" of sorts has emerged in the past two decades or so that is explicitly tied to the techniques and technologies of machine learning and its kin. This school has been called predictive processing (sometimes referred to as predictive coding). As Jakob Hohwy, an influential proponent, has concisely explained, predictive processing (PP) "is the theory that the brain is a sophisticated hypothesis-testing mechanism, which is constantly involved in minimizing the error of its predictions of the sensory input it receives from the world. This mechanism is meant to explain perception and action and everything mental in between."[3] The implication is that there is a singular "mechanism" (the logic of a Bayesian neural network) that underlies all cognition. As Hohwy writes, approvingly, "Though the description of the mechanism is statistical it is just a causal neuronal mechanism and the theory therefore sits well with a reductionist, materialist view of the mind."[4] Indeed, the research in both cognitive science and the functioning of the brain, especially the visual brain, in this mode has been impressive.[5] In particular, the way that predictive processing approaches make use of Hinton's internal generative models is striking and can help explain both inherited "hard-wired" input processing and network layers that have emerged from initially plastic conditions.[6] The brain, it seems, does not so much "analyze" data and use stored concepts and reasoning systems to make sense of the world. Rather, the brain uses past experiences to predict the future and pays cognitive attention only to those moments when prediction *fails*. And the "action" of an organism in this context can be understood as a way of making a "prediction" come true through a reorientation within, or transformation of, the environment.

It is quite clearly the case that our problem, the problem of intelligence and technogenesis (the origin of technology, which is to say, the origin of the modern *human*) remains a difficult issue. Hohwy's enthusiastic account of predictive processing is tempered by the admission that the focus on theories of perception "largely leave out higher cognitive phenomena such as thought, imagery, language, social cognition, and decision-making" (2). A similar caution appears in a recent collection of essays on *Philosophy and Predictive Processing*:

> in principle, one can use the language of PP across multiple domains of cognition and even across different disciplines. However, a gap remains between the domains of (relatively simpler) action-perception loops and (relatively more complex) higher cognitive abilities. The former have been characterized in formal and quantitative terms using PP, whereas explanations of the latter tend to appeal to the same PP concepts but often lack a comprehensive quantitative and computational characterization. Thus, it remains to be seen if PP really "scales up" to higher cognition domains.[7]

However, another major interpreter of predictive processing, the philosopher Andy Clark, is more than aware of the challenge these domains raise for the theory. Clark's early work in philosophy was on inactive and embodied perception, and he was also an innovator (with his collaborator David Chalmers) in the theory of the "extended mind," which argues that cognition is not limited by the mind but is distributed through a system of biological and technological support systems. Clark would go on to write a general interest book on the topic, *Natural Born Cyborgs*. The challenge for the predictive processing model would be integrating technology and distributed cognition, given that the very starting point of the model is a brain that must learn about the world and itself from within the highly constrained world of neural stimulation—what are known as "spike trains." How does a self-enclosed nervous system learn to extend itself, to "exteriorize" itself in new and artificial forms of perception and action?

In his own overview, Clark celebrates predictive processing as the "perfect neuro-computational partner" for the philosophical theories of enactive and embodied cognition, that is, research "that stresses the constant engagement of the world by cycles of perceptuo-motor activity. The predictive brain, if this is correct, is not an insulated inference engine so much as an action-oriented engagement machine. It is an engagement-machine."[8] For Clark, the brain is engaged with the environment and actively constructing its view of the world. Its predictions constitute a kind of "intuitive" judgment about the world. But this judgment is resolutely *unconscious*, which means that predictive processing fits nicely with so much ancillary work on unconscious cognition and intuition within cognitive science. As Clark admits in the opening of his book, conscious, directed thought is bracketed. The brain is a prediction *machine*, but in this prediction there is nothing like we would call, with Heidegger and Stiegler, *anticipation*. There is in fact no *gap* between "learning," "experience" and "prediction" here—which is to say, no genuine temporality, only a constantly shifting and updated "presence."[9]

But how can Clark reconcile the reductionism and limited scope of predictive processing with his prior work on extended minds, which dovetails to a certain degree with the philosophical claims that Bernard Stiegler was making on the technical origins of cognition in the same period? First, Clark will argue that what makes predictive processing so appealing is that it can explain both active inference in the perceptual sphere and what he calls "off-line" reasoning, or the imagination, since these all can be understood as the result of the internal generation of "models," just at different levels of the neural network system (3).

However, we are still left with the more difficult argument to make—namely, how does the brain (the human brain) exhibit the novel behavior of *technology*, and how does technology alter radically the cognitive potential of our species, in ways that have no parallel in the animal world? Clark is hardly one to avoid this question. He says very clearly:

> We humans—uniquely in the terrestrial natural order—build, and repeatedly rebuild, the social, linguistic, and technological worlds whose regularities then become reflected in the generative models making the predictions. It is because the brain itself is such a potent organ of unsupervised self-organization that our sociocultural immersions can be as efficacious

as they are. But it is only in the many complex and ill-understood interactions between these two fundamental forces (between complex self-organizing neural dynamics and the evolving swirl of social and material influence) that minds like ours emerge from the material flux. (270)

The answer is frustratingly vague. As Clark puts it, artifacts and symbols *do* enable new learning and new cognition. Instead of the brain looping and relooping generative models and prediction errors, the human mind makes use of "symbol-mediated loops" from social culture, such as notebooks, smartphones, and the like—what Stiegler calls "tertiary retentions" (277–78). The language deployed by Clark echoes Stiegler's own argument: "Once externalized, an idea or thought is thus able to participate in brand new webs of higher order and more abstract statistical correlation" (278). Humans must be constructing what Clark calls "designer environments" (art, culture, science), but there is no clear explanation for how, or why, human brains are capable of externalizing their thought in artificial technical forms of organization.

But this is a central problem in predictive processing if it is to be a model of human intelligence. As Regina Fabry recently explained, "Enculturation is a developmental process that is characterized by plastic changes to neural circuitry and motor profiles. It is strongly influenced by the structured interaction of human organisms with the cognitive niche."[10] For Fabry, culture provides "cognitive practices" that shape our interactions. In essence, Fabry—along with a number of scholars involved in what is called "cognitive archaeology"[11]—points to the importance of what Stiegler calls the *epiphylogenetic* essence of our cognitive capacity,[12] without directly acknowledging the radical question of origin. Human "niche construction" amounts to the unprecedented *anticipation* that is constituted by the appearance of the tool as a new (artificial, unnecessary) organization and therefore cannot simply be the result of a *prediction* based on the given organization of both brains and environments, as it is in the animal *Umwelt*.

The bridge between what we can call "natural" cognition and the artificial and "in-human" technicity of human thought may be found in another concept that emerges from neural network theory but that is not absolutely essential to the predictive processing paradigm. Taking up the idea of the Markov blanket, both Karl Friston and Andy Clark have sought to understand the boundaries between levels of the organism (Friston) or levels of cognition (Clark) by arguing that within neural networks there exist bounded zones where network nodes are protected, so to speak, by border nodes. In a Markov blanket internal nodes respond only to the bounding

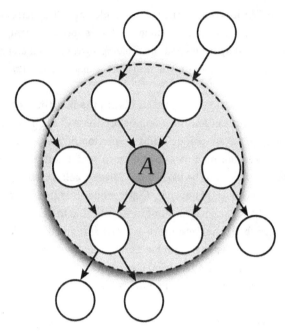

Figure 29.1. A Markov blanket.

nodes and other internal nodes and are never directly involved with the "outer" world of the network. (Figure 29.1.) There is a kind of homeostatic quality to the entity bounded by a Markov blanket, and indeed, as Friston and others have shown, the organismic cell can be described as a Markov blanket in line with the more statistical depiction of bounded areas within a network. The main point is that Clark sees in the Markov blanket a concept that unites neurophysiological systems, predictive inference, and, finally, the enculturated space of cognitive and symbolic practice that defines *human* intelligence. There is, to be sure, an unacknowledged spirit of the Leibnizian monad lurking here.[13]

In any case, what Clark argues is that humans transform themselves, in the sense that they "knit" their own Markov blankets, in the form of *artifacts* and *symbols*. These new organizations are self-enclosed and operate in the world—in our brains—as *systems*. For Clark, there is a slippage, a productive slippage, between the neural networks that harbor Markov blankets of relative stability (natural or "learned") and the *exteriorization* of an organization into the environment.

> Importantly, the predictive processing architecture itself provides a powerful mechanism enabling the flexible, repeated integration of capacities and

operations made available by the use of reliable bio-external resources. I won't rehearse those considerations here, since it is in any case evident—merely from observing our human capacity to become fluent users of a lifetime succession of new tools and technologies—that such fluidity is a crucial part of our heritage.[14]

Still, no account is offered (*can* be offered) of this critical leap across the threshold of interior and exterior, and thus no account of either the *différance* of thinking or the difference between animal and human cognition, the leap that is technology.

> The living being is very far from being a perfect machine and it's this imperfection which is both the cause and the means whereby poetry, resonances come about.
> Paul Valéry, *Cahiers*, 1927–28[15]

The paleoarchaeological and historical record shows clearly that the human at some point escaped natural determination and embarked on a new form of life that was superimposed on the biological—a political existence that navigated the inevitable conflicts between social forms of organization and the technical systems that enabled biological survival. This much is accepted by thinkers such as Clark and researchers in certain areas of evolutionary psychology. However, what this entails goes completely against the mainstream of contemporary cognitive neuroscience, namely, that we have to acknowledge that the human mind is *not* simply a product of the brain, but is instead the *site* for the operation of several different systems, systems that are often in competition even as they work together to maintain the cohesion of social groups and the survival of biological individuals.

What marks the human is this "invasion" of alien thought within our own minds—the thought of other minds, stored in technical forms of memory (like writing, for example) and reanimated in our own brains. Or the alien logic of cultural norms that dictate behaviors that will guarantee the survival of the group formation. Alongside this, the biological demands of the body also invade the mind, operating with their own insistent logics—both Freud and Damasio could be invoked in this context.

The point is that there are *no norms* governing these multiple systems that are instantiated by the individual human mind-brain. Or to put it another way, what we mean by "the mind" in this context is the continuing (and always *failing*) attempt to bring unity to the multiple systems in play.

As such, the mind is not itself a unity with given norms, and therefore an object of scientific study, but instead an evolving contingent effort to capture the varying logics of survival that stem from our biological, sociopolitical, and technical systems. The mind is both *defined* by these systems and is resolutely *not* these systems.

This is why the mind is always the site of crisis, of the unexpected, of the event. This is why the human mind is capable of genuine decision in crisis: the decision that there *is* a crisis (and not merely a prediction error) and the decision *for* a future that will relieve crisis—even if that means abandoning certain given norms of existence. One might imagine a computer programmed with conflicting norms and operations; however, there would never be a comprehensive unity in which this conflict could be recognized as a moment for genuine decision. To decide the conflict means taking seriously the possibility of an undetermined future. To imagine a future that exceeds past experience.

That is to say that the mind has *no norms of operation* itself but is only the site for the determination and configuration of *alien* norms. The mind is therefore not at all a biological entity conditioned by history, or a purely historical construction buffeted by biological imperatives. The mind appears in human existence as a novel entity, the space in which technical organization, social organization, and organismic organization, each with its own logic and its own demands, meet and can be constituted as codependent.

Human intelligence is constituted within this space—a mind that can escape its determinations and thereby open up an unforeseen, unpredictable future. To be sure, the imagination of a new path, a decision for a novel future and the institution of new norms is never an act of an actual entity. Like Carl Schmitt's sovereign, who decides the "exception," the mind is a kind of necessary fiction—the name for a unity that is not unitary, a unity that holds together difference while allowing a certain slippage, or leakage, that is to say, *gaps*, between the three spheres of life.

What escapes the machine, even the computer, even networks of computers, even the human mind in its "automatic" phases, is this capacity to *escape from its own determination*. As Stiegler wrote, human thought has "the power to *disrupt* and to *dis-automatize*, that is, to *change the rules*."[16] Only by leaving its own normative existence, through shock and decision, can a new norm ever be created, as a function of a novel future that is imagined—collectively—and not merely predicted.

Acknowledgments

It would be impossible to list everyone who helped me finish this project, so I will just give my sincere thanks to all the friends, colleagues, audiences, and, especially, students who gave me insights and sharpened my arguments. For material support, I would like to thank UC Berkeley; the Mellon Foundation; Tony Cascardi, former Dean of Arts and Humanities at UC Berkeley; Alan Tansman, former Director of the Townsend Center for the Humanities; and the Department of Rhetoric. At key moments of this project, I was grateful to those who offered exceptional support, intellectual and otherwise, as well as the spirit of collaboration; my thanks to Anne Alombert, Mitra Azar, Michael Dalebout, Nima Bassiri, Vicky Kahn, Gerald Moore, Luciana Parisi, Danilo Scholz, Yuk Hui, Catherine Malabou, Stefanos Geroulanos, Jonathan Sheehan, and Alan Tansman. Bernard Stiegler helped crystallize my argument while giving me new ways to think about technology and the mind; I dedicate the book to his memory. Finally, I thank my family: Heather, who urged me to begin writing during the first COVID shutdown; Graeme, particularly for his perspectives on computing and programming; and Ann, for being with me through this somewhat errant journey.

Portions of the text appeared in earlier publications, in the journal *Representations* and the following edited collections: Jessica Riskin, ed., *Genesis Redux: Essays in the History and Philosophy of Artificial Life* (Chicago: University of Chicago Press, 2007); Nitzan Lebovic and Andreas Killen, eds., *Catastrophes: A History and Theory of an Operative Concept* (Berlin: De Gruyter, 2014); David Bates and Nima Bassiri, eds., *Plasticity and Pathology: On the Formation of the Neural Subject* (New York: Fordham University Press, 2016); and Darrin McMahon and Joyce Chaplin, eds., *Genealogies of Genius* (London: Palgrave, 2017).

Notes

Chapter 1

1. See Jean-Pierre Dupuy, *The Mechanization of the Mind: On the Origin of Cognitive Science*, trans. M. B. DeBevoise (Princeton, NJ: Princeton University Press, 2009); and Paul N. Edwards, *The Closed World: Computers and the Politics of Discourse in Cold War America* (Cambridge, MA: MIT Press, 1996).
2. See Fernando Vidal, *Being Brains: Making the Cerebral Subject* (New York: Fordham University Press, 2017); David Bates and Nima Bassiri, eds., *Plasticity and Pathology: On the Formation of the Neural Subject* (New York: Fordham University Press 2016); and Tony D. Sampson, *The Assemblage Brain: Sense Making in Neuroculture* (Minneapolis: University of Minnesota Press, 2016).
3. Paul Merolla et al., "A Digital Neurosynaptic Core Using Embedded Crossbar Memory with 45pJ per Spike in 45nm," *IEEE Custom Integrated Circuits Conference* (Sept. 2011).
4. See, for one recent example, H. Clark Barrett, *The Shape of Thought: How Mental Adaptations Evolve* (Oxford: Oxford University Press, 2015).
5. J. A. Bargh and T. L. Chartrand, "The Unbearable Automaticity of Being," *American Psychologist* 54 (1999): 462–79.
6. Daniel M. Wegner, *The Illusion of Conscious Will* (Cambridge, MA: MIT Press, 2002).
7. Daniel M. Kahneman, *Thinking, Fast and Slow* (New York: Mcmillan, 2011).
8. Philip Johnson-Laird, *Mental Models* (Cambridge, MA: Harvard University Press, 1983).
9. Steven Pinker, "So How *Does* the Mind Work?," *Mind and Language* 20, no. 1 (2005): 2.
10. Andy Clark, Introduction to Andy Clark, Julien Kiverstein, and Tilmann Vierkant, eds., *Decomposing the Will* (Oxford: Oxford University Press, 2013), 5.
11. Yuval Harari, *21 Lessons for the 21st Century* (New York: Vintage, 2018), 29.
12. See, e.g., Cathy O'Neil, *Weapons of Math Destruction: How Big Data Increases Inequality and Threatens Democracy* (New York: Crown, 2016); and Safiya Noble, *Algorithms of Oppression: How Search Engines Reinforce Racism* (New York: New York University Press, 2018).

13 Harari, *21 Lessons*, 29.
14 Brian Christian and Tom Griffiths, *Algorithms to Live By: The Computer Science of Human Decisions* (New York: HarperCollins, 2015). "As computers become better tuned to real-world problems, they provide not only algorithms people can borrow for their own lives, but a better standard with which to compare human cognition itself" (5).
15 Daniel Kahneman, Olivier Sibony, and Cass R. Sunstein, *Noise: A Flaw in Judgment* (New York: Little, Brown, 2021). "We introduce several noise reduction techniques that we collect under the label of *decision hygiene*" (9).
16 Harari, *21 Lessons*, 56.
17 See the website for Stanford's Center for Human-Centered Computing. See as well Berkeley's Center for Human-Compatible AI and many other contemporary ventures along the same lines.
18 See Luciana Parisi, "Automatic Thinking and the Limits of Reason," *Cultural Studies – Critical Methodologies* 16, no. 5 (2016): 471–81.
19 Reinhart Koselleck, *Critique and Crisis: Enlightenment and the Pathogenesis of Modern Society* (Cambridge, MA: MIT Press, 1988).
20 Koselleck, *Critique*, 5.
21 Koselleck, *Critique*, 5.
22 Otto Bruner, Werner Conze, and Reinhart Koselleck, eds., *Geschichtliche Grundbegriffe: Historisches Lexicon zur politisch-sozialen Sprache in Deutschland* (Stuttgart: Klett Cotta, 1972), s.v. "Krise."
23 Carl Schmitt, "Dialogue on Power and Access to the Holder of Power," trans. Samuel Garrett Zeitlin, in Carl Schmitt, *Dialogues on Power and Space*, ed. Andreas Kalyvas and Federico Finchelstein (Cambridge: Polity, 2015), 45.
24 Carl Schmitt, "Dialogue on New Space" (1958), in Schmitt, *Dialogues*, 80.
25 Peter Thiel, "The Straussian Moment," in *Politics and Apocalypse*, ed. Robert Hamerton Kelly (East Lansing: Michigan State University Press, 2007): 189–218.
26 On the contemporary relevance of this idea, see Antoinette Rouvroy and Thomas Berns, "Algorithmic Governmentality and Prospects of Emancipation: Disparateness as a Precondition for Individuation through Relationships?," *Réseaux* 177 (2013): 163–96. The original concept is developed in Michel Foucault, "Governmentality," trans. Rosi Braidotti, rev. Colin Gordon, in *The Foucault Effect: Studies in Governmentality*, ed. Graham Burchell, Colin Gordon, and Peter Miller (Chicago: University of Chicago Press, 1991), 87–104.
27 See Fernando Vidal, *The Sciences of the Soul: The Early Modern Origins of Psychology*, trans. Saskia Brown (Chicago: University of Chicago Press, 2011).
28 Achille Mbembe, "Bodies as Borders," *From the European South* 4 (2019): 14. See as well Nick Land, *The Dark Enlightenment* (n.p.: Imperium Press, 2022); also available at http://www.thedarkenlightenment.com/the-dark-enlightenment-by-nick-land/. Accessed Nov. 24, 2022.
29 Achille Mbembe, *The Critique of Black Reason*, trans. Laurent Dubois (Durham, NC: Duke University Press, 2017), 5–6. Unless otherwise noted, emphasis in quotations is in the original.
30 Mbembe, "Bodies," 15.

Chapter 2

1 René Descartes, *Discours de la méthode*, in *The Philosophical Writings of Descartes*, vol. 1, trans. John Cottingham, Robert Stoothoff, and Dugald Murdoch (Cambridge: Cambridge University Press, 1985), 119.
2 Antonio Damasio, *Descartes' Error: Emotion, Reason, and the Brain* (New York: Penguin, 1995), 250.
3 Bernard Stiegler, *Technics and Time*, vol. 2: *Disorientation* (1996), trans. Stephen Barker (Stanford: Stanford University Press, 2009), 164. See as well Edwards, *The Closed World*, esp. chs. 6–8.
4 Olivier Houdé, "L'esprit, le cerveau et le corps: Descartes face aux sciences cognitives," in Annie Bitpol-Hespériès et al., *Descartes et son oeuvre aujourd'hui* (Sprimont: Mardaga, 1998).
5 For an overview, see the excellent essays on perception and thinking collected in Stephen Gaukroger, John Schuster, and John Sutton, eds., *Descartes' Natural Philosophy* (London: Routledge, 2000). Other recent works emphasizing Descartes's physiological approach to cognition include Daniel Garber, *Descartes Embodied: Reading Cartesian Philosophy though Cartesian Science* (Cambridge: Cambridge University Press, 2001); Dennis des Chene, *Spirits and Clocks: Machine and Organism in Descartes* (Ithaca, NY: Cornell University Press, 2001); Stephen Gaukroger, *Descartes: An Intellectual Biography* (Oxford: Oxford University Press, 1995); Dennis Sepper, *Descartes' Imagination: Proportion, Images, and the Activity of Thinking* (Berkeley: University of California Press, 1996); John Sutton, *Philosophy and Memory Traces: Descartes to Connectionism* (Cambridge: Cambridge University Press, 1998).
6 Vincente Aucante, *La philosophie médicale de Descartes* (Paris: Presses universitaires de France, 2006). See as well Richard B. Carter, *Descartes' Medical Philosophy: The Organic Solution to the Mind-Body Problem* (Baltimore: Johns Hopkins University Press, 1983).
7 Descartes, letter to Mersenne, 1632. AT 1:263; CSM 3:40. The following abbreviations will be used for Descartes's works: AT = *Oeuvres de Descartes*, ed. Charles Adam and Paul Tannery, 11 vols. (Paris: Vrin, 1973); CSM = *The Philosophical Writings of Descartes*, 3 vols., trans. John Cottingham, Robert Stoothoff, and Dugald Murdoch (vol. 3, including Anthony Kenny) (Cambridge: Cambridge University Press, 1988).
8 On the importance of this idea of information in Descartes's work, see Jacques Bouveresse, "La mécanique, la physiologie et l'âme," in Bitpol-Hespériès et al., *Descartes et son oeuvre aujourd'hui*, 99–100; Jean-Luc Marion, *Sur l'ontologie grise de Descartes: Science cartésienne et savoir aristotelicien dans les "Regulae"* (Paris: Vrin, 1981), 119, 123–25; Editors' Introduction to Gaukroger, Schuster, and Sutton, *Descartes' Natural Philosophy*, 16; Amélie Oksenberg Rorty, "Descartes on Thinking with the Body," in *Cambridge Companion to Descartes*, ed. John Cottingham (Cambridge: Cambridge University Press, 1992), 378.
9 On Descartes's conceptual move and its context, see Des Chene, *Spirits and Clocks*, ch. 6; and Gary Hatfield, "Mechanizing the Sensitive Soul," in *Matter and Form in Early Modern Science and Philosophy*, ed. Gideon Manning (Leiden: Brill, 2012), 151–86.
10 In general, see Otto Mayr, *Authority, Liberty, and Automatic Machinery in Early Modern Europe* (Baltimore: Johns Hopkins University Press, 1986); for religious

automata, see Jörg Jochen Berns, "Sakralautomaten: Automatisierungstendenzen in der mittelalterlichen und frühneuzeitlichen Frömmigkeitskulture," in *Automaten in Kunst und Literatur des Mittelalters und der Frühen Neuzeit*, ed. Klaus Grubmüller and Markus Stock (Wiesbaden: Harrassowitz, 2003), 197–22; and Jessica Riskin, *The Restless Clock: The Centuries-Long Argument over What Makes Living Things Tick* (Chicago: University of Chicago Press, 2016), ch. 1.

11 René Descartes, letter to Reneri for Pollot, April or May 1638, AT 2:39–40; CSM 3:99.
12 René Descartes, letter to More, Feb. 5, 1649, AT 5:277–78; CSM 3:366.
13 Descartes, letter to Reneri for Pollot, AT 2:40, CSM 3:99.
14 René Descartes, *Discourse on Method*, AT 6:56–57; CSM 1:140.
15 René Descartes, *Traité de l'homme*, AT 11:173, CSM 1:104.
16 John Cottingham's term "trialism" attempts to capture the importance for Descartes of this liminal zone where body and mind meet, a space rich with its own ontological possibilities, what Descartes might have called "native intelligence." See John Cottingham, "Cartesian Trialism," *Mind* 94 (1985): 218–30. See as well Fernando Vidal, *The Sciences of the Soul*, 77.
17 Gary Hatfield, "Descartes' Physiology and Psychology," in Cottingham, *Cambridge Companion*, 341.
18 Annie Bitpol-Hespériès, *Le principe de vie chez Descartes* (Paris: Vrin, 1990).
19 Stanley Finger, *Minds behind the Brains: A History of the Pioneers and Their Discoveries* (Oxford: Oxford University Press, 2000).
20 On Galen, see F. R. Freeman, "Galen's Ideas on Neurological Function," *Journal of the History of Neuroscience* 3, no. 4 (1992): 263–71.
21 On the transformation of "psychology" in the seventeenth century, from science of the soul as "animating principle of all living beings" to the study of "the rational human soul united with the body," see Vidal, *Sciences of the Soul*, 57. For the importance of the unity of mind and body in Descartes's physiology, see Annie Bitpol-Hespériès, Introduction to René Descartes, *Le monde, L'homme* (Paris: Seuil, 1996).
22 As Georges Canguilhem has forcefully argued, Descartes's mechanical understanding of the bodily organism merely displaces the Aristotelian question of the soul as form, and the telos of the formal organization, onto the figure of the Creator—whether it is divine or human in nature: "Mechanism can explain everything so long as we take machines as already granted, but it cannot account for the construction of machines." See Georges Canguilhem, "Machine and Organism" (1948), in Canguilhem, *Knowledge of Life*, ed. Paola Marati and Todd Meyers, trans. Stefanos Geroulanos and Daniela Ginsburg (New York: Fordham University Press, 2008), 87. I will return to the question of the unity of organism and machine in the discussion of Descartes's later work, the *Passions de l'âme*.
23 Descartes, *Traité de l'homme*, AT 11:131; CSM 1:100–101.
24 "And note by 'figures' I mean not only things that somehow represent the position of the edges and surfaces of objects, but also everything which, as indicated above, can cause the soul to sense movement, size, distance, colors, sounds, odors, and other such qualities; and even things that can make it sense titillation, pain, hunger, thirst, joy, sadness, and other such passions." Descartes, *Traité de l'homme*, AT 11:176; CSM 1:106.
25 "These shapes are not the literal analogues of perceived objects," writes one commentator, "rather they constitute a fully arbitrary code embedded in what we

now call a virtual function space, or, in Descartes' terms, the corporeal imagination." Paul Miers, "Descartes and the Algebra of the Soul," *MLN* 110, no. 4 (1995): 948.
26 Jean-Pierre Séris, "Language and Machine in the Philosophy of Descartes," in *Essays on the Philosophy and Science of René Descartes*, ed. Stephen Voss (New York: Oxford University Press, 1993), 183.
27 See Alison Simmons, "Representation," in *The Cambridge Descartes Lexicon*, ed. Larry Nolan (Cambridge: Cambridge University Press, 2016): 645–54. Thanks to Jacob Sheehan for alerting me to this thorough and sophisticated examination of Descartes's theory of perception and his version of "direct representation."
28 Descartes's theory of the pineal gland is often misunderstood. First, he was not the first to conjecture the cognitive function of the gland. Its peculiarity gave rise to various speculations and Descartes was relying heavily on prior anatomical work on the brain as well as his own animal dissections. Second, Descartes did not believe that only human brains had pineal glands; it was common knowledge that it existed in animals and Descartes himself often referred to them. See Bitpol-Hespériès, *Le principe de vie*; and Aucante, *La philosophie médicale*.
29 I am indebted to Nima Bassiri's lucid explication of the complex ontological status of the pineal gland in relation to both soul and body. See Nima Bassiri, "Material Translations in the Cartesian Brain," *Studies in History and Philosophy of Biological and Biomedical Sciences* 43, no. 1 (2012): 244–45.
30 *Objections and Replies*, AT 7:229–30; CSM 2:161
31 See Sutton, *Philosophy and Memory Traces*, ch. 3.
32 Sutton, *Philosophy and Memory Traces*, pt. 1.
33 Descartes, letter to Mersenne, Mar. 18, 1630, AT 1:134; CSM 3:20.
34 Descartes, *Traité de l'homme*, AT 11:185. (This passage is omitted in the translation in CSM.) Translations are my own unless otherwise cited.
35 It is worth noting that Aristotle himself preserved the explanatory power of causation in the higher animals with a similar argument: experience could be maintained in the body and therefore cause behaviors at a later moment. See Klaus Corcilius, *Streben und Bewegen: Aristoteles' Theorie der animalischen Ortsbewegung. Quellen und Studien zur Philosophie*, vol. 79 (Berlin: De Gruyter, 2008).
36 Descartes, *Traité de l'homme*, AT 11:132; CSM 1:101.
37 On this aspect of the *Regulae*, see Stephen Gaukroger, *Cartesian Logic: An Essay on Descartes's Conception of Inference* (Oxford: Oxford University Press, 1989), 25, 127–28; and Marion, *Sur l'ontologie grise*, 30.
38 René Descartes, *Rules for the Direction of the Mind*, AT 10:368, CSM 1:14.
39 René Descartes, *Meditations on First Philosophy*, AT 7:22; CSM 2:15.
40 Descartes, *Meditations on First Philosophy*, AT 7:31–32; CSM 2:20–21.
41 As Dennis Sepper explains, "There can be no thinking and knowing without the internal senses and their phantasms. All reasoning, conceiving, understanding, all science and truth, must come to us by way of and accompanied by phantasms." Sepper, *Descartes's Imagination*, 28.
42 Descartes, *Meditations on First Philosophy*, AT 7:31; CSM 2:21.
43 Descartes, *Meditations on First Philosophy*, AT 7:32; CSM 2:21.
44 René Descartes, *Passions de l'âme*, §6, AT 11:330; CSM 1:329.
45 Descartes, Conversation with Burman, AT 5:163–64; CSM 3:346; and *Passions de l'âme*, AT 5:179; CSM 3:354: "Even when we are ill, nature still remains the same." On Descartes's thinking on pathology, see Aucante, *La philosophie médicale*, ch. 8.

46   Descartes, *Passions de l'âme*, §16, AT 11:342; CSM 1:335.
47   Descartes, *Passions de l'âme*, §17, AT 11:342; CSM 1:335.
48   Descartes, letter to Regius, Jan. 1642, AT 3:493; CSM 3:206.
49   See Carter, *Descartes' Medical Philosophy*, 119–20, for a similar formulation.
50   Descartes, *Passions de l'âme*, §211, AT 11:487; CSM 1:403–4.
51   Descartes, *Passions de l'âme*, §30, AT 11:351; CSM 1:339.
52   Bassiri, "Material Translations," 249.
53   See, e.g., Deborah J. Brown, *Descartes and the Passionate Mind* (Cambridge: Cambridge University Press, 2006), 141.
54   Descartes, *Passions de l'âme*, §40, AT 11:359; CSM 1:343.
55   Descartes, *Passions de l'âme*, §52, AT 11:372; CSM 1:349.
56   Descartes, *Passions de l'âme*, §70, AT 11:380; CSM 1:353.
57   For Catherine Malabou and Adrian Johnston, the exception that is "wonder" constitutes the very possibility of the soul's own self-recognition. See Catherine Malabou and Adrian Johnston, *Self and Emotional Life: Philosophy, Psychoanalysis, and Neuroscience* (New York: Columbia University Press, 2013), ch. 1.
58   Descartes, *Passions de l'âme*, §70, AT 11:380; CSM 1:353.
59   Descartes, *Passions de l'âme*, §73, AT 11:382–83; CSM 1:354.
60   Christof Koch has framed this problem in ways that look back to Descartes's seminal effort to think the human from within the cybernetic automaton. As he writes, most actions of the higher animals depend on "zombie agents" that are automatically executed. "The hallmarks of a zombie system are stereotypical, limited sensorimotor behaviour, and immediate, rapid action. Its existence raises two questions. First, why aren't we just big bundles of unconscious zombie agents? Why bother with consciousness, which takes hundreds of milliseconds to set in?" For Koch, human thinking reveals best how nature has "evolved a powerful and flexible system whose primary responsibility is to deal with the unexpected and to plan for the future." "The function of consciousness ... is to handle those special situations for which no automatic procedures are available." See Christof Koch and Francis Crick, *The Quest for Consciousness: A Neurobiological Approach* (Englewood, CO: Roberts, 2004), 318–19.

Chapter 3

1   Thomas Hobbes, *Elements of Philosophy, the First Section, Concerning Body*, in *The English Works of Thomas Hobbes of Malmesbury*, vol. 1, ed. William Molesworth (London: John Bohn, 1839), 3. On Hobbes as a founding father of artificial intelligence, see, e.g., John Haugeland, *Artificial Intelligence: The Very Idea* (Cambridge, MA: MIT Press, 1985), ch. 1.
2   For a media history of certain kinds of "paper machines" that organize information, see Martin Krajewski, *Paper Machines: About Cards & Catalogs, 1548–1929* (Cambridge, MA: MIT Press, 2011); and Cornelia Vismann, *Files: Law and Media Technology*, trans. Geoffrey Winthrop-Young (Stanford: Stanford University Press, 2008).
3   Hobbes, *Elements*, 1:1–2.
4   Thomas Hobbes, *Leviathan*, ed. Richard Tuck (Cambridge: Cambridge University Press, 1996), ch. 4.
5   On the complexity of the figure of the automaton in Hobbes and lingering traces

of Aristotelianism, see Z. Bhorat, "Automata in Hobbes: Three Heads of Cerebus," *History of Political Thought* 42, no. 3 (2021): 441–63.
6   Thomas Hobbes, *Decameron Physiologicum*, in *English Works*, vol. 7 (London, 1845), 176.
7   Hobbes, *Leviathan*, 9.
8   Robert Boyle, *The Sceptical Chymist: or Chemico-Physical Doubts and Paradoxes, Touching the Spagyrist's Principles Commonly Call'd Hypostatical* . . . (1661), in *The Works of Robert Boyle*, 14 vols., ed. Michael Hunter and Edward B. Davis (London: Routledge, 1999) 2: 354–55.
9   Hobbes, *Leviathan*, 398. Subsequent citations are given parenthetically in the text.
10 Hobbes, *Elements*, 1:391.
11 Hobbes, *Elements*, 1:391.
12 Hobbes, *Elements*, 1:391–92.
13 Hobbes, *Elements*, 1:393.
14 Jakob Von Uexküll, "The Theory of Meaning," *Semiotica* 42, no. 1 (1982): 25–82, at 30.
15 Early users of microscopes, such as Pierre Borele, had found in samples of human blood what they took to be little worms—though they were most likely just seeing red blood cells.
16 Spinoza, letter to Henry Oldenburg, Nov. 1665, in Baruch de Spinoza, *Collected Works of Spinoza*, vol. 2, ed. and trans. Edwin Curley (Princeton, NJ: Princeton University Press, 1985), letter 32, p. 19.
17 Spinoza, letter to Oldenburg, 19.
18 Spinoza, letter to Oldenburg, 19–20.
19 On the complexities (and difficulties) of Spinoza's conception of reason, see Michael LeBuffe, *Spinoza on Reason* (Oxford: Oxford University Press, 2018). On the dependence of reason on the logic of the imagination, see Christof Ellsiepen, "The Types of Knowledge (2P38–47)," in Michael Hampe et al., *Spinoza's Ethics: A Collective Commentary* (Leiden: Brill, 2011), 133–34.
20 Spinoza, *Collected Works*, letter 76, 2:475.
21 Spinoza, letter to Simon de Vries, in *Collected Works*, 1:161. On essences and their access through mental representation, see Michael Della Rocca, *Spinoza* (New York: Routledge, 2008), ch. 3.
22 See Hans Jonas, "Spinoza and the Theory of Organism," *Journal of the History of Philosophy* 3, no. 1 (1965): 43–57.
23 Citations in the text are to Spinoza, *The Ethics*, in *A Spinoza Reader: The Ethics and Other Works*, ed. and trans. Edwin Curley (Princeton, NJ: Princeton University Press, 1994). Citations refer to book, axioms (a), propositions (p), scholia (s), corollaries (c), and demonstrations (d). When useful I have indicated the page number of Curley's edition.
24 Jonas, "Spinoza and the Theory of the Organism."
25 Gilles Deleuze, *Expressionism in Philosophy: Spinoza*, trans. Martin Joughin (New York: Zone, 1990).
26 Baruch de Spinoza, *Treatise on the Emendation of the Intellect*, in *Collected Works*, 1:36.
27 Gilles Deleuze, *Spinoza: Practical Philosophy*, trans. Robert Hurley (San Francisco: City Lights, 1988), 55.
28 Spinoza, *Emendation of the Intellect*: "But just as men, in the beginning, were able

to make the easiest things with the tools they were born with (however laboriously and imperfectly), and once these had been made, made other, more difficult things with less labor and more perfectly, and so, proceeding gradually from the simplest works to tools, and from tools to other works and tools, reached the point where they accomplished so many and so difficult things with little labor, in the same way the intellect, by its inborn power, makes intellectual tools for itself, but which it acquires other powers for other intellectual works, and from these works still other tools, or the power of searching further, and so proceeding by stages, until it reaches the pinnacle of wisdom" (17).

29 Spinoza, *Emendation of the Intellect*, 37.
30 An analysis of the spiritual automaton but not an account of its artificial origin can be found in Eugene Marshall, *The Spiritual Automaton: Spinoza's Science of the Mind* (Oxford: Oxford University Press, 2013).
31 Cf. Spinoza, *Ethics*: "The best thing, then, that we can do, so long as we do not have perfect knowledge of our affects, is to conceive a correct principle of living, or sure maxims of life, to commit them to memory, and to apply them constantly to the particular cases frequently encountered in life. In this way our imagination will be extensively affected by them, and we shall always have them ready" (5p10s).
32 Diane Steinberg, "Knowledge in Spinoza's Ethics," in *The Cambridge Companion to Spinoza's Ethics*, ed. Olli Koistinen (Cambridge: Cambridge University Press, 2009), 150–55.
33 On the possibility of intuiting essence despite the contingency of experience, see Kristin Primus, "*Scientia Intuitiva* in the Ethics," in *Spinoza's Ethics: A Critical Guide*, ed. Yitzhak Y. Melamed (Cambridge: Cambridge University Press, 2017), 185–86. See as well Sanem Soyarsian, "The Distinction between Reason and Intuitive Knowledge in Spinoza's Ethics," *European Journal of Philosophy* 24, no. 1 (2016): 27–64.

Chapter 4

1 Joseph Hoffman, quoted in François Duchesnau, "The Organism-Mechanism Relationship: An Issue in the Leibniz-Stahl Controversy," in *The Life Sciences in Early Modern Philosophy*, ed. Ohad Nachtomy and Justin E. H. Smith (Oxford: Oxford University Press, 2014), 103–4. Subsequent citations are given parenthetically in the text.
2 See François Duchesnau, "Physiology and Organic Bodies," in *Oxford Handbook of Leibniz*, ed. Maria Rosa Antognazza (Oxford: Oxford University Press, 2018), 466–84; Raphaële Andrault, *La vie selon la raison: Physiologie et métaphysique chez Spinoza et Leibniz* (Paris: Honoré Champion, 2014).
3 See François Duchesnau and Justin E. H. Smith, Introduction to *The Leibniz-Stahl Controversy*, trans. and ed. François Duschesneau and Justin E. H. Smith (New Haven, CT: Yale University Press, 2016); see as well Justin E. H. Smith, *Divine Machines: Leibniz and the Sciences of Life* (Princeton, NJ: Princeton University Press, 2011), ch. 2.
4 John Locke, *Essay concerning Human Understanding*, ed. Peter Nidditch (Oxford: Oxford University Press, 1975), bk. 2, ch. 27, §§4–5.

5   See Duchesnau, "The Organism-Mechanism Relation," 114; and Smith, *Divine Machines*, 100.
6   G. W. Leibniz, "Animadversions concerning Certain Assertions of the *True Medical Theory* [*Theoria Medica Vera*]" (1709), in Duchesnau and Smith, *The Leibniz-Stahl Controversy*, 23.
7   G. W. Leibniz, *Theodicy* (1710), trans. E. M. Huggard, ed. Austin Farrer (La Salle, IL: Open Court, 1985), 365.
8   Daniel Garber, *Leibniz: Body, Substance, Monad* (Oxford: Oxford University Press, 2009), 49.
9   Garber, *Leibniz*, 66. See as well Larry M. Jorgensen, *Leibniz's Naturalized Philosophy of Mind* (Oxford: Oxford University Press, 2019), ch. 4.
10 G. W. Leibniz, *New System of Nature*, in G. W. Leibniz, *Philosophical Essays*, ed. and trans. Roger Ariew and Daniel Garber (Indianapolis: Hackett), 139.
11 Leibniz, *New System*, 139.
12 On plastic nature in this context and a critique of Ralph Cudworth, see Leibniz, letter to Lady Masham (daughter of Cudworth), July 1705, in G. W. Leibniz, *Die philosophischen Schriften von Gottfried Wilhelm Leibniz*, vol. 3 (Berlin: Weidmann, 1887), 366–68; and G. W. Leibniz, "Considerations on the Principles of Life, and on Plastic Natures" (1705), in *The Philosophical Works of Leibnitz*, 2nd ed., trans. George Martin Duncan (New Haven, CT: Yale University Press, 1908).
13 Leibniz, *New System*, 140.
14 Leibniz, *New System*, 144.
15 Leibniz, *New System*, 142.
16 Leibniz, *New System*, 141.
17 Leibniz, *New System*, 143.
18 Leibniz, *New System*, 143.
19 See Adam Harmer, "Mind and Body," in Antognozza, *Oxford Handbook of Leibniz*, 393–409.
20 Smith, *Divine Machines*, 118–19.
21 Leibniz, *New System*, 144.
22 Garber, *Leibniz*, 205.
23 Garber, *Leibniz*, 217.
24 Leibniz, "Animadversions," in Duchesnau and Smith, *The Leibniz-Stahl Controversy*, 21. See Smith, *Divine Machines*, 118–19.
25 Leibniz, Letter to Lady Masham, undated, in *Philosophischen Schriften*, 3:374.
26 On these projects, see Matthew Jones, *Reckoning with Matter: Calculating Machines, Innovation, and Thinking about Thinking from Pascal to Babbage* (Chicago: University of Chicago Press, 2016).
27 G. W. Leibniz, *New Essays on Human Understanding*, ed. and trans. Peter Remnant and Jonathan Bennett (Cambridge: Cambridge University Press, 1996), 144; see as well Gilles Deleuze, *The Fold: Leibniz and the Baroque*, trans. Tom Conley (Minneapolis: University of Minnesota Press, 1993).
28 Leibniz, letter to Queen Sophie of Prussia, 1702, in *Philosophical Essays*, 188.
29 Leibniz, letter to Queen Sophie, 188.
30 Leibniz, letter to Queen Sophie, 191.
31 Leibniz, letter to Queen Sophie, 192.
32 Leibniz, letter to Queen Sophie, 192.
33 Leibniz, letter to Henry Oldenburg, 1675, in G. W. Leibniz, *Philosophical Papers*

and *Letters: A Selection*, 2nd ed., ed. and trans. Leroy M. Loemker (Dordrecht: Kluwer, 1969), 166.
34 G. W. Leibniz, "On the Universal Method," in Leibniz, *Selections*, ed. and trans. Philip P. Wiener (New York: Scribner, 2009), 4.
35 Leibniz, "On the General Characteristic," in *Philosophical Papers*, 224.
36 Leibniz, "Dialogue," in *Philosophical Papers*, 182–85.
37 Leibniz, *The Principles of Philosophy, or, the Monadology*, in *Philosophical Essays*, §17. Subsequent references to section number are given parenthetically in the text.
38 See Deleuze, *The Fold*, 119.
39 Leibniz, "Animadversions," in *The Leibniz-Stahl Controversy*, 22.

Chapter 5

1 Pascal, *Pensées*, trans. A. J. Krailsheimer (New York: Penguin, 1995), fragment 661.
2 Pascal, *Pensées*, fragment 661.
3 Nicolas Malebranche, *Search after Truth*, trans. T. Taylor (London: Bowyer, 1700), 1:47.
4 Malebranche, *Search after Truth*, 1:59.
5 John Locke, *Essay*, bk. 2, ch. 33, §6.
6 John Locke, *Of the Conduct of the Understanding*, 5th ed., ed. Thomas Fowler (Oxford: Oxford University Press, 1901), 66.
7 Locke, *Of the Conduct of the Understanding*, 67.
8 Jonathan Sheehan and Dror Wahrman, *Invisible Hands: Self-Organization and the Eighteenth Century* (Chicago: University of Chicago Press, 2015), ch. 5.
9 David Hume, *Dialogues Concerning Natural Religion*, ed. Dorothy Coleman (Cambridge: Cambridge University Press, 2007), 24.
10 Hume, *Dialogues*, 24.
11 Gilles Deleuze, *Empiricism and Subjectivity: An Essay on Hume's Theory of Human Nature*, trans. Constantin V. Boundas (New York: Columbia University Press, 1991), 22.
12 David Hume, *Treatise on Human Nature* (London: John Noon, 1739–40), 1.1.4. Subsequent citations of book, chapter, section, and paragraph are given parenthetically in the text.
13 Deleuze, *Empiricism*, 31.
14 Deleuze, *Empiricism*, 93.
15 David Hume, "The Epicurean," in David Hume, *Essays Moral, Political, and Literary*, 2 vols., ed. Eugene F. Miller (Indianapolis: Liberty Fund, 1987), 1:139–40.
16 Sheehan and Wahrman, *Invisible Hands*.
17 On habit and the default of the human in the Enlightenment, see Dorothea von Mücke, *The Practices of the Enlightenment: Aesthetics, Authorship, and the Public* (New York: Columbia University Press, 2015), ch. 2.
18 Charles Bonnet, *Essai de la psychologie* (London: n.p., 1755).
19 Denis Diderot, *Elémens de physiologie*, in *Oeuvres complètes de Diderot*, ed. Herbert Dieckmann (Paris: Hermann, 2003), 17:470.
20 David Hartley, *Observations on Man, His Frame, His Duties, and His Expectations* (London: S Richardson, 1749).

Chapter 6

1. Immanuel Kant, *Notes and Fragments*, ed. Paul Guyer, trans. Curtis Bowman, Frederick Rauscher, and Paul Guyer (Cambridge: Cambridge University Press, 2010), 331.
2. Immanuel Kant, *Natural Science*, ed. Eric Watkins (Cambridge: Cambridge University Press, 2015), 299.
3. Kant, *Natural Science*, 299.
4. Kant, *Natural Science*, 298.
5. Kant, *Natural Science*, 299.
6. On this topic, see Peter Szendy, *Kant in the Land of the Extraterrestrials: Cosmopolitical Philosofictions*, trans. Will Bishop (New York: Fordham University Press, 2013).
7. Kant, *Notes and Fragments*, 6.
8. Kant, *Notes and Fragments*, 39, 36.
9. Kant, *Notes and Fragments*: "Everything that happens is sufficiently determined, but not from appearances, rather in accordance with laws of appearance" (254).
10. Kant, *Notes and Fragments*, 254.
11. Theodor Adorno, *Kant's "Critique of Pure Reason"* (1959), ed. Rolf Tiedemann, trans. Rodney Livingstone (Stanford: Stanford University Press, 2001), 129.
12. Immanuel Kant, *Critique of Pure Reason*, ed. and trans. Paul Guyer and Allen W. Wood (Cambridge: Cambridge University Press, 1998). Citations in the text will be to the first (A) edition and the second (B) edition.
13. Jennifer Mensch, *Kant's Organicism: Epigenesis and the Development of Critical Philosophy* (Chicago: University of Chicago Press, 2013). I borrow the term "individuation" from Gilbert Simondon. See Simondon, *Individuation in Light of Notions of Form and Information*, trans. Taylor Adkins (Minneapolis: University of Minnesota Press, 2020).
14. As Hannah Ginsborg writes in *The Normativity of Nature: Essays on Kant's Critique of Judgement* (Oxford: Oxford University Press, 2014), "It is only by virtue of taking a normative attitude to one's mental activity that one can regard it as governed by rules, which in turn is required for recognizing the objects we perceive as falling under empirical concepts" (169).
15. See Catherine Malabou, *Before Tomorrow: Epigenesis and Rationality*, trans. Carolyn Shread (Cambridge: Polity, 2016), for a deep analysis of Kant on this question and for an overview of the debate on epigenesis among Kant scholars.
16. In his refutation of Idealism, Kant says, "Whether this or that putative experience is not mere imagination [or dream or delusion, etc.] must be ascertained according to its particular determinations and through its coherence with the criteria of all actual experience" (B279).
17. See Alison Laywine, "Problems and Postulates: Kant on Reason and Understanding," *Journal of the History of Philosophy* 36, no. 2 (1998): 279–309.
18. On the importance of unity for Kant, see, e.g., Susan Neiman, *The Unity of Reason: Rereading Kant* (Oxford: Oxford University Press, 1994); A. Nuzzo, *Kant and the Unity of Reason* (Oxford: Oxford University Press, 1997); and Paul Guyer, "The Unity of Reason: Pure Reason as Practical Reason in Kant's Early Conception of the Transcendental Dialectic," *Monist* 72 (1989): 139–67.
19. Immanuel Kant, *Critique of Practical Reason*, ed. and trans. Mary Gregor (Cambridge: Cambridge University Press, 1997), 85.

20 Kant, *Critique of Practical Reason*, 81–82.
21 Kant, *Critique of Practical Reason*, 85.
22 Kant, *Critique of Practical Reason*, n. 11.
23 Autonomous—but perhaps not fully self-enclosed? As Kant tells us, in the First Critique, some minds can *imitate* rational thought having learned and digested "a system of philosophy." As long as the principles and definitions are not disputed and left intact, this mind gives a simulation of active thought. However, this person "has formed himself from an alien reason" and was not self-generated; he is like "a plaster cast of a living human being" (A836/B865). As we see with earlier Enlightenment figures, the openness of the mind to "alien" thought is also the very possibility of education. What does it mean for Kant if the mind can be something it is not—a cognition of another mind? Of course, Kant dismissed the idea that there could be "a faculty of mind to stand in a community of thoughts with other men (no matter how distant they may be)" as "entirely groundless, because it cannot be grounded in experience and its known laws" (A222/B270). As we will see, this slippage will come back to haunt Kant's later reflections on the order and organization of nature.

Chapter 7

1 Immanuel Kant, *Critique of Judgment*, trans. Werner S. Pluhar (Indianapolis: Hackett, 1987), 265.
2 Kant, *Critique of Judgment*, 274.
3 See Ginsborg, *The Normativity of Nature*.
4 Kant, *Critique of Judgment*, 34.
5 Kant, *Critique of Judgment*, 255.
6 Kant, *Critique of Judgment*, 248–49.
7 Kant, *Critique of Judgment*, 271. On Nature's "technic," see Pheng Cheah, "Human Freedom and the Technic of Nature: Culture and Organic Life in Kant's Third Critique," *differences* 14, no. 2 (2003): 1–26.
8 Kant, *Critique of Judgment*, 248.
9 Kant, *Critique of Judgment*, 252.
10 Kant, *Critique of Judgment*, 252.
11 Immanuel Kant, *Opus Postumum*, ed. and trans. Eckart Förster and Michael Rosen (Cambridge: Cambridge University Press, 1995), 64.
12 Kant, *Critique of Judgment*, 254.
13 Kant, *Critique of Judgment*, 253.
14 Kant, *Critique of Judgment*, 253.
15 See James L. Larson, *Interpreting Nature: The Science of Living Form from Linnaeus to Kant* (Baltimore: Johns Hopkins University Press, 1994).
16 Kant, *Critique of Judgment*, 311.
17 Kant, *Critique of Judgment*, 306.
18 Kant, *Opus Postumum*, 85.
19 Immanuel Kant, *Anthropology from a Pragmatic Point of View*, in Kant, *Anthropology, History, and Education*, ed. Günter Zöller and Robert B. Louden, trans. Mary Gregor, Günter Zöller, and Robert B. Louden (Cambridge: Cambridge University Press, 2007), 306.
20 Kant, *Anthropology*, 293.

21 Kant, *Anthropology*, 247.
22 Kant, *Anthropology* 283.
23 Kant, *Anthropology*, 417–18.
24 Kant, *Anthropology*, 418.

Chapter 8

1. Karl Marx, *Capital*, 1:ch. 15. Available at https://www.marxists.org/archive/marx/works/1867-c1/ch15.htm. Accessed Jan. 6, 2023.
2. Karl Marx, "Fragment on Machines," from the *Grundrisse*, available at https://thenewobjectivity.com/pdf/marx.pdf. Accessed 1/6/2023.
3. Charles Babbage, *Science and Reform: Selected Works of Charles Babbage* (Cambridge: Cambridge University Press, 1989), 39.
4. See Simon Schaffer, "Babbage's Intelligence: Calculating Engines and the Factory System," *Critical Inquiry* 21, no. 1 (1994): 203–27.
5. William J. Ashworth, "Memory, Efficiency, and Symbolic Analysis: Charles Babbage, John Herschel, and the Industrial Mind," *Isis* 97, no. 4 (1986): 629–53.
6. Ashworth, "Memory," 648.
7. Charles Babbage and John Herschel, preface to *Memoirs of the Analytical Society 1813* (Cambridge: Cambridge University Press, 1813), i.
8. Charles Babbage, *The Exposition of 1851*, ch. 13, "Calculating Engines," in *The Works of Charles Babbage*, ed. Martin Campbell-Kelly (New York: New York University Press, 1989), 10:104.
9. Charles Babbage, *Charles Babbage and His Calculating Engines: Selected Writings* (New York: Dover, 1961), 55.
10. L. F. Manabrea, *Sketch of the Analytical Engine Invented by Charles Babbage, Esq., with Notes by the Translator*, trans. Ada Lovelace (London: Taylor, 1843), note A, 694.
11. Ada Lovelace, letter to Woronzow Greig, Nov. 15, 1844, printed in Betty Alexandra Toole, *Ada, Enchantress of Numbers: Poetical Science* (Sausalito, CA: Critical Connection, 2010), ch. 18, Apple Books.

Chapter 9

1. Alfred Smee, *The Process of Thought Adapted to Words and Language, Together with a Description of the Relational and Differential Machines* (London: Longman, 1851), ix.
2. Hermann von Helmholtz, *Vorträge und Reden*, 4th ed. (Braunschweig: Vieweg, 1896), 15.
3. Smee, *Process*, 30. Subsequent citations of this work are given parenthetically in the text.
4. George Boole, *The Laws of Thought: An Investigation of the Laws of Thought on Which Are Founded the Mathematical Theories of Logic and Probabilities* (London: Macmillan, 1854), 24.
5. On the historical development of the idea of the *coup d'ceil*, see Lorraine Daston, "The *Coup d'Oeil*: On a Mode of Understanding," *Critical Inquiry* 4, no. 2 (2019): 307–31.

6   John Stuart Mill, *A System of Logic Ratiocinative and Inductive, Being a Connected View of the Principles of Evidence and the Methods of Scientific Investigation* (London: Longmans, 1891), 124a.
7   William Thomson, *An Outline of the Necessary Laws of Thought: A Treatise on Pure and Applied Logic*, 2nd ed. (London: William Pickering, 1849), 309.
8   Thompson, *An Outline*, 90.
9   W. Stanley Jevons, *The Principles of Science: A Treatise on Logic and Scientific Method* (London: Macmillan, 1887), 20. Subsequent citations of this work are given parenthetically in the text.
10 "It is the capability of mutual substitution which gives such great value to the modern methods of mechanical construction, according to which all the parts of a machine are exact facsimiles of a fixed pattern. The rifles used in the British army are constructed on the American interchangeable system, so that any part of any rifle can be substituted for the same part of another. A bullet fitting one rifle will fit all other of the same bore. Whitworth has extended the same system to the screws and screw-bolts used in connecting together the parts of machines, by establishing a series of standard screws." Jevons, *Principles*, 20–21.
11 Georg Wilhelm Friedrich Hegel, *The Science of Logic*, ed. and trans. George Di Giovanni (Cambridge: Cambridge University Press, 2010), 181.
12 Charles Sanders Peirce, "Deduction, Induction, and Hypothesis," in *Chance, Love, and Logic: Philosophical Essays*, ed. Morris Cohen (London: Kegan Paul, 1923), 131–53.
13 On abduction and conjecture, see Jaako Hintikka, "What Is Abduction? The Fundamental Problem of Contemporary Epistemology," in Hintikka, *Jaakko Hintikka Selected Papers*, vol. 5, *Inquiry as Inquiry: A Logic of Scientific Discovery* (Dordrecht: Springer, 1999).
14 Peirce, "Deduction," 142.
15 Ernest Naville, *La logique de l'hypothèse* (Paris: Germer Baillière, 1880), 253.
16 Charles Sanders Peirce, *The Collected Papers of Charles Sanders Peirce*, vol. 5, ed. Charles Hartshorne and Paul Weiss (Cambridge, MA: Harvard University Press, 1974), 181.
17 Charles Sanders Peirce, "Logical Machines" (1887), in *The Writings of Charles Sanders Peirce*, vol. 5, ed. Max H. Fisch (Bloomington: Indiana University Press, 1993), 65.
18 See K. L. Ketner and A. F. Stewart, "The Early History of Computer Design: Charles Sanders Peirce and Marquand's Logical Machines," *Princeton University Library Chronicle* 45, no. 3 (1984): 187–224.
19 Peirce, *Collected Papers*, vol. 2, §59.
20 Peirce, "Deduction," 152.
21 Peirce, *Collected Papers*, vol. 2, §56.
22 See David W. Bates, "Unstable Brains and Ordered Societies: On the Conceptual Origins of Plasticity, circa 1900," in *Distributed Perception: Resonances and Axiologies*, ed. Natasha Lushetich and Iain Campbell (London: Routledge, 2021), 122–24. Cf. Peter Skagestad, "Thinking with Machines: Intelligence Augmentation, Evolutionary Epistemology, and Semiotic," *Journal of Social and Evolutionary Systems* 16, no. 2 (1993), 163–68: "People are like machines, in that the unaided individual is as limited by his/her design as is the machine. But people do not think unaided, and therein lies the difference" (172).
23 Pierre Janet, *L'automatisme psychologique: Essai de psychologie expérimentale sur*

les formes inférieures de l'activité humaine (Paris: Alcan, 1889), 473. Subsequent citations are given parenthetically in the text.

Chapter 10

1. Hermann von Helmholtz, "On the Interaction of Natural Forces," London, Edinburgh, and Dublin Philosophical Magazine and Journal of Science 11, no. 75 (1856): 490.
2. Helmholtz, "On the Interaction of Natural Forces," 509.
3. On the French context, see John Tresch's wonderful study, The Romantic Machine: Science and Technology after Napoleon (Chicago: University of Chicago Press, 2016).
4. Balfour Stewart, The Conservation of Energy; with an Appendix, Treatise of the Vital and Mental Applications of the Doctrine (New York: D. Appleton, 1874), appendix. Subsequent citations are given parenthetically in the text.
5. James Clerk Maxwell, "Paradoxical Philosophy," Nature 19, no. 477 (1878): 141–43.
6. James Clerk Maxwell, "Essay for the Apostles of Nature on 'Analogies in Nature'" (Feb. 1856), in The Scientific Letters and Papers of James Clerk Maxwell, 3 vols., ed. P. M. Harman (Cambridge: Cambridge University Press, 1990–2002), 2:379–80.
7. James Clerk Maxwell, letter to Francis Galton, Feb. 26, 1879, in Maxwell, Scientific Letters and Papers, 3:757.
8. Maxwell, "Essay for the Apostles of Nature," 380.
9. Maxwell, "Essay for the Apostles of Nature," 380.
10. James Clerk Maxwell, "Essay for the Eranus club on Science and Free Will" (1873), in Maxwell, Scientific Letters and Papers, 2:817.
11. A. A. Cournot, Matérialisme, vitalisme, rationalisme: Etudes sur l'emploi des données de la science en philosophie (1875), ed. Claire Salomon-Bayet, in A. A. Cournot, Oeuvres complètes, 2nd ed., 11 vols. (Paris: Vrin, 1987), 5:53, 71–72.
12. Cournot, Matérialisme, 5:74.
13. Cournot, Matérialisme, 5:74.
14. Cournot, Matérialisme, 5:75.

Chapter 11

1. See, e.g., Barbara Tizard, "Theories of Brain Localization from Flourens to Lashley," Medical History 3, no. 2 (1959): 132–45; and S. Zola-Morgan, "Localization of Brain Function: The Legacy of Franz Joseph Gall (1758–1828)," Annual Review of Neuroscience 18 (1995): 359–83.
2. Marie-Jean-Pierre Flourens, Recherches expérimentales sur les propriétés et les fonctions du système nerveux dans les animaux vertébrés, 2nd ed. (Paris: J. B. Baillière, 1842), preface, xv. Subsequent citations are given parenthetically in the text.
3. Geert-Jan Rutten, "Broca-Wernicke Theories: A Historical Perspective," Handbook of Clinical Neurology 185 (2022): 25–34.
4. Karl Zilles, "Brodmann: A Pioneer of Human Brain Mapping—His Impact on Concepts of Cortical Organization," Brain 141, no. 11 (2018): 3262–78.

5   H. Beaunis, *Nouveaux éléments sur la physiologie humaine* (Paris: Baillière, 1876), 1017.
6   Beaunis, *Nouveaux éléments*, 1017.
7   Charles-Édouard Brown-Séquard, "Aphasia as an Effect of Brain-Disease," *Dublin Journal of Medical Science* 63 (1877): 212.
8   Brown-Séquard, "Aphasia," 219–20. See as well Charles-Édouard Brown-Séquard, "The Localisation of the Functions of the Brain Applied to the Use of the Trephine," *The Lancet* 110, no. 2812 (1877): 107–8.
9   Quoted in Stanley Finger, *Origins of Neuroscience: A History of Explorations into Brain Function* (Oxford: Oxford University Press, 2001), 53.
10 David Ferrier, *Functions of the Brain* (London: Smith, 1886), 435.
11 See Ferrier, *Functions*, 451, for the example of the right brain controlling speech but the left brain controlling writing.
12 Ferrier, *Functions*, 467.
13 On Hughlings Jackson's topological theory of the brain, see Nima Bassiri, "Dislocations of the Brain: Subjectivity and Cerebral Topology from Descartes to Nineteenth-Century Neuroscience" (PhD diss., University of California, Berkeley, 2010).
14 John Hughlings Jackson, *Evolution and Dissolution of the Nervous System*, in *Selected Writings of John Hughlings Jackson*, vol. 2 (London: Hodder and Stoughton, 1932), 67.
15 Hughlings Jackson, *Evolution and Dissolution*, 78.
16 Hughlings Jackson, *Evolution and Dissolution*, 68.
17 Charles Scott Sherrington, *The Integrative Action of the Nervous System* (Oxford: Oxford University Press, 1906). Subsequent citations are given parenthetically in the text.
18 C. Judson Herrick, "The Evolution of Intelligence and Its Organs," *Science*, n.s., 31, no. 784 (1910): 16.
19 Ralph Lillie, "What Is Purposive and Intelligent Behavior from the Physiological Point of View?," *Journal of Philosophy, Psychology and Scientific Methods* 12, no. 22 (1915): 592–93.
20 Lillie, "What Is Purposive and Intelligent Behavior," 609.

Chapter 12

1   Charles Mercier, *The Nervous System and the Mind: A Treatise on the Dynamics of the Human Organism* (London: Macmillan, 1888), 130.
2   George Lewes, *Life and Mind*, 3rd ser. (Boston: Houghton, 1879), 153. Subsequent citations are given parenthetically in the text.
3   Hermann von Helmholtz, *Handbuch der physiologischen Optik* (Leipzig: Voss, 1867), 431.
4   It is worth noting that Comte suggested, near the end of his life, in the so-called "Letters on Illness" (1854–85), that we should see the brain as a site of mediation and not as a machine generating thought. The brain, he wrote, was a "permanent double placenta between the individual and Humanity." Jean François Eugène Robinet, *Notice sur l'oeuvre et sur la vie d'Auguste Comte* (Paris: Dunod, 1860), 533. Which also meant that the social organization was always going to be an

essential intermediary between the individual mind and its external world, its "milieu."

5 Charles Darwin, *Expression of the Emotions in Man and Animals* (London: Murray, 1873), 15. Subsequent citations are given parenthetically in the text.
6 Charles Darwin, *The Descent of Man, and Selection in Relation to Sex*, 2nd ed. (London: John Murray, 1874), 610. Subsequent citations are given parenthetically in the text.
7 On the importance of the technological concept of invention throughout Darwin's work on evolution, see Giuliano Pancaldi, "Darwin's Technology of Life," *Isis* 110, no. 4 (2019): 680–700.
8 George Campbell, Duke of Argyll, *Primeval Man: An Examination of Some Recent Speculations* (New York: Boies, 1884). Originally published 1869. Subsequent citations are given parenthetically in the text.
9 "The man who first lifted a stone and threw it, practiced an art which not one of the lower animals is capable of practising." Argyll, *Primeval Man*, 48.
10 Lazarus Geiger, *Contributions to the History of the Development of the Human Race: Lectures and Dissertations*, trans. David Asher (London: Trübner and Co., 1880). Subsequent citations are given parenthetically in the text.
11 Eduard Reich, *"Der" Mensch und die Seele: Studien zur physiologischen und philosophischen Anthropologie und zur Physik des täglichen Lebens* (Berlin: Nicolai, 1872), 178.
12 Ernst Kapp, *Grundlinien einer Philosophie der Technik: Zur Entstehungsgeschichte der Kultur aus neuen Gesichtspunkten* (Braunschweig: Westermann, 1877); in English translation as Kapp, *Elements of a Philosophy of Technology: On the Evolutionary History of Culture*, ed. Jeffrey West Kirkwood and Leif Weatherby, trans. Lauren K. Wolfe (Minneapolis: University of Minnesota Press, 2018). Subsequent citations of the English translation are given parenthetically in the text.
13 The intimate connection between communication technologies and experimental neurology in this period is brilliantly demonstrated in works such as Laura Otis, *Networking: Communicating with Bodies and Machines in the Nineteenth Century* (Ann Arbor: University of Michigan Press, 2001), esp. ch. 1; and Friedrich Kittler, *Gramaphone, Film, Typewriter*, trans. Geoffrey Winthrop-Young and Michael Wutz (Stanford: Stanford University Press, 1999), originally published 1986.

Chapter 13

1 Charles Richet, "Défense (fonctions de)," in *Dictionnaire de physiologie*, 10 vols. (Paris: Alcan, 1895), 4:721.
2 Énile Boutroux, *De la contingence des lois de la nature* (Paris: Alcan, 1874). In English as *The Contingency of the Laws of Nature*, trans. Fred Rothwell (Chicago: Open Court, 1916), 69–70. Subsequent citations are given parenthetically in the text.
3 James Mark Baldwin, *Mental Development in the Child and the Race: Methods and Processes*, 2nd ed. (New York: Macmillan, 1900), 174–75.
4 Baldwin, *Mental Development*, 175–76.
5 James Mark Baldwin, "A New Factor in Evolution," *American Naturalist* 30, no. 354 (1896): 445. Subsequent citations are given parenthetically in the text.

6   James Mark Baldwin, *The Individual and Society; or, Psychology and Sociology* (Boston: Badger, 1911).
7   Charles Sanders Peirce, "Logic as Semiotic: The Theory of Signs," in *Philosophical Writings of Peirce* (New York: Dover, 1940).
8   Edward Bradford Titchener, *A Text-Book of Psychology* (New York: Macmillan, 1910), 456.
9   Charles Hubbard Judd, *Psychology: General Introduction* (New York: Scribner, 1907), 366.
10 C. H. Judd, "Evolution and Consciousness," *Psychological Review* 17, no. 2 (1910): 80.
11 Judd, "Evolution," 85.
12 Judd, "Evolution," 87.
13 Judd, "Evolution," 87.
14 Henri Bergson, *Creative Evolution*, trans. Arthur Mitchell (New York: Henry Holt, 1911), 141. Subsequent citations are given parenthetically in the text.
15 Henri Bergson, "The Soul and the Body," in *Mind-Energy: Lectures and Essays*, trans. H. Wildon Carr (London: Macmillan, 1920), 45. Not insignificantly, Bergson was very interested in parapsychological phenomena, although with a critical eye. This exteriorization of thought beyond the body—what was called telepathy—may well be explained, he suggested, by action at a distance, as in the case of electricity. Perhaps our minds are always being influenced by other minds, he speculated, but below our threshold of conscious awareness. See Bergson, "Phantasms of the Living" and "Psychical Research," in *Mind-Energy*.
16 Werner Sombart, "Technology and Culture," in *Sociological Beginnings*, ed. and trans. Christopher Adair-Toteff (Liverpool: Liverpool University Press, 2005), 97.
17 Sombart, "Technology," 102.

Chapter 14

1   Filippo Marinetti, "Extended Man and the Kingdom of the Machine" (1910), in *Critical Writings*, ed. Günter Berghaus, trans. Doug Thompson (New York: Farrar, 2006), 86.

Chapter 15

1   Samuel Taylor Coleridge, *Lectures on Literature*, 2 vols., ed. R. A. Foakes (London: Routledge & Kegan Paul, 1987), 2:221–22.
2   Arthur Schopenhauer, "The Art of Literature," in *The Essays of Arthur Schopenhauer*, trans. T. Bailey Saunders (New York: A. L. Burt, 1902).
3   Henri F. Ellenberger, *The Discovery of the Unconscious: The History and Evolution of Dynamic Psychiatry* (New York: Basic Books, 1970).
4   Herbert Spencer, *The Principles of Psychology* (London: Longman, 1855), 495.
5   Eduard von Hartmann, *Philosophie des Unbewussten* (Berlin: C. Duncker, 1869).
6   William B. Carpenter, *Principles of Mental Physiology with Their Applications to the Training and Discipline of the Mind, and the Study of Its Morbid Conditions* (London: Henry S. King, 1874), 515.
7   L. Dumont, "Une philosophie nouvelle en Allemagne" (1872), in *Un débat sur l'inconscient avant Freud: La reception de Eduard von Hartmann chez les psychologues*

et philosophes français, ed. Serge Nicolas and Laurent Fedi (Paris: L'Harmattan, 2008), 73.
8   Otis, *Networking*.
9   Charles Mercier, *The Nervous System and the Mind: A Treatise on the Dynamics of the Human Organism* (London: Macmillan, 1888), 190, 370.
10 Joseph John Murphy, *Habit and Intelligence, in Their Connexion with the Laws of Matter and Force: A Series of Scientific Essays*, 2 vols. (London: Macmillan, 1869), 2:47.
11 Henry Maudsley, *On the Method of the Study of Mind: An Introductory Chapter to a Physiology and Pathology of the Mind* (London: John Churchill, 1865), 19.
12 Francis Galton, "Psychometric Facts," *Popular Science Monthly* 14 (1879): 779–80.
13 Thomas Huxley, "On the Hypothesis That Animals Are Automata," *Fortnightly Review* 95 (1874): 550–80.
14 Simon Schaffer, "Enlightened Automata," in *The Sciences in Enlightened Europe*, ed. William Clark, Jan Golinski, and Simon Schaffer (Chicago: University of Chicago Press, 1999), 126–65.
15 Arthur Rothenberg, *The Emerging Goddess: The Creative Process in Art, Science, and Other Fields* (Chicago: University of Chicago Press, 1979), 395–96.
16 Helmholtz, *Vorträge und Reden*, 15.
17 Henri Poincaré, "L'invention mathématique" (1908), reprinted in Jacques Hadamard, *Essai sur la psychologie de l'invention dans la domaine mathématique* (Paris: Jacques Gabay, 2007), 144.
18 Graham Wallas, *The Art of Thought* (London: Jonathan Cape, 1931), 80 ff.
19 This research is summarized in Karl Lashley, *Brain Mechanisms and Intelligence: A Quantitative Study of Injuries to the Brain* (Chicago: University of Chicago Press, 1929).
20 Wallas, *Art*, 36–45, at 44–45.
21 William James, "Are We Automata?," *Mind* 4 (1879): 5.
22 William James, "Great Men, Great Thoughts, and the Environment," *Atlantic Monthly* 46 (1880): 457.
23 See Catherine Malabou, *The Future of Hegel: Plasticity, Temporality and Dialectic*, trans. Lisabeth During (London: Routledge, 2005).
24 William James, *Principles of Psychology*, 2 vols. (New York: Holt, 1890), 1:105.
25 See Donna Haraway, *Primate Visions: Gender, Race, and Nature in the World of Modern Science* (London: Routledge, 1990), although there is only a brief reference to Köhler in the introduction, on the colonial siting of German primate research in the Canary Islands (21).
26 Wolfgang Köhler, *The Mentality of Apes*, trans. Ella Winter (London: Kegan Paul, 1925), 2.
27 "The genuine achievement takes place as a single continuous occurrence, a unity, as it were, in space as well as in time. . . . A successful chance solution consists of an agglomeration of separate movements, which start, finish, start again, remain independent of one another in direction and speed, and only in a geometrical summation start at the starting-point, and finish at the objective." Köhler, *Mentality*, 16–17.
28 Köhler, *Mentality*, 125–27.
29 "When a man or animal takes a roundabout way (in the ordinary sense of the word) to his objective the beginning—considered by itself and regardless of the further course of the experiment—contains at least one component which must

seem irrelevant; in very indirect routes there are usually some parts of the way which, when considered alone, seem in contradiction to the purpose of the task, because they lead away from the goal." Köhler, *Mentality*, 103–4.

30 Max Wertheimer, *Productive Thinking* (New York: Harper, 1945), 24, 34, 41.
31 Karl Duncker, *Zur Psychologie des produktive Denken* (Berlin: Springer, 1935). In English as *On Problem-Solving*, trans. Lynne S. Lees, Psychological Monographs, no. 270 (Washington, DC: American Psychological Association, 1945), 4–6, 8, 29.
32 Kurt Koffka, *Principles of Gestalt Psychology* (New York: Harcourt, Brace, 1935), 684.
33 Wolfgang Köhler, "Some Gestalt Problems," in *A Sourcebook of Gestalt Psychology*, ed. Willis D. Ellis (London: Kegan Paul, 1938), 66.
34 Wolgang Köhler, *Gestalt Psychology* (New York: Liveright, 1947), 114, 193.
35 Koffka, *Principles*, 105.
36 Wolfgang Köhler, *Die physischen Gestalten in Ruhe und im stationären Zustand* (Braunschweig: Vieweg, 1920).
37 Maurice Merleau-Ponty, *The Structure of Behavior*, trans. A. L. Fisher (London: Methuen, 1942), 39.
38 E. C. Tolman and C. H. Honzik, "Insight in Rats," *University of California Publications in Psychology* 4, no. 14 (1930): 230.
39 See Kurt Koffka, *The Growth of the Mind: An Introduction to Child Psychology* (New York: Harcourt, 1925): "It would then follow that plasticity must be something more than memory—something more than the retention of an achievement by the simple means of effecting a new combination of reaction—pathways which the organism already possesses—and we could then rightfully say that by virtue of his plasticity man is superior to all other living creatures" (125).
40 Wertheimer, *Productive Thinking*, 24, 34, 41.
41 Joshua Gregory, "Neo-Realism and the Origin of Consciousness," *Philosophical Review* 29, no. 3 (1920): 243.
42 Gregory, "Neo-Realism," 244.
43 Hugh A. Reyburn, "A Functional Theory of Knowledge II," *Journal of Philosophical Studies* 2, no. 8 (1927): 463–76, at 463.
44 Henri Wallon, "La conscience et la vie subconsciente," in George Dumas, *Nouveau traité de psychologie* vol. 7, fascicule 1 (Paris: Alcan, 1942), 4.
45 Achille Urbain, *Psychologie des animaux sauvages: Instinct-Intelligence* (Paris: Flammarion, 1940), 91.
46 George Boas, "Insight, Habituation, and Enjoyment," with a response by Robert Francis Creegan, *Journal of Philosophy* 36, no. 26 (1939): 709–715, at 710.
47 "Précisément grace à ce qu'elle est l'acte exceptionnel, l'invention peut être un guide précieux dans l'exploration de ce monde si complexe et encore si mystérieux qu'est la pensée humaine." Jacques Hadamard, "Subconscient, intuition, et logique dan la recherche scientifique," *Conference faite au Palais de la Découverte*, Dec. 3, 1945 (Paris: Palais de la Découverte, 1945), 21.
48 Emile Guyénot, "La vie comme invention," in Fondation "Pour la science," Centre International de Synthèse, *Neuvième semaine internationale de synthèse: L'invention* (Paris: Alcan, 1938), 186, 188, 192.
49 Tournay, "Physiologie spéciale du système nerveux," in Georges Dumas, *Nouveau traité de psychologie*, vol. 1, fascicule 3 (Paris: Presses universitaires de France, 1938), 288. See as well Henri Delacroix, *Les grandes formes de la vie mentale* (Paris: Alcan, 1934): "La plasticité totale de la matière vivante asservirait complètement

l'être au milieu. C'est parce que l'organisme résiste aux changements incessants du milieu que ses variations, difficilement acquises, persistent et deviennent à leur tour le point de départ d'autres variations" (89).
50 Lev Vygotsky, *Mind in Society: The Development of Higher Psychological Processes*, ed. Michael Cole, Vera John-Steiner, Sylvia Scribner, and Ellen Souberman (Cambridge, MA: Harvard University Press, 1979), 37. Subsequent citations are given parenthetically in the text.
51 Ernst Cassirer, *The Philosophy of Symbolic Forms*, vol. 3: *The Phenomenology of Knowledge*, trans. Ralph Manheim (New Haven, CT: Yale University Press, 1955), 275.
52 Cassirer, *The Philosophy of Symbolic Forms*, 3:276.
53 Cassirer, *The Philosophy of Symbolic Forms*, 3:277.
54 Cassirer, *The Philosophy of Symbolic Forms*, 3:277 n. 117.
55 Koffka, *Growth of the Mind*, 2. See also Koffka, *Principles*, 422.
56 Henri Delacroix, *Les grandes formes de la vie mentale* (Paris: Alcan, 1934): "L'homme est surtout adapté à ce qui n'est pas" (6).
57 Henri Delacroix, *Le langage et la pensée*, 2nd ed. (Paris: Alcan, 1930), 115.
58 Eugenio Rignano, *The Psychology of Reasoning*, trans. Winifred A. Holl (London: Kegan Paul, 1923), 7.
59 Alfred North Whitehead, *Science and the Modern World* (London: Mentor, 1925), 207.

Chapter 16

1 Shepherd Ivory Franz, "Discussion: Cerebral Adaptation vs. Cerebral Organology," *Psychological Bulletin* 14, no. 4 (1917): 137–38.
2 Franz, "Discussion," 138.
3 Franz, "Discussion," 140; see also Shepherd Ivory Franz, "Cerebral-Mental Relations," *Psychological Review* 28, no. 2 (1921): 81–95.
4 Shepherd Ivory Franz and Karl Lashley, "The Retention of Habits by the Rat after Destruction of the Frontal Portion of the Cerebrum," *Psychobiology* 1, no. 1 (1917): 17.
5 Shepherd Ivory Franz, "Conceptions of Cerebral Functions," *Psychological Review* 30, no. 6 (1923): 444–45.
6 Franz, "Conceptions," 445.
7 Henry Head, *Aphasia and Kindred Disorders of Speech* (Cambridge: Cambridge University Press, 1926), 445. Subsequent citations are given parenthetically in the text.
8 Stefanos Geroulanos and Todd Meyers, *The Human Body in the Age of Catastrophe: Brittleness, Integration, Science, and the Great War* (Chicago: University of Chicago Press, 2018).
9 Walter B. Cannon et al., *The Nature and Treatment of Wound Shock and Allied Conditions* (Chicago: American Medical Association, 1918).
10 Walter B. Cannon, *The Wisdom of the Body* (New York: Norton, 1932), 25.
11 J. B. S. Haldane, *The New Physiology and Other Addresses* (London: Charles Griffin, 1919). Subsequent citations are given parenthetically in the text.
12 On Lashley's work, see Nadine Weidman, *Constructing Scientific Psychology: Karl Lashley's Mind-Brain Debates* (Cambridge: Cambridge University Press, 1999).

As Weidman demonstrates, Lashley was a profound racist, and it is therefore astonishing that one cannot easily find traces of this in his research publications. There is, it seems, no hint of a racialized nervous system in Lashley's own thinking.
13 Lashley, *Brain Mechanisms*, 176.
14 Lashley, *Brain Mechanisms*, 25.
15 Lashley, *Brain Mechanisms*, 154.
16 Karl Lashley, "Basic Neurological Mechanisms in Behavior," *Psychological Review*, 37, no. 1 (1930): 1–24.
17 Constantin von Monakow, *Die Lokalisation im Grosshirn und der Abbau der Funktion durch Kortikale Herde* (Wiesbaden: J. F. Bergmann, 1914), 30.
18 Constantin von Monakow and R. Mourgue, *Introduction biologique à l'étude de la neurologie et désintégration de la fonction* (Paris: Félix Alcan, 1928), 29.
19 Kurt Goldstein, *Der Aufbau des Organismus: Einführing in die Biologie unter besonderer Berücksichtigung der Ehfahrungen am kranken Menschen* (The Hague: Martinus Nijhoff, 1934); in English as *The Organism: A Holistic Approach to Biology derived from Pathological Data in Man* (New York: Zone, 1995). Subsequent citations of the translation are given parenthetically in the text. An excellent account of Goldstein's work and the German contexts of theoretical biology can be found in Anne Harrington, *Reenchanted Science: Holism in German Culture from Wilhelm II to Hitler* (Princeton, NJ: Princeton University Press, 1996), ch. 5. The complex conceptual landscape framing Goldstein's neurological work is laid out brilliantly in Geroulanos and Meyers, *The Human Body*.
20 Kurt Goldstein, *Human Nature in the Light of Psychopathology* (Cambridge, MA: Harvard University Press, 1940), 112.

Chapter 17

1 Jakob Von Uexküll, *A Foray into the Worlds of Animals and Humans*, trans. Joseph D. O'Neil (Minneapolis: University of Minnesota Press, 2010), 41.
2 Jakob Von Uexküll, *Leitfaden in das Studium der experimentellen Biologie der Wassertiere* (Wiesbaden, 1905), 66. On the development of these ideas in Von Uexküll's early work, see Carlo Brentari, *Jakob Von Uexküll: The Discovery of the Umwelt between Biosemiotics and Theoretical Biology* (Dordrecht: Springer, 2015), ch. 3.
3 Jakob Von Uexküll, *Theoretische Biologie* (Berlin: Paetel, 1920); in English as *Theoretical Biology* (New York: Harcourt, Brace, 1926). Subsequent citations of the English translation are given in the text.
4 On the *Umwelt*, see Giorgio Agamben, *The Open: Man and Animal*, trans. Kevin Attell (Stanford: Stanford University Press, 2004), §§ 10–11.

Chapter 18

1 Erwin Schrödinger, *What Is Life? With Mind and Matter and Autobiographical Sketches* (Cambridge: Cambridge University Press, 2012), 71.
2 Alfred J. Lotka, *Principles of Physical Biology* (Baltimore: Wilkins and Wilkins, 1925), 15. Subsequent citations are given parenthetically in the text.

3   Alfred J. Lotka, "The Intervention of Consciousness in Mechanics," *Science Progress in the Twentieth Century* 18 (1924): 415.
4   Lotka, *Physical Biology*, 364–65.
5   Alfred J. Lotka, "Population Analysis as a Chapter in the Mathematical Theory of Evolution," in *Essays on Growth and Form*, ed. W. E. Le Gros Clark, and P. B. Medavar (Oxford: Oxford University Press, 1945), 381.
6   Alfred J. Lotka, "The Spread of Generations," *Human Biology* 1, no. 3 (1929): 317–18.
7   Lotka, "Spread," 319–20.
8   Lotka, "Population Analysis," 384.
9   Alfred J. Lotka, "The Law of Evolution as a Maximal Principle," *Human Biology* 17, no. 3 (1945): 167–94. Subsequent citations are given parenthetically in the text.

Chapter 19

1   William McDougall, "Men or Robots?," *Pedagogical Seminary and Journal of Genetic Psychology* 33 (1926): 71–89. Subsequent citations are given parenthetically in the text.
2   C. Judson Herrick, *The Thinking Machine* (Chicago: University of Chicago Press, 1929). Subsequent citations are given parenthetically in the text.
3   Preface to the second edition of Herrick, *The Thinking Machine* (Chicago: University of Chicago Press, 1932).

Chapter 20

1   Jacques Lafitte, "Sur la science des machines," *Revue de Synthèse* 6, no. 10 (1933): 155. Subsequent citations are given parenthetically in the text.
2   On these kind of machines, see David A. Mindell, *Between Human and Machine: Feedback, Control, and Computing before Cybernetics* (Baltimore: Johns Hopkins University Press, 2000).
3   José Ortega y Gasset, "Thoughts on Technology" (1939), in *Philosophy and Technology: Readings in the Philosophical Problems of Technology*, ed. Carl Mitcham and Robert Mackey (New York: Free Press, 1971), 292.
4   Ortega y Gasset, "Thoughts," 292.
5   Ortega y Gasset, "Thoughts," 292, 294.
6   Ortega y Gasset, "Thoughts," 299.
7   Ortega y Gasset, "Thoughts," 306.

Chapter 21

1   Charles Hubbard Judd, *The Psychology of Social Institutions* (New York: Macmillan, 1926). Subsequent citations are given parenthetically in the text.
2   Thomas Morgan, *The Scientific Basis of Evolution* (New York: Norton, 1932), 203–4.
3   Morgan, *The Scientific Basis of Evolution*, 208.
4   Morgan, *The Scientific Basis of Evolution*, 214.
5   André Leroi-Gourhan, *Evolution et techniques: L'homme et la matière* (Paris: Albin Michel, 1943), 472.

6   Ernst Cassirer, *The Philosophy of Symbolic Forms*, vol. 4: *The Metaphysics of Symbolic Forms*, trans. John Michael Krois (New Haven, CT: Yale University Press, 1996). The fourth volume is a collection of texts related to the uncompleted last book. Subsequent citations are given parenthetically in the text.
7   See Peter Gordon's excellent contextualized account, *Continental Divide: Heidegger, Cassirer, Davos* (Cambridge, MA: Harvard University Press, 2010).
8   As Leslie White argued in his comparison of tool-use in primates and humans, what technology demonstrates in human life is not tool use per se but rather the *transmission of knowledge* via cultural life. Leslie White, "On the Use of Tools by Primates," *Journal of Comparative Psychology* 34 (1942): 371.
9   Ernst Cassirer, "Form and Technology" (1930), in *Ernst Cassirer on Form and Technology: Contemporary Readings*, ed. Aud Sissel Hoel and Ingvild Folkvord (London: Palgrave Macmillan, 2012), 304.
10 Cassirer, "Form and Technology," 372.
11 Hans Freyer, *Theorie des Objectiven Geistes: Eine Einleitung in die Kulturphilosophie* (Leipzig: B. G. Teubner, 1923).
12 Freyer, *Theorie des Objectiven Geistes*, 13.
13 Freyer, *Theorie des Objectiven Geistes*, 66.
14 Freyer, *Theorie des Objectiven Geistes*, 137.
15 Max Scheler, *The Human Place in the Cosmos*, trans. Manfred S. Frings (Evanston, IL: Northwestern University Press, 2009). Subsequent citations are given parenthetically in the text.
16 Martin Heidegger, *Being and Time*, trans. John Macquarrie and Edward Robinson (New York: Harper, 1962), 30.
17 Martin Heidegger, *The Fundamental Concepts of Metaphysics: World, Finitude, Solitude*, trans. William McNeill and Nicholas Walker (Bloomington: Indiana University Press, 1995).
18 Heidegger, *Fundamental Concepts*, 213.
19 Heidegger, *Fundamental Concepts*, 214.
20 Heidegger, *Fundamental Concepts*, 215.
21 Heidegger, *Fundamental Concepts*, 215. Subsequent citations are given parenthetically in the text.
22 See Heidegger, *Being and Time*: "When, for instance, a man wears a pair of spectacles which are so close to him distantially that they are 'sitting on his nose,' they are environmentally more remote from him than the picture on the opposite wall. Such equipment has so little closeness that often it is proximally quite impossible to find. Equipment for seeing—and likewise for hearing, such as the telephone receiver—has what we have designated as the inconspicuousness of the proximally ready-to-hand. So too, for instance, does the street, as equipment for walking" (141).
23 Martin Heidegger, *Contributions to Philosophy (of the Event)*, trans. Daniela Vallega-Neu and Richard Rojcewicz (Bloomington: Indiana University Press, 2012), 37.
24 This is the critique Bernard Stiegler will make in *Technics and Time*, vol. 2: *Disorientation*, trans. Stephen Barker (Stanford: Stanford University Press, 2008).
25 Heidegger, *Contributions*, 38.
26 Heidegger, *Contributions*, 69–71.
27 Helmuth Plessner, *Levels of Organic Life and the Human: An Introduction to Philosophical Anthropology*, trans. Millay Hyatt (New York: Fordham University

Press, 2019). Subsequent citations given in the text are to the English edition; the German text is Helmuth Plessner, *Die Stufen des Organischen und Der Mensch: Einleitung in die philosophische Anthropologie*, 3rd ed. (Berlin: De Gruyter, 1975).
28   Plessner here cites the work of Buytendijk. See, e.g., F. Buytendijk, *Psychologie des animaux*, trans. R. Bredo, preface Eduard Claparède (Paris: Payot, 1928).
29   Catherine Malabou, *Ontology of the Accident: An Essay on Destructive Plasticity*, trans. Carolyn Shread (Cambridge: Polity, 2012).
30   See the section, "Selbstregulierbarkeit des lebendigen Einzeldinges und harmonische Äquipotentialität der Teile." Plessner, *Die Stufen*, 160.

Chapter 22

1   Ludwig Wittgenstein, *Zettel*, ed. G. E. M. Anscombe and G. H. von Wright, trans. G. E. M. Anscombe (Berkeley: University of California Press, 1967), editors' preface. This is a bilingual edition. Subsequent citations are given in the text and are to the numbers assigned to the individual fragments.

Chapter 23

1   W. Ross Ashby, "Homeostasis," in *Cybernetics: Circular Causal and Feedback Mechanisms in Biological and Social Systems*, ed. Heinz von Foerster (New York: Josiah Macy Jr. Foundation, 1952), 73.
2   Wolfgang Köhler, review of *Cybernetics, or Control and Communication in the Animal and the Machine*, by Norbert Wiener, *Social Research* 18, no. 1 (1951): 128; my emphasis.
3   Warren McCulloch, "What Is a Number, That a Man May Know It, and a Man, That He May Know a Number?," in *Embodiments of Mind* (Cambridge, MA: MIT Press, 1965), 14.
4   Steve Joshua Heims, *The Cybernetic Group* (Cambridge, MA: MIT Press, 1991).
5   McCulloch, "What's in the Brain That Ink May Character?" in McCulloch, *Embodiments*, esp. 389; and see "Where Is Fancy Bred?," another essay in the same volume that examines the insight problem in relation to AI.
6   Warren McCulloch and Walter Pitts, "A Logical Calculus of the Ideas Immanent in Nervous Activity," in McCulloch, *Embodiments*, 38. Although Turing's iconic 1950 essay, "Computing Machinery and Intelligence," is famous for introducing the "Turing test" for machine intelligence, one of the more disturbing (almost Nietzschean) undercurrents of the argument was the idea that human beings are no more than complex machines, whose "unpredictability" and "creativity" are just consequences of our own ignorance concerning the processes unraveling within. I take this up in the next section.
7   Warren McCulloch, "Why the Mind Is in the Head," in McCulloch, *Embodiments*, 73.
8   George Canguilhem, "Machine and Organism," 90.
9   Georges Canguilhem, *Essai sur quelques problèmes concernant le normal et le pathologique* (Clermont-Ferrand: La Montagne, 1943).
10   See George Canguilhem, *The Normal and the Pathological*, trans. Carolyn R. Fawcett and Robert S. Cohen (New York: Zone, 1991), ch. 4.

11 Canguilhem, "Machine and Organism," 79.
12 Ludwig von Bertalanffy, *Robots, Men, and Minds: Psychology in the Modern World* (New York: George Braziller, 1967), 68.
13 Ludwig von Bertalanffy, *Das Biologische Weltbild* (Bern: Francke, 1949): "ein gänzlich mechanisierter Organismus wäre unfähig zur Regulation nach Störungen" (29).
14 Ludwig von Bertalanffy, *General System Theory: Foundation, Development, Applications* (New York: George Braziller, 1968), 213.
15 See Ludwig von Bertalanffy, "The Theory of Open Systems in Physics and Biology," *Science* 111, no. 2872 (1950): 23–29.
16 Lewis Mumford, "The Automation of Knowledge: Are We Becoming Robots?," *Audio-Visual Communication Review* 12 (1964): 275.
17 Gregory Bateson, *Steps to an Ecology of Mind* (Chicago: University of Chicago Press, 1972), 286.
18 Raymond Ruyer, *Paradoxes de la conscience et limites de l'automatisme* (Paris: Albin Michel, 1966): "Ses erreurs doivent précéder ses essais, puisqu'elles en constituent la cause motrice" (232).
19 Norbert Wiener, *Cybernetics, or Control and Communication in the Animal and Machine* (Paris and Cambridge, MA: Hermann and MIT Press, 1948), 172.
20 Jurgen Ruesch and Gregory Bateson, *Communication: The Social Matrix of Psychiatry* (New York: Norton, 1951), 91.
21 Bateson, *Ecology of Mind*, 143.
22 Wiener, *Cybernetics*, 15.
23 W. Ross Ashby, *An Introduction to Cybernetics* (London: Chapman and Hall, 1961), 5–6.
24 W. Ross Ashby and Mark R. Gardner, "Connectance of Large Dynamic (Cybernetic) Systems: Critical Values for Stability," *Nature* 228, no. 784 (1970): 784. On catastrophe, see Wiener, *Cybernetics*, 113; and Norbert Wiener, *The Human Use of Human Beings* (Boston: Houghton Mifflin, 1950), 49.
25 Michael Polanyi, *Personal Knowledge: Towards a Post-Critical Philosophy* (Chicago: University of Chicago Press, 1958), 336. Subsequent citations are given parenthetically in the text.
26 "L'hypothèse cybernétique," *Tiqqun* 2 (2001): 40–83.
27 Peter Galison, "Ontology of the Enemy: Norbert Wiener and the Cybernetic Vision," *Critical Inquiry* 21, no. 1 (1994): 228–66.
28 Andrew Pickering, *The Cybernetic Brain: Sketches of Another Future* (Chicago: University of Chicago Press, 2010), 132.

Chapter 24

1 Nicholas Georgescu-Roegen, *The Entropy Law and the Economic Process* (Cambridge MA: Harvard University Press, 1971), 90.
2 Wiener, *Cybernetics*, 40.
3 C. E. Shannon and J. McCarthy, eds., *Automata Studies* (Princeton, NJ: Princeton University Press, 1956), v.
4 Arturo Rosenblueth, Norbert Wiener, and Julian Bigelow, "Behavior, Purpose, and Teleology," *Philosophy of Science* 10, no. 1 (1953): 18–24.
5 See Anatol Rapoport, "Technological Models of the Nervous System" (1955), in

The Modeling of Mind: Computers and Intelligence, ed. Kenneth M. Sayre and Frederick J. Crosson (New York: Simon and Schuster, 1963), 32.
6 Warren McCulloch and John Pfeiffer, "Of Digital Computers Called Brains," *Scientific Monthly* 69, no. 6 (1949): 368–76.
7 See the iconic essay, Rosenblueth, Wiener, and Bigelow, "Behavior, Purpose, and Teleology."
8 G. A. Miller, "The Magical Number Seven, Plus or Minus Two: Some Limits on our Capacity for Processing Information," *Psychological Review* 63, no. 2 (1956): 81–97.
9 G. A. Miller, "Encoding the Unexpected," *Proceedings of the Royal Society of London. Series B, Biological Sciences* 171, no. 1024 (1968): 361–75. Subsequent citations are given parenthetically in the text.
10 Wiener, *Cybernetics*, 135.
11 Ashby, "Homeostasis," 73.
12 Ashby, *Introduction*, 197.
13 W. Ross Ashby, *Journal*, 1523–24. Available at the W. Ross Ashby Digital Archive, rossashby.info/index.html, accessed July 24, 2013.
14 Ashby, *Journal*, 1906–9.
15 Ashby, *Journal*, 1054.
16 W. Ross Ashby, "The Nervous System as Physical Machine: With Special Reference to the Origin of Adaptive Behaviour," *Mind* 56 (1947): 50.
17 Ashby, *Journal*, 1054.
18 Ashby, "Nervous System," 55.
19 W. Grey Walter, *The Living Brain* (New York: Norton, 1953), 184; Grey Walter is commenting on the conditional reflex analogue (CORA) and its relationship to theories in psychopathology.
20 Grey Walter, *The Living Brain*, 25, 31.
21 Grey Walter, *The Living Brain*, 27.
22 W. Ross Ashby, *Design for a Brain*, 2nd. ed. (New York: John Wiley & Sons, 1960), 95.
23 Ashby, *Introduction*, 37.
24 Ashby, *Introduction*, 24.
25 For a critical view, see, e.g., Paul Cossa, *La Cybernétique: Du cerveau humain aux cerveaux artificiels* (Paris: Masson, 1955), 56.
26 Wiener, *Cybernetics*, 168.
27 Wiener, *Cybernetics*, 171.
28 Wiener, *Cybernetics*, 168.
29 Wiener, *Cybernetics*, 172.
30 Wiener, *Cybernetics*, 178–79.
31 Norbert Wiener, "Problems of Organization," *Bulletin of the Menninger Clinic* 17 (1953): 132.
32 Wiener, *Cybernetics*, 178.
33 Wiener, "Problems," 134.
34 Alfred Fessard, "Points de contact entre Neurophysiologie et Cybernétique," *S.E.T.* 5, nos. 35–36 (July 1953–Jan. 1954): 32–33.
35 Fessard, "Points," 33.
36 See John von Neumann and H. H. Goldstine, "On the Principles of Large-Scale Computing Machines," in John von Neumann, *Collected Works*, vol. 5: *Design of Computers, Theory of Automata, and Numerical Analysis* (Oxford: Pergamon Press, 1963), 2.

37 See William Asprey, "The Origins of John von Neumann's Theory of Automata," in *The Legacy of John von Neumann*, ed. James Glimm, John Impagliazzo, and Isadore Singer (Providence, RI: American Mathematical Society, 1990), 289–90.
38 John von Neumann, *Theory of Self-Reproducing Automata*, ed. Arthur W. Burks (Urbana: University of Illinois Press, 1966), 35.
39 A good example is John von Neumann, "The Mathematician," in *Collected Works*, vol. 1: *Logic, Theory of Sets, and Quantum Mechanics* (Oxford: Pergamon Press, 1961).
40 Von Neumann, *Self-Reproducing Automata*, 33–35. See also the editor's introduction to this text, esp. p. 3. For a broader perspective, see Peter Galison, "Computer Simulations and the Trading Zone," in *The Disunity of Science: Boundaries, Contexts, and Power*, ed. Peter Galison and David J. Stumpf (Stanford: Stanford University Press, 1996).
41 Von Neumann, *Self-Reproducing Automata*, 45.
42 Von Neumann, *Self-Reproducing Automata*, 46.
43 Von Neumann, *Self-Reproducing Automata*, 47.
44 Von Neumann, *Self-Reproducing Automata*, 49.
45 On this point, see Claude E. Shannon, "Von Neumann's Contributions to Automata Theory," in "John von Neumann, 1903–1957," special issue of *Bulletin of the American Mathematical Society* (May 1958): 123–25.
46 Von Neumann, *Self-Reproducing Automata*, 57.
47 John von Neumann, "The General and Logical Theory of Automata," in *Collected Works*, 5:322. This point is made by von Neumann in the discussion period following this, his Hixon Symposium paper.
48 John von Neumann, "Probabilistic Logics and the Synthesis of Reliable Organisms from Unreliable Components," in *Collected Works*, 5:329. This paper was edited from the notes for a 1952 lecture given at the California Institute of Technology. On von Neumann and the role of error in his thought, see Giora Hon, "Living Extremely Flat: The Life of an Automaton; John von Neuman's Conception of Error of (In)animate Systems," in *Going Amiss in Experimental Research*, ed. Giora Hon, Jutta Schickore, and Friedrich Steinle, *Boston Studies in the Philosophy of Science*, no. 267 (2009): 55–71.
49 Von Neumann, *Self-Reproducing Automata*, 57, 71–73.
50 Von Neumann, "General and Logical Theory," 307–8. See also John von Neumann, *The Computer and the Brain* (New Haven, CT: Yale University Press, 1958), 76–79.
51 Von Neumann, *Self-Reproducing Automata*, 71; my emphasis.
52 Ibid., 73.
53 Von Neumann, "General and Logical Theory," 289.
54 Von Neumann, *Computer and the Brain*, 81–82.
55 On artificial plasticity and cybernetics, see Catherine Malabou, *Morphing Intelligence: From IQ Measurement to Artificial Brains* (New York: Columbia University Press, 2019).

Chapter 25

1 Alan Turing, "Computing Machinery and Intelligence," *Mind* 49 (1950): 442.
2 John McCarthy, Marvin Minksy, N. Rochester, and Claude E. Shannon, "A Pro-

posal for the Dartmouth Summer Research Project on Artificial Intelligence" (1955) Available at http://jmc.stanford.edu/articles/dartmouth/dartmouth.pdf. Accessed Jan. 3, 2023.
3   Edward Tenney Brewster, *Natural Wonders Every Child Should Know* (New York: Grosset and Dunlap, 1912), 239.
4   Arthur Eddington, *The Nature of the Physical World* (London: Macmillan, 1928), 332.
5   Eddington, *Nature of the Physical World*, 321.
6   Eddington, *Nature of the Physical World*, 343.
7   Alan Turing, "The Nature of Spirit" (ca. 1932), cited in Andrew Hodges, *Alan Turing: The Enigma* (New York: Simon and Schuster, 1983), 64. Scan of original manuscript text available at https://turingarchive.kings.cam.ac.uk/unpublished-manuscripts-and-drafts-amtc/amt-c-29. Accessed Oct. 30, 2022.
8   Turing, "Computing Machinery," 447.
9   Alan Turing, "Systems of Logic Based on Ordinals," in *The Essential Turing: Seminal Writings in Computing, Logic, Philosophy, Artificial Intelligence, and Artificial Life, plus The Secrets of Enigma*, ed. B. Jack Copeland (Oxford: Oxford University Press, 2004), 166.
10 Alan Turing, "On Computable Numbers, with an Application to the *Entscheidungsproblem*" (1936), in Copeland, *The Essential Turing*, 60.
11 Turing, "Computable Numbers," 59.
12 On Gödel numbering, see Ernest Nagel and James R. Newman, *Gödel's Proof*, rev. ed (New York: New York University Press, 2001), 68–79.
13 Turing, "Computable Numbers," 75–77.
14 Turing, "Computable Numbers," 60.
15 Turing, "Systems of Logic," 192.
16 Turing, "Systems of Logic," 156.
17 See Paulo Mancuso, ed., *From Brouwer to Hilbert: The Debate on the Foundations of Mathematics in the 1920s* (Oxford: Oxford University Press, 1997); and Dennis E. Hesseling, *Gnomes in the Fog: The Reception of Brouwer's Intuitionism in the 1920s* (Dordrecht: Springer, 2003).
18 Alan Turing, "Solvable and Unsolvable Problems" (1954), in *The Essential Turing*, 595.
19 Turing, "Computing Machinery," 459; my emphasis.
20 Alan Turing, "Lecture on the Automatic Computing Engine" (1947), in *The Essential Turing*, 393.
21 Alan Turing, "Intelligent Machinery" (1948), in *The Essential Turing*, 411. "It is related that the infant Gauss was asked at school to do the addition 15 + 18 + 21 + . . . + 54 (or something of the kind) and that he immediately wrote down 483, presumably having calculated it as (15 + 54)(54 -12)/ 2. One can imagine circumstances where a foolish master told the child that he ought instead to have added 18 to 15 obtaining 33, then added 21, etc. From some points of view this would be a 'mistake,' in spite of the obvious intelligence involved."
22 Turing, "Intelligent Machinery," 429.
23 Turing, "Intelligent Machinery," 422.
24 Turing, "Intelligent Machinery," 429.
25 Turing, "Intelligent Machinery," 430; my emphasis.
26 Alan Turing, letter to W. Ross Ashby, ca. Nov. 19, 1946. Image of letter available at the Ashby Digital Archive, http://www.rossashby.info/letters/turing.html. Accessed Oct. 30, 2022.

27 Alan Turing, "Proposal for Development in the Mathematics Division of an Automatic Computing Engine (ACE)" (1945), in *Collected Works*, vol. 3: *Mechanical Intelligence* (Amsterdam: North Holland, 1992), 19–20.
28 Turing, "Proposal for Development," 22.
29 Turing, "Intelligent Machinery," 419.
30 Turing, "Intelligent Machinery," 424. Subsequent citations are given parenthetically in the text.
31 Quoted in Sara Turing, *Alan M. Turing* (Cambridge: Heffer, 1959), xx.
32 M. H. A. Newman, "A Note on Electric Computing Machines," *British Medical Journal* 1, no. 4616 (1949): 1133.
33 Geoffrey Jefferson, "The Mind of Mechanical Man," *British Medical Journal* 1, no. 4616 (1949): 1105–10. Subsequent citations of this work are given parenthetically in the text.
34 On this point, see Stuart Shanker, *Wittgenstein's Remarks on the Foundations of AI* (London: Routledge, 1998), ch. 2.
35 The most brilliant of these readings is, I believe, Jean Lassègue's kaleidoscopic analysis of Turing's essay as symptoms of various important episodes in his life—including, most controversially perhaps, Turing's circumcision. Lassègue, "What Kind of Turing Test Did Turing Have in Mind?," *Tekhnema: Journal of Philosophy and Technology* (Spring 1996), available at http://tekhnema.free.fr/3Lasseguearticle.htm. Accessed Jan. 3, 2023.
36 Turing, "Computing Machinery," 435. Subsequent citations are given parenthetically in the text.
37 Max Black, *Critical Thinking: An Introduction to Logic and Scientific Method* (New York: Prentice Hall, 1946), 348.
38 It seems Douglas Hofstadter was the first to notice this mistake, according to Jean Lassègue.
39 Blaise Pascal, *Pensées*, ed. Philippe Sellier (Paris: Mercure de France, 1976), no. 78, p. 55.
40 I have argued that this was how the Enlightenment understood error, in David W. Bates, *Enlightenment Aberrations: Error and Revolution in France* (Ithaca, NY: Cornell University Press, 2002).
41 Martin Heidegger, "On the Essence of Truth" (1930), in *Basic Writings*, rev. ed., ed. David F. Krell (San Francisco: Harper, 1993), 133.
42 Martin Heidegger, *Aus der Erfahrung des Denkens* (Pfullingen: Günther Neske, 1954), 17.
43 Turing, "Computing Machinery," 456.
44 Turing, "Computing Machinery," 456.
45 Turing, "Computing Machinery," 448.
46 Turing, "Intelligent Machinery," 421.
47 Turing, "Computing Machinery," 459.
48 Turing, "Computing Machinery," 459.
49 Alan Turing, "Intelligent Machinery, a Heretical Theory," in Copeland, *The Essential Turing*, 472.
50 McCarthy et al., "A Proposal," 7.

Chapter 26

1 Sherwood L. Washburn, "Speculations on the Interrelations of the History of Tools and Biological Evolution," *Human Biology* 31, no. 1 (1959): 21–31, at 24.
2 Julian S. Huxley, "Evolution, Cultural and Biological," *Yearbook of Anthropology* (1955): 10–11. Subsequent citations are given parenthetically in the text.
3 Peter Medawar, *Uniqueness of the Individual* (New York: Basic Books, 1958), 135–36. Subsequent citations are given parenthetically in the text.
4 As Medawar writes, "[Niko] Tinbergen and [Konrad] Lorenz have given us reasons for believing that many kinds of behaviour which seem to us to be peculiarly human are part of a very ancient heritage—'showing off,' for example; playing with dolls; sexual rivalry; and many kinds of 'displacement activity,' in which a thwarted instinctive impulse vents itself in actions of an apparently quite irrelevant kind" (135–36).
5 J. Z. Young, *A Model of the Brain* (Oxford: Oxford University Press, 1964), 13.
6 See Karl Popper, *The Logic of Scientific Discovery* (1934) (New York: Basic Books, 1959).
7 On Popper's evolutionary epistemology and its relation to technology, see Skagestad, "Thinking with Machines."
8 Karl Popper, *Conjectures and Refutations: The Growth of Scientific Knowledge* (London: Routledge, 1963), 64.
9 Karl Popper, *Objective Knowledge: An Evolutionary Approach* (Oxford: Oxford University Press, 1972).
10 Karl Popper, "Epistemology without a Knowing Subject," in *Objective Knowledge*, 115.
11 Popper, "Epistemology," 116.
12 Karl Popper, "Of Clouds and Clocks," in *Objective Knowledge*, 222–23.
13 Popper, "Of Clouds," 230.
14 Popper, "Epistemology," 118.
15 Popper, "Epistemology," 145.
16 Popper, "Epistemology," 135.
17 Popper, "Epistemology," 145.
18 Popper "Of Clouds," 238.
19 Skagestad, "Thinking with Machines," 166.
20 Popper, "Of Clouds," 238–39.
21 Popper, "Of Clouds," 239.
22 Popper, "Of Clouds," 253–54.
23 Bateson, *Ecology of Mind*, 331–33. Subsequent citations are given parenthetically in the text.
24 Vannevar Bush, "As We May Think," *Atlantic Monthly* 176, no. 1 (1945): 101. Subsequent citations are given parenthetically in the text.
25 Gilbert Simondon, *On the Mode of Existence of Technical Objects* (1958), trans. Cecile Malaspina and John Rogove (Minneapolis: Univocal Press, 2017), 105.
26 J. C. R. Licklider, "Man-Computer Symbiosis," *IRE Transactions on Human Factors in Electronics* (Mar. 1960): 4. Subsequent citations are given parenthetically in the text.
27 On the trajectory of intelligence augmentation, from Bush to Licklider to Engelbart, see Skagestad, "Thinking with Machines," 158–63. Skagestad also notices the key conceptual relationship between Engelbart and Popper, although there seems to be no direct connection between the two thinkers.

28 Douglas C. Engelbart, "Augmented Man, and a Search for Perspective" (1960). This is an abstract for a potential conference paper. Available at https://web.stanford.edu/dept/SUL/library/extra4/sloan/mousesite/EngelbartPapers/B5_F15_AugmPap1.html. Accessed July 13, 2020.

29 Douglas C. Engelbart, *Augmented Human Intellect Study*, Prepared for the Headquarters of Air Force Office of Scientific Research Washington 25, D.C. Computer Techniques Laboratory (1961). Available at https://web.stanford.edu/dept/SUL/library/extra4/sloan/mousesite/EngelbartPapers/B6_F1_AugmProp2.html. Accessed July 13, 2020.

30 Engelbart, letter to V. Bush, May 24, 1962, asking for permission to quote at length from "As We May Think," in Engelbart's SRI report.

31 Douglas C. Engelbart, *Augmenting Human Intellect: A Conceptual Framework*, Prepared for Director of Information Sciences, Air Force Office of Scientific Research, Washington 25, D.C. Contract AF49(638)-1024 (1962), 8. Available at https://web.stanford.edu/dept/SUL/library/extra4/sloan/mousesite/EngelbartPapers/B5_F18_ConceptFrameworkInd.html. Accessed July 13, 2020

32 Engelbart, *Augmenting*, 9 Subsequent citations are given parenthetically in the text.

33 Douglas C. Engelbart, "Intellectual Implications of Multi-Access Computer Networks," in *Proceedings of the Interdisciplinary Conference on Multi-Access Computer Networks*, Austin, Texas, Apr. 1970. Available at https://web.stanford.edu/dept/SUL/library/extra4/sloan/mousesite/Archive/Post68/augment-5255.htm. Accessed July 13, 2020.

34 Garrett Hardin, "An Evolutionist Looks at Computers with Some Alarm," *Datamation* 15 (1969): 98–109.

35 Hardin, "An Evolutionist," 98.

Chapter 27

1 Washburn, "Speculations on the Interrelations of the History of Tools and Biological Evolution," 31.

2 Washburn, "Speculations," 25.

3 André Leroi-Gourhan, "Technique et société chez l'animal et chez l'homme" (1957), in *Le fils du temps: Ethnologie et préhistoire 1935–1970* (Paris: Fayard, 1983), 11. Subsequent citations are given parenthetically in the text.

4 André Leroi-Gourhan, *Gesture and Speech* (1964), trans. Anna Bostock Berger (Cambridge, MA: MIT Press, 1993), 18. Subsequent citations are given parenthetically in the text.

5 "There is probably no reason, in the case of the earliest anthropoids, to separate the level of language from that of toolmaking: Throughout history up to the present time, technical progress has gone hand in hand with progress in the development of technical language symbols" (114).

6 As Leroi-Gourhan explains, in a way reminiscent of Vannevar Bush's depiction of the imagined Memex machine, "A punched card index is a memory-collecting machine. It works like a brain memory of unlimited capacity that is endowed with the ability—not present in the human brain—of correlating every recollection with all others" (264).

7 See Edwards, *The Closed World*.

8   In a passage that sounds a lot like theorists of the "Singularity" (such as Ray Kurzweil), Leroi-Gourhan also will write: "Putting it more positively, we could say that if humans are to take the greatest possible advantage of the freedom they gained by evading the risk of organic overspecialization, they must eventually go even further in exteriorizing their faculties" (265).
9   Bruce Mazlish, "The Fourth Discontinuity," *Technology and Culture* 8, no. 1 (Jan. 1967): 1–15.
10 Friedrich Georg Jünger, *The Failures of Technology: Perfection without Purpose*, trans. F. D. Wieck (Hinsdale, IL: Henry Regnery, 1949), 64.
11 S. Demczynski, *Automation and the Future of Man* (London: George Allen Unwin, 1964), 143.
12 Hannah Arendt, "Cybernetics" (1964), Hannah Arendt Papers at the Library of Congress (Series: Speeches and Writings File, 1923–1975, n.d.), [1].
13 Hannah Arendt, "The Conquest of Space and the Stature of Man" (1963), in Arendt, *Between Past and Future: Six Exercises in Political Thought* (New York: Faber, 1961), 264.
14 Hannah Arendt, *The Human Condition* (Chicago: University of Chicago Press, 1958), 172.
15 Arendt, *The Human Condition*, 191–92.
16 Martin Heidegger, preface to *Wegmarken* (Frankfurt: Klostermann, 1967), x.
17 See Erich Hörl, *Sacred Channels: The Archaic Illusion of Communication* (Amsterdam: Amsterdam University Press, 2018), appendix.
18 Leroi-Gourhan, *Gesture and Speech*, 266.
19 Leroi-Gourhan, *Gesture and Speech*, 269.

Chapter 28

1   Lambros Malafouris, "Mind and Material Engagement," *Phenomenology and the Cognitive Sciences* 18, nos. 1–2 (2019): 1–17, at 3.
2   See Bruno Latour, "On Actor-Network Theory: A Few Clarifications," *Soziale Welt* 47, no. 4 (1996): 369–81; and Edward Hutchins, *Cognition in the Wild* (Cambridge, MA: MIT Press, 1995).
3   Gilles Deleuze, "Postscript on the Societies of Control," *October* 59 (1992): 3–7.
4   Andy Clark and David Chalmers, "The Extended Mind," *Analysis* 58, no. 1 (1998): 7–19; and Andy Clark, *Natural Born Cyborgs: Minds, Technologies, and the Future of Human Intelligence* (Oxford: Oxford University Press, 2004).
5   Humberto Maturana and Francisco Varela, *Autopoiesis and Cognition: The Realization of the Living* (Dordrecht: Reidl, 1980); and Francesco Varela, Evelyn Rosch, and Evan Thompson, *The Embodied Mind: Cognitive Science and Human Experience* (Cambridge, MA: MIT Press, 1991).
6   Donna Haraway, "A Cyborg Manifesto" (1985), in Haraway, *Simians, Cyborgs and Women: The Reinvention of Nature* (London: Routledge, 1991), 149–82.
7   See, e.g., John Tooby and Leda Cosmides, "Evolutionary Psychology and the Generation of Culture, Part 1: Theoretical Considerations," *Ethology and Sociobiology* 10 (1989): 29–49.
8   Maturana and Varela, *Autopoiesis and Cognition*, 36.
9   Maturana and Varela, *Autopoiesis and Cognition*, 135.
10 Frank Rosenblatt, "The Perceptron: A Probabilistic Model for Information Stor-

age and Organization in the Brain," *Psychological Review* 65, no. 6 (1958): 407–8. Subsequent citations are given parenthetically in the text.
11 Geoffrey Hinton and David Touretzky, "Symbols among the Neurons: Details of a Connectionist Inference Architecture," *IJCAI'85: Proceedings of the 9th International Joint Conference on Artificial Intelligence* 1 (1985): 238.
12 David Rumelhart, Paul Smolensky, Jay McClelland, and Geoffrey Hinton, "Schemata and Sequential Thought Processes in PDP Models," in *Parallel Distributed Processing*, vol. 2: *Psychological and Biological Models* (Cambridge, MA: MIT Press, 1986), 7–57. Subsequent citations are given parenthetically in the text.
13 Clifford Geertz, "The Impact of the Concept of Culture on the Concept of Man," in *The Interpretation of Cultures* (New York: Basic Books, 1973), 44.
14 Merlin Donald, *Origins of the Modern Mind: Three Stages in the Evolution of Culture and Cognition* (Cambridge, MA: Harvard University Press, 1991).
15 Merlin Donald, "The Exographic Revolution: Neuropsychological Sequelae," in *The Cognitive Life of Things: Recasting the Boundaries of the Mind*, ed. L. Malafouris and C. Renfrew (Cambridge, MA: McDonald Institute for Archaeological Research, 2010), 71–79.
16 Franz M. Wuketits, "Evolution as a Cognitive Process: Toward an Evolutionary Epistemology," *Biology and Philosophy* 1 (1986): 191–206.
17 Wuketits, "Evolution," 200.
18 Franz M. Wuketits, *Evolutionary Epistemology and Its Implications for Humankind* (Albany: SUNY Press, 1990), 135.
19 Tom Stonier, "The Natural History of Humanity: Past, Present and Future," *International Journal of Man-Machine Studies* 14, no. 1 (1981): 122.
20 Tom Stonier, *Beyond Information: The Natural History of Intelligence* (London: Springer, 1992), 6.
21 Peter C. Reynolds, "The Complementation Theory of Language and Tool Use," in *Tools, Language and Cognition in Human Evolution*, ed. Kathleen R. Gibson and Tim Ingold (Cambridge: Cambridge University Press, 1993), 423.
22 Pierre Lévy, *L'Intelligence collective: Pour une anthropologie du cyberespace* (Paris: La Découverte, 1994).
23 Terrence Deacon, "The Co-Evolution of Language and the Brain," by David Boulton, KCSM (PBS) Television (Sept. 5, 2003). Available at https://childrenofthecode.org/interviews/deacon.htm. Accessed Mar. 11, 2023.
24 Merlin Donald, "Human Cognitive Evolution: What We Were, What We Are Becoming," *Social Research* 60 (1993): 143–34.
25 Andy Clark, "Symbolic Invention: The Missing (Computational) Link?," comment on Merlin Donald, "Précis of *Origins of the Modern Mind: Three Stages in the Evolution of Culture and Cognition*," *Behavioral and Brain Sciences* 16 (1993): 753–54.
26 Terrence Deacon, *The Symbolic Species: The Co-evolution of Language and the Brain* (New York: Norton, 1997). Subsequent citations are given parenthetically in the text.
27 We note that Deacon will refer to Lev Vygotsky and the idea that "psychological processes could be understood as internalized versions of processes that are inherently social in nature" (450).
28 François Lyotard, *The Inhuman: Reflections on Time*, trans. Geoffrey Bennington and Rachel Bowlby (Stanford: Stanford University Press, 1992), 12.
29 Bernard Stiegler, *Technics and Time*, vol. 1: *The Fault of Epimetheus*, trans. George

Beardsworth and Richard Collins (Stanford: Stanford University Press, 1998), 151. Subsequent citations are given parenthetically in the text.
30 "Its body and brain are defined by the existence of the tool, and they thereby become indissociable. It would be artificial to consider them separately, and it will therefore be necessary to study technics and its evolution just as one would study the evolution of living organisms. The technical object in its evolution is at once inorganic matter, inert, and organization of matter. The latter must operate according to the constraints to which organisms are submitted" (150).
31 Recently, Fiona Coward has argued, "The incorporation of non-human *things* into social networks allows for their other-than-human qualities—notably, in the case of material culture, the qualities of *durability, persistence* and *divisibility* . . .—to be co-opted (or exapted) for the purposes of constructing and maintaining ever-greater social networks." Coward, "Small Worlds, Material Culture and Ancient Near Eastern Social Networks," in *Social Brain, Distributed Mind*, ed. Robin Dunbar, Clive Gamble, and John Goblet (Oxford: Oxford University Press, 2010), 468.
32 Bernard Stiegler, *Symbolic Misery*, vol. 1: *The Hyperindustrial Epoch*, trans. Barnaby Norman (Cambridge: Polity, 2014).
33 Bernard Stiegler, *Automatic Society*, vol. 1: *The Future of Work*, trans. Daniel Ross (London: Wiley, 2016); and Bernard Stiegler, *The Age of Disruption: Technology and Madness in Computational Capitalism*, trans. Daniel Ross (Cambridge: Polity, 2019).

Chapter 29

1 Yann LeCun (@ylecun), Twitter, May 16, 2017, https://twitter.com/ylecun/status/864463163135324165.
2 As Jeff Hawkins wrote in the early 2000s, "Prediction is not just one of the things your brain does. It is the *primary function* of the neocortex, and the foundation of intelligence. The cortex is an organ of prediction." Jeff Hawkins with Sandra Blakeslee, *Intelligence: How a New Understanding of the Brain Will Lead to the Creation of Truly Intelligent Machines* (New York: Henry Holt, 2004).
3 Jakob Hohwy, *The Predictive Mind* (Oxford: Oxford University Press, 2013), 1.
4 Hohwy, *The Predictive Mind*, 2.
5 As the brain theorist Karl Friston writes, "Predictive coding offered a compelling process theory that lent notions like the Bayesian brain a mechanistic substance. The Bayesian brain captured a growing consensus that one could understand the brain as a statistical organ, engaging in an abductive inference of an ampliative nature. Predictive coding articulated plausible neuronal processes that were exactly consistent with the imperative to optimize Bayesian model evidence." Friston, "A Free Energy Principle for Biological Systems," *Entropy* 14, no. 11 (2012): 2119.
6 See Chris Frith, *Making up the Mind* (Oxford: Blackwell, 2007). 41.
7 Giovanni Pezzulo, "Tracing the Roots of Cognition in Predictive Processing," in *Philosophy and Predictive Processing*, ed. Thomas Metzinger and Wanja Wiese (Frankfurt am Main: MIND group, 2021), no. 20, p. 2. Available at https://predictive-mind.net/papers/tracing-the-roots-of-cognition-in-predictive-processing. Accessed Jan. 4, 2023.

8   Andy Clark, *Surfing Uncertainty: Prediction, Action, and the Embodied Mind* (Oxford: Oxford University Press, 2015), 1. Subsequent citations are given parenthetically in the text.

9   For a critique of machine learning from the perspective of error, see Matteo Pasquinelli, "How a Machine Learns and Fails: A Grammar of Error for Artificial Intelligence," *Spheres* 5 (2019): 1–17.

10 Regina E. Fabry, "Betwixt and Between: The Enculturated Predictive Processing Approach to Cognition," *Synthese* 195 (2018): 2483–2518, at 2484.

11 See Lambros Malafouris, *How Things Shape the Mind: A Theory of Material Engagement* (Cambridge MA: MIT Press, 2013); Thomas Wynn and Frederick L. Coolidge, eds., *Cognitive Models in Paleolithic Archaeology* (Oxford: Oxford University Press, 2016); and Dunbar, Gamble, and Goblet, eds., *Social Brain, Distributed Mind*.

12 As Fabry writes, "An important consequence of cognitive niche construction is that both the phylogenetic and ontogenetic development of human cognitive capacities cannot be understood independently from considerations of the structured environment. Thus, we need to take the interaction and the mutual dependence of human cognitive systems and their cognitive niche into account." "Betwixt," 2487.

13 "Because the internal states of each cell are sequestered behind their respective Markov blankets, all interactions between cells must be mediated by their Markov blankets. This means that we can describe the self-organisation of the cellular ensemble purely in terms of transactions among the (sensory and active) states of Markov blankets. However, these exchanges will themselves have a sparsity structure that induces a Markov blanket of Markov blankets. For example, one subset of the ensemble could be an organ that is surrounded by epithelia that provide a Markov blanket for all of the cells that constitute the internal states of the organ. However, this still follows exactly the same (statistical) structure—an ensemble of Markov blankets. We can then repeat this process, at increasingly larger scales of self-organisation, to create a series of hierarchically nested systems (e.g., the body)." Karl Friston, Maxwell Ramstead and Paul Badcock, "Answering Schrödinger's Question: A Free-Energy Formulation," *Physics of Life Reviews* 24 (2018): 1–16, at 8.

14 Andy Clark, "How to Knit Your Own Markov Blanket: Resisting the Second Law with Metamorphic Minds," in Metzinger and Wiese, *Philosophy and Predictive Processing*, no. 3, p. 13. Available at https://predictive-mind.net/papers/how-to-knit-your-own-markov-blanket. Accessed Jan. 4, 2023.

15 Paul Valéry, *Notebooks*, trans. Paul Gifford, ed. Brian Stimpson (Franfurt am Main: Peter Lang, 2000), 402.

16 Bernard Stiegler, *Nanjing Lectures 2016—2019*, ed. and trans. Daniel Ross (London: Open Humanities Press, 2020), 60.

Index

abduction (logic), 121
actor network theory, 5, 329
algorithms, 4, 6, 114, 243, 261, 266, 279, 284, 330
apes, 169–72, 177, 208
Arendt, Hannah, 323–25
Aristotle, 359n35
artificial intelligence, 2, 4–5, 7, 11, 16, 33, 110, 232, 250, 322–23, 328, 330
Ashby, W. Ross, 240–41, 247–51, 299
Ashworth, William, 110
automata, 8, 18–19, 20, 22, 27, 31, 35, 41, 44, 49, 51, 60, 63, 65–66, 83, 93, 126, 167, 196, 231

Babbage, Charles: Analytical Engine, 109–10, 114, 285, 294; on analytics, 11; Difference Engine, 109–11, 112–13, 167; *Ninth Bridgewater Treatise*, 111–13; on printing, 113–14
Baldwin, James Mark, 153–55
Bassiri, Nima, 28
Bateson, Gregory, 239–40, 297–99
Beaunis, Henry, 133, 135
Bergson, Henri, 158–59
Bernard, Claude, 125, 152
Bertalanffy, Ludwig von, 238, 249
bifurcation, 128
Bonnet, Charles, 77
Boole, Georges, 118
Boutroux, Émile, 152–53
Boyle, Robert, 34–35
brain, 2, 3, 16, 20–25, 28, 30–31, 66, 73–74, 81, 117–18, 123, 131–40, 141, 142, 181–87, 215, 232, 256, 272–73, 313–15, 318, 322, 324, 339–40, 346–47, 359n28; localization, 131–32, 135, 181; plasticity, 132, 138, 142, 152, 155–56, 168, 176, 184, 204, 238, 243, 252, 256, 286
Broca, Paul, 132
Brodmann, Kordanian, 133
Brown-Séquard, Charles-Éduard, 135

Bush, Vannevar, 299–302; on the Memex, 300–301

Campbell, George, Duke of Argyll, 146–47
Canguilhem, Georges, 237–38, 358n22
Cannon, Walter, 182, 240
Carpenter, William, 165
Cassirer, Ernst, 178–79, 211–13; on technology, 212–13
Center for Human-Centered Computing (Stanford University), 5
Chalmers, David, 330
Clark, Andy, 3, 330, 338, 347–51
cognitive science, 3, 16, 114, 260, 336, 346, 348
Cold War, 6, 243, 329
Coleridge, Samuel Taylor, 164
computers, 2, 4, 11–12, 114, 236, 254–59, 277–78, 279–84, 297, 308, 312, 324, 333, 346, 352
consciousness, 3, 123, 139, 144, 151, 153–54, 157, 158, 165–66, 168, 176, 190, 197, 209, 239, 247, 262, 281, 297, 299, 341, 360n60
Cosmides, Leda, 330
Cournot, A. A., 129–30
Coward, Fiona, 389n31
Craik, Kenneth, 247
Cudworth, Ralph, 39
cybernetics, 10, 11, 16, 17, 22, 29, 139–40, 200, 234–59, 316, 321

Damasio, Antonio, 16
Darwin, Charles, 141, 143–45; on technology, 145
Deacon, Terrence, 338–40
decision, 4, 5, 6–7, 8, 28, 31, 124, 185, 221, 272, 352
decision problem, 263
deep learning, 2, 330, 345
Delacroix, Henri, 179
Deleuze, Gilles, 68–70, 71, 329–30
Derrida, Jacques, 242

Descartes, René, 9, 10, 39, 59, 166, 245, 276; on animals, 18–19; on animal spirits, 20–21, 22, 31; on automata, 18–19, 20, 22, 27, 31; on information, 23, 25; on memory, 24–25; on the nervous system and the brain, 16, 20–25, 28, 30–31, 359n28; *Passions of the Soul*, 27–31, 42; on reason, 19, 25, 26–27; on wonder, 30–31
Diderot, Denis, 77
Donald, Merlin, 336–37, 338
Driesch, Hans, 174, 188
Duncker, Karl, 173–74

Eddington, Arthur, 262
EDVAC, 254
Ellenberger, Henri, 164
Engelbart, Douglas C., 304–10
epigenesis, 89, 93–94
error, 10, 27, 34, 60, 66–67, 82–83, 90, 111, 114, 225, 237, 238–39, 242, 246–47, 254, 256–59, 270, 282–85, 287–88, 308–9, 322, 345–46. *See also* error correction
error correction, 114, 256–57, 346
exosomatic evolution, 199–200, 296, 310

Fabry, Regina, 349
Facebook, 8
feedback, negative, 235, 237, 240, 246–47
Ferrier, David, 135–36
Fessard, Alfred, 253–54
Flourens, Jean-Pierre, 131–32, 133
Fodor, J. A., 330
Foucault, Michel, 8, 293, 329
Franz, Shepherd Ivory, 181
Freyer, Hans, 213–14
Friston, Karl, 389n5
Fukiyama, Francis, 329

Galison, Peter, 243
Galton, Francis, 166
Gates, Bill, 7
Geiger, Lazarus, 147–48
Gestalt theory, 172–75, 176, 179, 184, 185, 222–23, 241
Gibson, William, 338
Ginsborg, Hannah, 88
Girard, René, 8
Gödel, Kurt, 264
Goldstein, Kurt, 184–87, 222–23, 237–38, 332

Haldane, J. B. S., 182–83
Harari, Yuval, 3–4
Haraway, Donna, 329–30
Hardin, Garrett, 310
Hartley, David, 77
Hartmann, Eduard von, 165
Hawking, Stephen, 7
Hawkins, Jeff, 389n2

Head, Henry, 182
Heidegger, Martin, 217–22, 284–85, 324, 325, 343; on the organism, 217–19; on technology, 220
Helmholtz, Hermann von, 126, 142, 168
Herrick, C. Judson, 201–4
Herschel, William, 110, 11
Hinton, Geoffrey, 333–34, 347
Hixon Symposium (California Institute of Technology), 235, 236, 242, 382n47
Hobbes, Thomas, 10, 39, 42, 44, 60, 61; on the organism, 34, 36; on reason, 33–34; on sovereignty, 35
Hoffmann, Josef, 49–50
Hohwy, Jakob, 346, 347–48
homeostasis, 216, 234, 240, 241, 244, 246, 247, 248, 250, 251, 298, 350
Hume, David, 10–11, 81, 83, 293; on anticipation, 71, 72; on association of ideas, 70, 72, 75; *Dialogues on Natural Religion*, 68; on the nervous system and brain, 73–74; on reason, 71–72, 74
Hutchins, Edward, 329
Huxley, Julian, 289–90
Huxley, Thomas, 166
hylomorphism, 52
hypothetical reasoning, 20, 116, 119–20, 121–22, 230, 282, 292–93, 346

information, 2–3, 17, 22–26, 28, 30, 31, 33, 42, 59, 85, 17, 125, 127, 138, 151, 152, 157, 166, 196, 198, 207, 221, 235, 236, 239, 245–46, 251–53, 254, 258, 260, 261, 263, 269, 270, 273, 292, 298, 299–302, 303, 308, 315, 325, 332, 334, 337–38, 339

Jackson, John Hughlings, 136–38
James, William, 168, 249
Janet, Pierre, 123–24
Jefferson, Geoffrey, 273–79
Jevons, William, 116, 119–20, 122
Jonas, Hans, 40
Judd, Charles, 156–58, 209–10

Kahneman, Daniel, 3
Kant, Immanuel, 10–11, 77, 293; on artifice, 80; on automata, 93; on the categories, 86, 89–90; on empirical concepts, 88, 90; on extraterrestrials, 82; on the hand, 105; on intuitions, 85; on judgment, 88–89, 90–91, 95–99, 103; on logic, 83; on the manifold, 86, 104; on the nervous system and brain, 81; on the organism, 100–102; on pathology, 102, 103; on reason, 88, 89–90, 92, 93–94; on technology, 103–5; on the transcendental apperception of unity, 92
Kapp, Ernst, 149–51, 211–12
Kekulé, August, 168
Koch, Christof, 360n60

Koffka, Kurt, 174, 175, 179
Köhler, Wolfgang, 169–72, 174–75, 179, 208, 235
Koselleck, Reinhart, 6–7

Lafitte, Jacques, 205–7
Land, Nick, 9
Lashley, Karl, 168, 181–82, 183–84
Lassègue, Jean, 384n35
Latour, Bruno, 328–29
Leibniz, Gottfried Willhelm, 10, 87; and Hoffmann, Josef, 49–50; on the human soul, 52, 53, 55–56, 57–58; on monads, 56–57, 60, 62, 63; on the organism, 50, 51, 51–52, 53, 55, 57; on perception, 57, 61–62; on preestablished harmony, 51, 54, 55, 63; on reason, 61–63; and Stahl, Georg Ernst, 50; on technology, 60–61
Leroi-Gourhan, André, 311–26, 336, 341–42; on the brain, 313–15, 318, 322; on memory, 312
Lévy, Pierre, 338
Lewes, George, 142–43; on technology, 143
Licklider, J. C. R., 302–4
Lillie, Ralph, 139–40
Locke, John, 51, 66–67, 82, 87
logic machines, 116, 118, 119–20, 122
Lotka, Alfred J., 203; on exosomatic evolution, 199–200; on technology, 196, 198–99; on thermodynamics, 194–95, 200
Lovelace, Ada, 114–15, 285

machine learning, 345–47
Malabou, Catherine, 223
Malebranche, Nicolas, 66
Marinetti, Filippo, 162
Markov blanket, 349–51
Marquand, Alan, 116, 122
Marx, Karl, 108–9, 321
Maxwell, James Clerk, 126–29
Mbembe, Achille, 9–10
McCarthy, John, 245, 260, 288
McCulloch, Warren, 236, 240
McDougall, William, 201
McTaggart, John, 262
mechanical philosophy, 16, 17, 34, 51–52
mechanism, 8, 21, 33–34, 36, 65, 113, 117, 123, 139–40, 203, 217–18, 234, 236, 263–64, 286, 346, 358n22, 389n5. *See also* mechanical philosophy
Medawar, Peter, 290–92
Mensch, Jennifer, 88
Merleau-Ponty, Maurice, 175
Mill, John Stuart, 118–19
Miller, George A., 247
Morgan, Thomas, 210–11
morphogenesis, 88, 174, 188, 244
Mumford, Lewis, 238
Murphy, Joseph J., 166
Musk, Elon, 7

neural networks, 328, 331–35, 346
neuroscience, 2, 3–4, 11, 141, 204, 258, 260, 329, 330, 346, 351
niche construction, 158, 349

Oldenburg, Henry, 37, 60
Ortega y Gasset, José, 207–8

Pappert, Seymour, 332
Pascal, Blaise, 57, 65–66, 284
Peirce, Charles Sanders, 121–23, 155–56
perceptron. *See* Rosenblatt, Frank
Pickering, Andrew, 244
Pitts, Walter, 236
plastic nature, 57
Plessner, Helmuth, 222–26
Pinker, Steven, 3
plasticity. *See* brain: plasticity
Poincaré, Henri, 168
Polanyi, Michael, 240–43
Popper, Karl, 292–97
predictive processing, 346–49

rats, 175, 177
rector (Roman law), 63
Reich, Eduard, 148–49
Rignano, Eugenio, 180
Rosenblatt, Frank, 331–33
Rosenblueth, Arturo, 240
Rousseau, Jean-Jacques, 81
Ruyer, Raymond, 239

Schaffer, Simon, 110
Scheler, Max, 214–17
Schmitt, Carl, 7–8, 352
Schopenhauer, Arthur, 164
self-organization, 67, 77
Shannon, Claude, 122, 245
Sherrington, Charles, 138–39, 181, 197, 249
singularities, 127–29
Smee, Alfred, 116, 117–18
Sombart, Werner, 159
Spencer, Herbert, 165
Spinoza, Baruch, 10, 51, 53, 63; on the bloodworm, 37–38; on the human body, 40–41, 46; on intuition, 45–48; on method, 43–44; on the passions, 42–43; on reason, 38, 39–40, 43, 44, 45
Stahl, Georg Ernst, 50
Stewart, Balfour, 126
Stiegler, Bernard, 12, 340–44
Stonier, Tom, 337–38
Strauss, Leo, 8

telegraph, 127, 151, 159
Thiel, Peter, 7–8
Thompson, William, 119
Thorndike, Edmund, 170

Tichener, Edward Bradford, 156
Tolman, E. C., 175
Tooby, John, 330
Touretsky, David, 333
Turing, Alan, 260–75, 279–88, 295; on intelligence, 265–69, 271–73, 282–84; on the Oracle, 265; on unorganized machines, 268–69, 271

Umwelt, 37–38, 190, 194, 212, 213, 218–19
unconscious, 3, 57, 65, 116–17, 151, 164–66, 167–68, 176, 348, 360n60

Varela, Francisco, 330
Vaucanson, Jacques, 93, 126
von Monakow, Constantin, 182, 184
Von Neumann, John, 254–59
Von Uexküll, Jakob, 194, 212–13, 216, 218–19, 224; on the nervous system and brain, 190–91; on technology, 189, 192–93
Vygotsky, Lev, 177–78

Wallas, Graham, 168–69
Walter, W. Grey, 239, 276
Washburn, Sherwood, 311
Whitehead, Alfred North, 180
Whorf, Benjamin Lee, 305
Wiener, Norbert, 235, 239, 245, 251–53
Wittgenstein, Ludwig, 228–32; on slavery, 228, 231
Wuketits, Franz, 337

Young, J. Z., 273, 291–92

zombies, 3, 360n60